# World Sustainability Series

**Series editor**

Walter Leal Filho, Hamburg, Germany

More information about this series at http://www.springer.com/series/13384

Paula Castro · Ulisses M. Azeiteiro
Paula Bacelar-Nicolau · Walter Leal Filho
Anabela Marisa Azul
Editors

# Biodiversity and Education for Sustainable Development

Springer

*Editors*
Paula Castro
Department of Life Sciences, Centre for Functional Ecology
University of Coimbra
Coimbra
Portugal

Ulisses M. Azeiteiro
Department of Science and Technology
Universidade Aberta and Centre for Functional Ecology
University of Coimbra
Coimbra
Portugal

Paula Bacelar-Nicolau
Department of Science and Technology
Universidade Aberta and Centre for Functional Ecology
University of Coimbra
Coimbra
Portugal

Walter Leal Filho
HAW Hamburg
Hamburg
Germany

and

Manchester Metropolitan University
Manchester
UK

Anabela Marisa Azul
Center for Neuroscience and Cell Biology
University of Coimbra
Coimbra
Portugal

ISSN 2199-7373 ISSN 2199-7381 (electronic)
World Sustainability Series
ISBN 978-3-319-32317-6 ISBN 978-3-319-32318-3 (eBook)
DOI 10.1007/978-3-319-32318-3

Library of Congress Control Number: 2016935961



Printed on acid-free paper

This Springer imprint is published by Springer Nature
The registered company is Springer International Publishing AG Switzerland

# Editorial

This book joins the global debate on education for sustainable development (ESD) and explores its contribution to biodiversity conservation. This book collects reflections from various areas of knowledge and links biological and ecological principles and pedagogical practices. In parallel, it reports on a remarkable wide range of formal, non-formal and informal learning and participatory research and awareness actions, as well as educational proposals and resources for critical thinking and for fostering practices and attitudes, reconciling education with principles of human behaviour and nature.

Along with the challenge of the *United Nations Decade on Biodiversity* (2011–2020), this book aims to bring a scientific and pedagogical contribution targeting the *Strategic Plan for Biodiversity* 2011–2020, including the *Aichi Biodiversity Targets* and also the *Transforming our world: the 2030 Agenda for Sustainable Development* that states in its *Goal 14. Conserve and sustainably use the oceans, seas and marine resources for sustainable development* and *Goal 15. Protect, restore and promote sustainable use of terrestrial ecosystems, sustainably manage forests, combat desertification, and halt and reverse land degradation and halt biodiversity loss,* biodiversity conservation related goals.

The contents of this book are divided into five parts:

Part I Education Research on Biodiversity
Part II Teaching About Biological Invasions and Threats to Biodiversity
Part III Engaging Society in Biodiversity Conservation and Sustainability
Part IV Designing Sustainable Futures
Part V Online Education to Biodiversity and Sustainability Awareness

The first part gathers papers concerning education research on biodiversity as well as conservation of biological resources. The second part focuses on the biological invasions that threaten native biodiversity. The third part reports on study cases that engage society in the conservation of biodiversity and sustainable development. The fourth part focuses on novel approaches leading to more sustainable futures, and the final part discusses the use of e-learning in the progresses towards conservation of biodiversity and sustainability awareness. Using e-learning is becoming widely accepted in both formal and non-formal education, proving to be effective in expanding education for sustainability (EfS).

Readers will find variety of topics, covering from urban biodiversity, island ecosystems, polar regions biodiversity, endemisms and protected species to biological invaders/invasive alien species (IAS), from fungal biodiversity and teaching to insects biodiversity, in the context of endemic biodiversity in the Azores Islands (North Atlantic) and soil biodiversity. These topics are addressed in a scientific and investigative perspective, educational context (education for sustainable development) and incorporate social awareness/raising awareness into the broader socio-environmental topic of biodiversity conservation and communication strategies for public engagement. Biodiversity and rural communities and biodiversity and indigenous communities are addressed together with the discussion of biodiversity policies in national contexts.

Teaching biodiversity conservation and sustainable use of resources, ecosystem services and sustainable development may also make wider use of, and profit from, the new information and communication technologies, as also addressed in this book by the study case of online teaching (*e*-learning and *b*-learning) for biodiversity conservation.

A total of 20 double-blind peer-reviewed chapters from Europe (contributors from Portugal, Germany, United Kingdom, Norway, Spain and Italy), South America (contributors from Brazil and Mexico) and Africa (contributors from South Africa) cover the different subjects related to the above-mentioned topics of this book.

An introductory chapter from the editors Walter Leal Filho, Paula Castro, Paula Bacelar-Nicolau, Anabela Marisa Azul and Ulisses M. Azeiteiro entitled "Biodiversity and Education for Sustainable Development (ESD): Tendencies and Perspectives" draws some theoretical concepts with reference to the links between biodiversity research, biodiversity loss, biodiversity in education for sustainable development, education and sustainable development, tendencies and perspectives in the context of the *United Nations Decade on Biodiversity* (2011–2020).

Part I of the book includes five chapters relating to education research on biodiversity and conservation of biological resources:

Paula Castro, Anabela Marisa Azul, and Jorge Paiva in the chapter "Conservation of Biological Resources: Why Does It Matter?" emphasise on the relevance of biodiversity in people's life. The authors address the importance of forests, the unknown biodiversity and the extinction of species in order to alert the readers and the general public, students, teachers and other stakeholders on the importance of all biological resources. This is the viewpoint of one of the lasting researchers of a generation of naturalists—Jorge Paiva.

As we are increasingly becoming aware, "urban growth and human competition for land have led to deep structural changes in the composition and dynamics of the landscape, significantly affecting the fragile rural/urban equilibrium, and migration from rural to urban areas has forced cities to expand into the surrounding environments which created substantial environmental impacts on the functioning of natural ecosystems, affecting land use of the rural-urban interface". In this context, the chapter untitled "Urban Biodiversity and Cities' Sustainability Development", by Lurdes Barrico and Paula Castro, looks at understanding the human causes and consequences of land use change in urban environments. Authors state that "with a correct model of urban development we may preserve native biodiversity, ecosystem services, and diminish natural hazards in urban environment" and "highlight the importance of greening cities and engaging residents and other stakeholders in the planning process and decision-making, as well as the importance of teaching, training or raising awareness as key actions to achieve these goals".

"The Polar Regions are famously associated with extreme temperatures, ice, snow, legendary explorers, indigenous people, polar bears, penguins and other impressive fauna and flora. The past decades have witnessed a revolution in the amount of data collected in the Polar Regions, with considerable advances in the knowledge of numerous areas, including polar biodiversity. Educationally, the Polar Regions can be perfect vehicles to transfer educational concepts related to biodiversity, but unfortunately, the evaluation of the impact of educational activities related to polar biodiversity is scarce". In this fourth chapter, untitled "Education on Biodiversity in the Polar Regions", José C. Xavier, Gerlis Fugmann, Inga Beck, Louise Huffman and Eric Jensen provide a general review of the importance of the polar regions, the increasing status of polar education in the last decade, examples of polar educational activities on biodiversity and a resource to stakeholders interested in polar science and education.

María P. Martín and Roy Watling in "Teaching Mycology Worldwide" write about taxonomy and fungi, stating the difficulty "to delimit a species concept which is common to all fungal groups. In general, taxonomists continue using morphological characters; however, DNA sequence analysis (barcoding) is now essential to discover the true identity of new fungal species". In this chapter, some aspects of teaching mycology are summarised, including well-documented fungi websites from different countries, and scientific databases for various groups of fungi and the application of new technologies. Some teaching experiences, while training taxonomists all around the world, are shared with the reader, such as sampling in the rain forest with Malaysian and Thai students, teaching general and master courses in Australia, Brazil, Ecuador and India and *online* tools for the effective training of students and teachers, particularly when these participate at distance.

"The negative impact of biodiversity loss on ecosystem functioning and services, and ultimately on human well-being, has been unequivocally established; however, despite all efforts, biodiversity is still declining worldwide. It is widely accepted that biodiversity awareness is crucial for its conservation. Nevertheless, after many initiatives to alert society about the consequences of losing biodiversity, biodiversity loss is still perceived as a minor environmental risk compared to others, such as climate change. Thus far, most communication strategies have involved conventional venues, targeting people who are already "environmentally-aware", and have not incorporated societal idiosyncrasies and cultural backgrounds". In this context, Ana Moura Arroz, Rosalina Gabriel, Isabel R. Amorim, Rita São Marcos and Paulo A.V. Borges in the chapter "Bugs and Society I: Raising Awareness About

Endemic Biodiversity" review the role of communication in risk governance, and the principles and strategic options of the Azorean intervention.

The second part of the book discusses biological invasions and their threats to native biodiversity and includes chapters "Science and Education at the Centre for Invasion Biology" and "Engaging Society to Fight invasive Alien Plants in Portugal—One of the Main Threats to Biodiversity":

"South Africa has severe problems caused by biological invasions in terrestrial, freshwater and marine ecosystems, and a long history of managing biological invasions. However, appreciation and systematic study of the problems associated with invasive species are relatively recent. In 2004, the Centre for Invasion Biology (CIB) was established as one of the first six national Centres of Excellence funded by the South African government". Sarah J. Davies, G. John Measey, Dorette du Plessis and David M. Richardson in "Science and Education at the Centre for Invasion Biology" present the readers the aims, structure and activities of Centres of Excellence. Researchers at these centres are involved in citizen science programmes on invasive species, in collaborative work with partner organisations which implement invasive species management programmes and provide employment opportunities for graduates at the centre.

In the chapter "Engaging Society to Fight invasive Alien Plants in Portugal—One of the Main Threats to Biodiversity", Elizabete Marchante and Hélia Marchante write about invasive alien species (IAS) in Portugal. A list of 32 animal and plant species has been recognised as invasive species by the Portuguese legislation since 1999. Nevertheless, a large portion of the population is still unaware of the problem. This paper reviews the various strategies developed by the researchers in engaging and raising public awareness with the topic of IAS, namely plant species, since 2003, including the website *invasoras.pt*. Overall, awareness about IAS is increasing amongst the Portuguese population and citizens are contributing more to the prevention and control of IAS. Still, the challenge to reach public other than the peers or professionals related to the topic is still daunting.

In Part III of this book, readers can find three study cases that engage society in biodiversity conservation and sustainable development, Chaps. "Bugs and Society II: Testing Two Communication Strategies for Public Engagement in the Azores" to "Education for Sustainability in the Context of Community Forestry":

Following from chapter 6 above, Isabel R. Amorim, Ana Moura Arroz, Rita São Marcos, Paulo A.V. Borges and Rosalina Gabriel, in "Bugs and Society II: Testing Two Communication Strategies for Public Engagement in the Azores", introduce the communicational programme "Bugs and society" and test two initiatives to raise biodiversity awareness relating to Azorean endemic species—an outdoor exhibition *Açorianos há milhões de anos* (Azoreans for millions of years) and a web contest to name insects *Chama-lhe Nomes*! (Pick a name!). Both communicational strategies targeted non-traditional audiences and relied on the Portuguese and Azorean cultural identity and on anthropic verisimilitude of situations involving insects. The assessment of impacts produced in the knowledge, attitudes and behaviours of viewers and users are discussed. Communication devices like the ones herein presented are expected to raise biodiversity awareness and empower people regarding its preservation in the Azores.

"Dialogue and mutual learning between civil society and researchers involved in natural resource management have been increasingly advocated as a means of improving public understanding of science, biodiversity conservation, and local well-being. In rural areas in developing countries, however, science communication and environmental education strategies for disseminating biodiversity conservation research have traditionally used methods based on top–down, one-way approaches that have limited local engagement in research and undermined feedback generation between local people and researchers". In the Chap. "Communicating Biodiversity Conservation Research Through Dialogue and Mutual Learning in Rural and Indigenous Communities" Isabel Ruiz Mallen examines a participatory process of developing a communication strategy for an environmental conservation project in Southeastern Mexico. Such participatory approach increases the social relevance of the research and improves both research results and dissemination products.

"Community forestry in Portugal is emerging as a promising form of multifunctional forestry that combines scientific and technical knowledge with the participation of the local residents in decision-making. These forests are governed by collective property arrangements (*baldios*) based on millenarian traditional usufruct rights of a local community of commoners (*compartes*). Participation is open to all the new residents regardless of their gender, activity or status. This connection between the commoners and the commons lands was severely disrupted during the twentieth century, by the national Forest Services, by compulsory afforesting the lands with tree species unknown to local populations and causing the decline of collective agro-pastoral practices". In "Education for Sustainability in the Context of Community Forestry" Rita Serra, Patrícia Ferreira, Iryna Skulska, Mayrén Alavez-Vargas, Anailton Salgado, João Arriscado Nunes and Raúl García-Barrios describe the study case of a community of *compartes* that recently gained back control of its common lands and initiated a project to regenerate a degraded mountain forest. This includes the challenges and the educational activities jointly developed to activate meaningful engagement in collective practices, intergenerational responsibility and active citizenship.

Part IV focuses on novel approaches leading to more sustainable futures in the scope of education for biodiversity conservation and sustainability, with Chaps. "Urban Forest Governance: FUTURE—The 100,000 Trees Project in Porto Metropolitan Area" to "Interdisciplinary and Participatory Research at Early Childhood to Biodiversity Education and Sustainable Development":

Marta Pinto, Conceição Almeida, Ana Maria Pereira and Margarida Silva in "Urban Forest Governance: FUTURE—The 100,000 Trees Project in the Porto Metropolitan Area" present the reader the case study FUTURE, describing its context, scale, institutional framework, actors and partnerships, resources and processes and governance model. The Porto Metropolitan Area is a region in northern Portugal with approximately 2,000 km$^2$ and 16 % of the Portuguese population. The region is a jigsaw puzzle of urban, agricultural and forest areas. A broad participatory regional planning process, conducted from 2003 to 2008, concluded that major challenges ahead included the improvement of the green infrastructure (forest, riverside areas and natural corridors), the need for education and training for sustainability, as well as improved interinstitutional coordination. The *FUTURE – the 100,000 trees project in the Porto Metropolitan Area* is the outcome of this process.

"Sustainable development (SD) is a controversial concept informed by conflictive narratives which reshape the way we envision the earth, the sea and the stars. Its integration in international policies and national strategy plans for development influences the ways we now know the past, our understanding of the present, and our paths to the future. It influences our lives through policies that regulate daily practices, such as the European Common Fisheries Policy which focuses its strategies for SD in trade and education". Chapter "Reflexive Research and Education for Sustainable Development with Coastal Fishing Communities in the Azores Islands: A theatre for Questions", by Alison Laurie Neilson and Irina Castro, raises the questions of environmental justice that challenges one to look critically at research and education norms for SD, and also questions how the deficit-model of the research is built on the assumption that failures of SD are due to lack of knowledge. The authors bring together the research experience on education and research practices, with the Azores archipelago in Portugal, as a background to their reflexive and educational practices. The participation of Augusto Boal's Theatre of the Oppressed helps exploring the potential of multi-directional learning via aesthetic practices and action-based research to engage people in research, and in SD policy development which are environmentally just and sustainable.

Chapter "Public Policies and Education for Biodiversity: Brazilian Challenges in a New Global Context", by Klautau de Araújo, analyses Brazilian education and public policies for environment and biodiversity, through the (i) legal, (ii) organisational and administrative and (iii) participatory points of view. This review allows us to understand the potential of environmental education (formal and informal), in the Brazilian context, and the challenges of reforming the present paradigm. The study case of the *Escola Bosque* (Forest School) illustrates a good example of an integrated environmental education system, which has been internationally awarded for its positive impact on education for biodiversity and the environment.

"Economic progress has been noteworthy in almost all fields in the last 70 years. However, the unsustainable use of non-renewable resources, the destruction of biological diversity and greenhouse gas emissions accelerated the environmental crisis and highlighted social inequalities. The accountability over this civilizational crisis is diffuse, but environmental education in Brazil is mostly focused on those who cause the least environmental damage". In the Chap. "Education for Sustainable Development in Brazil: Challenges for Inclusive, Differentiated and Multicultural Education", Andreia Setti and Ulisses Miranda Azeiteiro analyse the interactions between the determinants of the current environmental crisis and the contribution of education to sustainable development, critically considering the sustainability of the current means of production and consumption, and as a strategy for the promotion of autonomy and equity in the Brazilian reality. Authors approach three main aspects: (i) the original, foundational and practical constitution of environmental education, (ii) understanding the challenges of institutionalising environmental education (the Brazilian context) and (iii) the paths of political action needed to attain the sustainable development goals (SDGs).

"Soil is fundamental for human life as we know it. The top layer of Earth's crust, essentially composed of minerals, water and air, also harbours an immense variety of organisms, from plants to microorganisms, which qualifies it as an actual living system. On account of soil biodiversity and its functioning, soil can deliver services essential for regulating, providing and supporting human life". Sara Mendes, Anabela Marisa Azul, Paula Castro, Jöerg Röembke and José Paulo Sousa in the Chap. "Protecting Soil Biodiversity and Soil functions: Current Status and Future Challenges" review the main advances in soil research, knowledge and monitoring and discuss the status of current strategies towards soil protection and sustainability. They also present a three vector plan for effectively contributing for soil protection based on monitoring, experimenting and raising awareness towards soil issues, which hopefully can change the way people use soil.

The Chap. "Interdisciplinary and Participatory Research at Early Childhood to Biodiversity Education and Sustainable Development" by João Miranda, Raquel Maricato, Joana Vila Nova, Joana Margarida Baptista, João Lourenço Monteiro, Nuno Freitas, Odete Gonçalves, Vera Vale and Anabela Marisa Azul explored how biological understanding of the biodiversity and ecological processes at early childhood may contribute to biodiversity education and sustainable development (BESD) awareness. The participatory research undertook a constructive programme, with the active collaboration of researchers from life sciences, humanities, science education, kindergarten teachers, children and artists. The perceptions, evaluation and validation of the approach are emphasised in the ateliers progressively designed in the drawings by the children, the documentation by the kindergarten teachers and the interviews to the children and findings focused attention on the understanding of biological. Focus was given to the ecological interactions, the adaptations to climate, the food and the products of the Mediterranean forests and the biodiversity legacy in the Mediterranean region, where children were actors/authors of knowledge that resulted from the mutual learning and the active collaboration.

The fifth and the last part of the book focuses on information and communication technologies (ICT) that lead to *e*-learning's (and *b*-learning) significant role within learning and educational processes, namely in Biodiversity Conservation and Education for Sustainable Development (Chaps. "Engaging ODL Students with Biodiversity Issues: A South African Case Study on the Role of ESD" to "Knowledge Dissemination and Best Practice Transfer on Biosafety, Biosecurity and Biorisk Management Through a Sustainable and Effective Education and Awareness System"):

"Biodiversity is regarded as a key asset in safeguarding the well-being of future generations. The threat to biodiversity through indiscretionary human activities is increasingly gaining attention, from local to global scales. Biodiversity forms part of the agenda for Education for Sustainable Development through addressing inter-linkages between the various components and systems comprising the environment. This has been illustrated throughout the Decade of Education for Sustainable Development, which ended in 2014. As a result, significant progress has been made at various levels on the understanding of the impact of human consumption on biodiversity, together with the sensitization of students in terms of their potential roles to curb habitat and species loss as well as environmental degradation". Rudi W. Pretorius, Mathilda E. Brand and Leslie R. Brown in the paper "Engaging ODL Students with Biodiversity Issues: A South African Case Study on the Role of ESD" reflect on the way biodiversity is dealt with through the *b*-learning approach to ESD in the Diploma in Nature Conservation offered by the University of South Africa (UNISA). A review of the flexibility of a *b*-learning approach to open distance learning (ODL), the challenges that were experienced, the means through which these were addressed and a future perspective concludes this chapter.

Ulisses Miranda Azeiteiro and Paula Bacelar-Nicolau in the Chaps. "Online Teaching for Biodiversity Conservation" reviews the *e*-learning experience of biodiversity conservation courses at Diploma and Master's level, offered by the Universidade Aberta, the Portuguese distance learning university. This review includes the *e*-learning methodologies, their learning effectiveness, challenges (throughout nine course editions') and discusses future course's improvements. "Education is the first step to create a society that respects the others and the environment and that works to design and build a peaceful future. An effective and sustainable education system could rely on a tailored methodology that synergizes self-evaluation, gap-analysis, and train-the-trainers methods. This combination allows gathering information about real needs and expectations of training targets, elaborating a made-to-measure educational programme and training future educators on topics of interest, making education and awareness system sustainable".

In the last paper of the book the Chaps. "Knowledge Dissemination and Best Practice Transfer on Biosafety, Biosecurity and Biorisk Management Through a Sustainable and Effective Education and Awareness System", Carola Argiolas, Veronica Baldo and Maurizio Martellini review the methodologies applied to develop a sustainable education and awareness system and the training contents related to biosafety/biosecurity/biorisk management, and describe how the authors set up a knowledge development and transfer of best practice system on biosafety/biosecurity/biorisk management, in order to spread awareness and know-how on these topics. Twenty-two countries, in four different regions (South-east Europe, South-east Asia, North and West Africa) were involved in the project titled "Knowledge development and transfer of best practice in biosafety, biosecurity and biorisk management".

This book offers science research on biodiversity and educational approaches for biodiversity valuation and conservation within a geographical coverage. In addition, the chapters address important challenges and future developments, also giving insights into how education for sustainability may be pursued in a conservation biology context.

We would like to take this opportunity to thank all authors who submitted their manuscripts for consideration of inclusion in this book. And since the peer review was a double-blind process, we also thank the reviewers who have taken time to provide timely feedback to the authors, thereby helping the authors to improve their manuscripts, and ultimately the quality of this book.

Paula Castro
Ulisses M. Azeiteiro
Paula Bacelar-Nicolau
Walter Leal Filho
Anabela Marisa Azul

# Contents

# Biodiversity and Education for Sustainable Development (ESD): Tendencies and Perspectives

Walter Leal Filho, Paula Castro, Paula Bacelar-Nicolau, Anabela Marisa Azul and Ulisses Miranda Azeiteiro

**Abstract**

This introductory paper outlines some of the areas where research and action is needed, so as to allow a more systematic development of Education for Sustainable Development (ESD) in the Biodiversity Conservation context. Apart from presenting the concepts of ecosystem services and biodiversity, this chapter indicates the main initiatives needed to be developed to incorporate ESD in the curricula as a multiple-perspective approach. Additionally, it is discussed how biodiversity conservation through education action fit within the actual premises and programmes of sustainability as the Convention on Biological Diversity and the 2010 Biodiversity Target or under the scope of the Strategic Plan for Biodiversity 2011–2020. Tendencies and perspectives on Education for Biodiversity Conservation in the coming decades is also a subject of debate in the present chapter.

W.L. Filho
HAW Hamburg, Hamburg, Germany

W.L. Filho
Manchester Metropolitan University, Manchester, UK

P. Castro
Department of Life Sciences, Centre for Functional Ecology,
University of Coimbra, Coimbra, Portugal

P. Bacelar-Nicolau · U.M. Azeiteiro
Department of Science and Technology,
Universidade Aberta and Centre for Functional Ecology,
University of Coimbra, Coimbra, Portugal

A.M. Azul
Center for Neuroscience and Cell Biology, University of Coimbra,
Coimbra, Portugal

 
P. Castro et al. (eds.), *Biodiversity and Education for Sustainable Development*,
World Sustainability Series, DOI 10.1007/978-3-319-32318-3_1

## 1 Introduction

The incorporation of Education for Sustainable Development (ESD) in the curriculum in a multiple-perspective approach is usually used for teaching and learning in primary and secondary schools, to promote interdisciplinary and intercultural competencies to address ESD (United Nations Educational, Scientific and Cultural Organization [UNESCO] 2012). There are many recommended ways to do it: the scientific perspective way of knowing about the world around us; the historical perspective of changes in the world over time; the geographic perspective of events, problems and issues take on different complexities; the human rights perspective; the gender equality perspective; the values perspective of individuals, cultures and countries; the cultural diversity perspective, and the sustainability perspective (sustainability balances environmental, social, and economic concerns, as well as taking into account the well-being of future generations) (UNESCO Teaching and Learning for a Sustainable Future programme: http://www.unesco.org/education/tlsf/). For Higher Education Institutions (HEIs) approaches to incorporate ESD in the curricula include: coverage of some environmental issues and material in an existing course; a separate module, or chapter in a traditional course, tailored to the nature of each specific training; finally, a specific sustainable development course and renewal of the curricula (formal or non-formal programs, undergraduate or post-graduate programmes) (Lozano et al. 2015a).

ESD will require the development of initiatives needed to:

(a) reorient and train teachers (linking theory and practice, interdisciplinarity, transdisciplinary, informal learning, transformative learning, leadership approaches);
(b) make use of holistic, collaborative learning methods, approaches and tools (e.g. problem based learning, Triple Bottom Line, life cycle analysis, games, on-line collaborative learning);
(c) fulfil student learning outcomes and competences for ESD (Aktas et al. 2015; Amador et al. 2015; Barth and Rieckmann 2012; Dlouha and Burandt 2015; Lozano and Lozano 2014) and key competences for ESD as described by Disterheft et al. (2013).

Besides training and education of human resources and physical changes, organizations have to institutionalize sustainability into their systems and cultures (through values, visions, philosophies, policies, employee empowerment, and change management practices) (Lozano et al. 2015b), together with community outreach activities (Müller-Christ et al. 2014).

This chapter aims, through a whole-institution approach, multi-stakeholder interactions and lifelong learning, to provide a description of the achievements in the field of biodiversity and Education for Sustainable Development. Furthermore, this chapter will put in perspective future tendencies and point out some of the key issues which will guide biodiversity and conservation teaching, education for sustainable development, and sustainable development.

## 2 Ecosystem Services and Biodiversity

Ecosystem services may be defined as the benefits that humans obtain from ecosystems that support, directly or indirectly, their survival and quality of life (Millennium Ecosystem Assessment [MEA] 2003). However, human modification of Earth's biological resources—its species and genetically distinct populations—is substantial and growing. The overall risks associated with modification of landscapes and soil functions, are seriously threatening biodiversity conservation and the services provided by ecosystems (European Environment Agency [EEA] 2011). Human socio-economic activities as well as their quality of life and well-being are highly dependent on the function and associated services that ecosystems provide. These include services such as providing food, water, wool, fertile soils, timber, and other products; they regulate ecosystem processes such as climate, land degradation, disease as well as ensuring the flow of clean water and protection from flooding or other hazards like soil erosion and landslides. They are also important in supporting services like soil formation, nutrient cycling, and primary production. Finally, they provide cultural services through recreation, education, spiritual well-being, and other nonmaterial benefits (MEA 2003, 2005; Pittock et al. 2012).

The connection between ecosystems and human well-being is well known (Fig. 1). Therefore, it is crucial to understand human dependence on how ecosystems work so to better understand their resilience capacity.

Biodiversity and ecosystems are closely related as biodiversity is essential for the functioning and sustainability of an ecosystem. Biodiversity can be described as the sum of the total biotic variation, from genes to ecosystems, and supports ecosystem functioning which ultimately affects human well-being (EEA 2015;

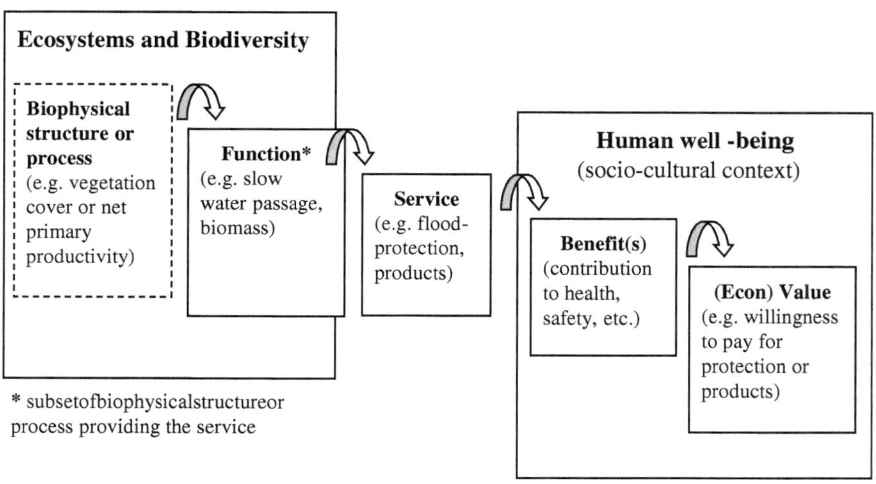


**Fig. 1** The relationship between biodiversity, ecosystem function, and human well-being (adapted from de Groot et al. 2002, after modification by Haines-Young and Potschin 2010)

MEA 2005). Biodiversity conservation emerged as a field in international policies in the 2nd half of the 20th century, culminating in the United Nations Convention on Biological Diversity [CBD], which entered into force in 1992.

Changes in drivers that indirectly affect biodiversity, such as population, technology, and lifestyle, can lead to changes in drivers directly affecting biodiversity, such as the catchment of fish, the application of fertilizers to increase food production, or urbanization, which in turn affect ecosystem's function by changing their species composition and species richness, and communities' structures (Alberti 2005; Leung 2015). At the same time, human activities have homogenized the Earth's biota, introducing many species into new areas where they can disrupt both natural and human systems (Ceballos et al. 2015; Vitousek et al. 1997).

The loss of biodiversity is the most critical global environmental threat. More than a third of the global species are facing extinction and an estimated 60 % of the Earth's ecosystems have been degraded in the last 50 years, with consequences for the services that depend on them (Joppa et al. 2015; European Commission [EC] 2010). Marine biodiversity is also under pressure, a worrying fact given that approximately 90 % of the planet's biomass lives in the ocean (EC 2010).

## 3 2010 Biodiversity Target—The Convention on Biological Diversity

Since the 2nd half of the 20th century, concerns about the use of natural resources and the recognition of the importance of biological diversity, have grown. In May 1992, the Nairobi Conference resulted in the adoption of the Agreed Text of the CBD (www.cbd.int/), which was open for signature at the Rio "Earth Summit" (United Nations Conference on Environment and Development) in the same year.

The CBD text's (1992) main objective (https://www.cbd.int/doc/legal/cbd-en.pdf) was to preserve biological diversity through its sustainable use. Biological diversity was then defined as "the variability among living organisms from all sources including, inter alia, terrestrial, marine and other aquatic ecosystems and the ecological complexes of which they are part: this includes diversity within species, between species and of ecosystems". This includes all populations, organisms, and genetic resources or "any other biotic component of ecosystems with actual or potential use or value for humanity".

Since then, several periodic meetings—The Conference of the Parties [COP]—have occurred as to monitor and discuss advances and new strategies about the CBD commitments. In COP 6 (Decision VI/26, April 2002) a new challenge for scientists, policy-makers, and society was undertaken (www.cbd.int/doc/decisions/cop-06/full/cop-06-dec-en.pdf). All the parties, at the time, should comply with the Convention objectives and work towards achieving "by 2010 a significant reduction of the current rate of biodiversity loss at the global, regional and national level as a contribution to poverty alleviation and to the benefit of all life on earth". This agreement became known as the **2010 Biodiversity Target** or the Strategic Plan for the Convention on Biological Diversity, a global target incorporated as a new target under the Millennium Development Goals.

To measure the progress towards the target of reducing biodiversity loss in Europe by 2010, a set of 26 environmental indicators were presented in 2007 (EEA 2007). The process was however, initiated in 2005 by the Streamlining European Biodiversity Indicators [SEBI] (http://biodiversity.europa.eu/topics/sebi-indicators). Their goal was to build on current monitoring available data, to avoid duplication of efforts and to complement—and not replace—other activities to describe, model, and understand biodiversity and the pressures upon it.

The 2010 Biodiversity Target provides the scientific community with the challenge to engage in stimulating fundamental science and to participate in what is likely to be the most significant conservation agreement of the early 21st century.

Despite some local successes and increasing responses (including the extent and biodiversity coverage of protected areas, sustainable forest management, policy responses to invasive alien species, and biodiversity-related aid), the rate of biodiversity loss did not appear to be slowing down (Butchart et al. 2010). Some concern regarding the design of the indicators and communication to different and relevant audiences was also demonstrated (Mace and Baillie 2007). Reports on the assessment of biodiversity in Europe (EEA 2010) and the Global Biodiversity Outlook 3 (http://gbo3.cbd.int), both launched in the 2010 International Year of Biodiversity, demonstrated that despite some progresses, there was a continuing decline in biodiversity at genes, species, and ecosystems levels, with the most severe threats affecting several types of ecosystems (e.g. freshwater, coastal and marine, forests, grassland, and urban ecosystems). It became clear that the original global target had not been met and the loss of biodiversity in Europe was still a fact!

## 4 Strategic Plan for Biodiversity 2011–2020

Failing to meet the 2010 targets forced the Conference of the Parties to rethink and update its strategy to halt the continuous loss of biodiversity. Thus, a new decision (X/2) at COP 10 (2010) was adopted and a revised global Strategic Plan for Biodiversity for the period 2011–2020 was approved to safeguard biodiversity and the benefits it provides to people (https://www.cbd.int/sp/), under the flag "Living in harmony with nature". This decision, among other outputs, adopted the 20 Aichi Biodiversity Targets (https://www.cbd.int/sp/targets/), which are grouped under 5 strategic goals: "(a) address the underlying causes of biodiversity loss by mainstreaming biodiversity across government and society; (b) reduce the direct pressures on biodiversity and promote sustainable use; (c) improve the status of biodiversity by safeguarding ecosystems, species and genetic diversity; (d) enhance the benefits to all from biodiversity and ecosystem services; and (e) enhance implementation through participatory planning, knowledge management and capacity building".

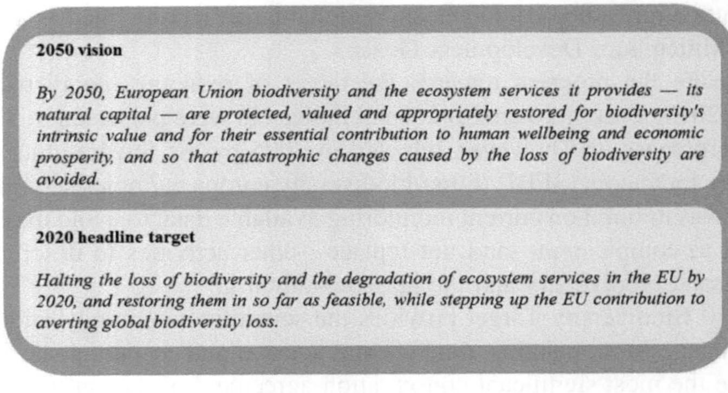


**Fig. 2** EU biodiversity strategy vision for 2050 and target for 2020 (EU 2011)

The Strategic Plan for Biodiversity 2011–2020 aims to provide a flexible broad-based framework for the establishment of national and regional targets by the parties and other stakeholders, for the implementation of the CBD requirements. The plan focused on the necessity of developing effective communication tools to engage all stakeholders in complying with the Aichi Targets, as well as streamlining the mainstreaming of biodiversity into broader national and global agendas.

An ambitious global vision is transmitted in the plan, where "Living in harmony with nature" will serve as a pillar to ensure that: "By 2050, biodiversity is valued, conserved, restored and wisely used, maintaining ecosystem services, sustaining a healthy planet and delivering benefits essential for all people".

The European Union [EU] also shared the same long-term 2050 vision and the 2020 headline targets proposed by the Strategic Plan for Biodiversity 2011–2020 and a new EU biodiversity strategy—"Our life insurance, our natural capital: an EU biodiversity strategy to 2020" was agreed by the EC in 2011 (COM/2011) (Fig. 2).

Recognizing the importance of the Strategic Plan for Biodiversity 2011–2020, as well as the Aichi Targets and the need to reduce the rate of global biodiversity loss, the United Nations [UN] Secretary-General, Mr. Ban Ki-moon declared the period 2011–2020 the "Decade on Biodiversity" (Resolution 65/161—https://www.cbd.int/2011-2020/), reinforcing the urgent need for the global implementation of the Plan.

## 5 Education for Biodiversity Conservation in the 21st Century: Beyond 2015

The 2014 UNESCO World Conference on Education for Sustainable Development marked the end of the United Nations [UN] Decade of ESD (2005–2014) and launched the Global Action Programme [GAP] on ESD. Under the banner of "Learning Today for a Sustainable Future", the Conference celebrated the achievements of the Decade, identified lessons learnt while setting the stage for the future of ESD. It also showcased initiatives, key players, networks, and ideas that the decade has stimulated. Such examples from all over the world will help to generate future action under the GAP. The outcomes of the World Conference will inform the deliberations of the World Education Forum to held from 19 to 22 May 2015 in Incheon, Republic of Korea. The GAP on ESD, as a follow-up to the UN Decade of ESD after 2014, in its principles stated, "ESD relates to the environmental, social and economic pillars of sustainable development in an integrated, balanced and holistic manner. It equally relates to a comprehensive sustainable development agenda as contained in the outcome document of Rio+20, which includes, among others, the interrelated issues of poverty reduction, climate change, disaster risk reduction, ***biodiversity***, and sustainable consumption and production. It responds to local specificities and respects cultural diversity".

It is also noteworthy to refer to the UN 2030 Agenda for Sustainable Development, launched in October 2015, entitled "Transforming our world: the 2030 Agenda for Sustainable Development" (Resolution 70/1, 2015) (https://sustainabledevelopment.un.org/post2015/transformingourworld), a plan of action for people, planet, and prosperity. The plan identifies as extremely important, the quality of an inclusive and equitable education to promote lifelong learning opportunities for all, as the way to achieve the sustainable development goals set for 2030. The vision expressed in this agenda embraces a world where equitable and universal access to quality education at all levels is mandatory.

## 6 Education for Biodiversity: Tendencies and Perspectives

In 2015, Europe stands roughly halfway between the initiation of the European environmental policy in the early 1970s and the General Union Environment Action Programme to (7th Environment Action Programme) "Living well within the limits of the planet" (Decision N.º 1386/2013/EU, 2013). Access to ecosystem services will become an even more critical factor for economic success and resilience in the 21st century.

A survey carried out by the Gallup Organization (2010) entitled "Attitudes of Europeans towards the issue of biodiversity" showed that two-thirds of European citizens were familiar with the term "biodiversity". Moreover, 38 % of interviewees said they knew the meaning of the term and 28 % stated they had heard of the term but did not know its meaning. Approximately one-third (34 %) of respondents claimed they had never heard of the term biodiversity, let alone understand what the threats and challenges to its conservation are. Respondents with the lowest level of education, manual workers and non-working respondents, were the most likely to claim that they had never heard of the term. This study also demonstrated that citizens were unaware of what Europe was doing to save biodiversity. However, they were conscious of environmental matters. When the issue was explained to them, over two-thirds considered the loss of biodiversity a serious problem,

**Table 1** Key findings and trends of the decade of education for sustainable development (DESD) (UNESCO 2014)

| Main subject | Findings/trends |
|---|---|
| ESD, an enabler for sustainable development | Education systems are addressing sustainability issues |
| | Sustainable development agendas and education agendas are converging |
| Importance of stakeholder engagement for ESD | Political leadership has proven instrumental |
| | Multi-stakeholder partnerships are particularly effective |
| | Local commitments are growing |
| ESD is galvanizing pedagogical innovation | Whole-institution approaches practise ESD |
| | ESD facilitates interactive, learner-driven pedagogies |
| ESD has spread across all levels and areas of education | ESD is being integrated into formal education |
| | Non-formal and informal ESD is increasing |
| | Technical and vocational education and training advances sustainable development |

principally at global level. These results clearly indicate that the level of understanding of the problem is still inadequate.

Despite these worrying numbers, the final report on the UN Decade of Education for Sustainable Development (2005–2014), showed that during this time, some interesting findings and trends could be observed regarding the integration of all features of education and learning with principles and practices of sustainable development (UNESCO 2014) (Table 1).

Despite these advances in education for sustainable development, societies still face several important challenges. According to UNESCO (2014), additional alignment between education and sustainable development sectors is needed, more efforts are required to improve the institutionalization of ESD, and tools to monitor and evaluate programmes and action should also be perfected. Bennett et al. (2015) argued that in order to achieve a better management of natural resources, it is essential to improve communication with different stakeholders (e.g. private actors) and for inspiring advances in policy.

Thus it is essential to identify and clarify trends and assess the effectiveness of current and future policies aimed at raising public awareness and participation with regards to biodiversity, as far as communication to the public is concerned. In addition, biodiversity research and intervention should integrate new perspectives, bringing together biological sciences, social sciences and local knowledge (Alves et al. 2013).

## References

Aktas, B., Whelan, R., Stoffer, H., Todd, E., & Kern, C. L. (2015). Developing a university-wide course on sustainability: A critical evaluation of planning and implementation. *Journal of Cleaner Production, 106*, 216–221.

Alberti, M. (2005). The effects of urban patterns on ecosystem function. *International Regional Science Review, 28*(2), 168–192.

Alves, F., Filho, W. L., Araújo, M. J., & Azeiteiro, U. M. (2013). Crossing borders and linking plural knowledge: Biodiversity conservation, ecosystem services and human well–being. *International Journal of Innovation and Sustainable Development, 7*(2), 111–125.

Amador, F., Martinho, A. P., Bacelar-Nicolau, P., Caeiro, S., & Oliveira, C. P. (2015). Education for sustainable development in higher education: Evaluating coherence between theory and praxis. *Assessment and Evaluation in Higher Education, 40*(6), 867–882.

Barth, M., & Rieckmann, M. (2012). Academic staff development as a catalyst for curriculum change towards education for sustainable development: An output perspective. *Journal of Cleaner Production, 26*, 28–36.

Bennett, E. M., Cramer, W., Begossi, A., Cundill, G., Díaz, S., Egoh, B. N., et al. (2015). Linking biodiversity, ecosystem services, and human well-being: Three challenges for designing research for sustainability. *Current Opinion in Environmental Sustainability, 14*, 76–85.

Butchart, S. H., Walpole, M., Collen, B., Van Strien, A., Scharlemann, J. P., Almond, R. E., et al. (2010). Global biodiversity: Indicators of recent declines. *Science, 328*(5982), 1164–1168.

Ceballos, G., Ehrlich, P. R., Barnosky, A. D., García, A., Pringle, R. M., & Palmer, T. M. (2015). Accelerated modern human–induced species losses: Entering the sixth mass extinction. *Science Advances, 1*, e1400253.

de Groot, R. S., Wilson, M. A., & Boumans, R. M. J. (2002). A typology for the classification, description and valuation of ecosystem functions, goods and services. *Ecological Economics, 41*(3), 393–408.

De Vos, J. M., Joppa, L. N., Gittleman, J. L., Stephens, P. R., & Pimm, S. L. (2015). Estimating the normal background rate of species extinction. *Conservation Biology, 29*(2), 452–462.

Disterheft, A., Caeiro, S., Azeiteiro, U. M., & Leal Filho, W. (2013). Sustainability science and education for sustainable development in universities: a way for transition. In S. Caeiro, W. Leal Filho, C. Jabbour & U. M. Azeiteiro (Eds.), *Sustainability Assessment Tools in Higher Education Institutions* (pp. 3–27). Springer International Publishing.

Dlouha, J., & Burandt, S. (2015). Design and evaluation of learning processes in an international sustainability oriented study programme. In search of a new educational quality and assessment method. *Journal of Cleaner Production, 106*, 247–258.

European Commission. (2010). *Options for an EU vision and target for biodiversity beyond 2010 (COM (2010) 4 final)*. Brussels: European Commission.

European Commission. (2011). *Our life insurance, our natural capital: An EU biodiversity strategy to 2020 (COM (2011) 244 final)*. Brussels: European Commission.

European Environment Agency. (2007). Halting the loss of biodiversity by 2010: Proposal for a first set of indicators to monitor progress in Europe. (Techical report N.° 11/2007). Copenhagen: European Environment Agency.

European Environment Agency. (2010). Assessing biodiversity in Europe—the 2010 report. (Report N.° 5/2010) Copenhagen: European Environment Agency.

European Environment Agency. (2011). Landscape fragmentation in Europe. (Report N.° 2/2011) Copenhagen: European Environment Agency.

European Environment Agency. (2015). *The European environment—state and outlook 2015: Synthesis report*. Copenhagen: European Environment Agency.

Gallup Organisation. (2010). Attitudes of Europeans towards the issue of biodiversity. (Analytical report, wave 2). Brussels: European Commission.

Haines-Young, R., & Potschin, M. (2010). The links between biodiversity, ecosystem services and human well-being. In D. G. Raffaelli & C. L. J. Frid (Eds.), *Ecosystem ecology. A new synthesis* (pp. 110–139). Cambridge: Cambridge University Press.

Leung, J. Y. S. (2015). Habitat heterogeneity affects ecological functions of macrobenthic communities in a mangrove: Implication for the impact of restoration and afforestation. *Global Ecology and Conservation, 4*, 423–433.

Lozano, F., & Lozano, R. (2014). Developing the curriculum for a new Bachelor's degree in Engineering for sustainable development. *Journal of Cleaner Production, 64*, 136–146.

Lozano, R., Ceulemans, K., & Seatter, C. S. (2015a). Teaching organizational change management for sustainability: Designing and delivering a course at the University of Leeds to better prepare future sustainability change agents. *Journal of Cleaner Production, 106*, 205–215.

Lozano, R., Ceulemans, K., Alonso-Almeida, M., Huisingh, D., Lozano, F. J., Waas, T., et al. (2015b). A review of commitment and implementation of sustainable development in higher education: Results from a worldwide survey. *Journal of Cleaner Production, 108*, 1–18.

Mace, G. M., & Baillie, J. E. (2007). The 2010 biodiversity indicators: Challenges for science and policy. *Conservation Biology, 21*(6), 1406–1413.

Millennium Ecosystem Assessment. (2003). *Ecosystems and human well-being: A framework for assessment*. Washington, DC: World Resources Institute.

Millennium Ecosystem Assessment. (2005). *Ecosystems and human well-being: Biodiversity synthesis*. Washington, DC: World Resources Institute.

Müller-Christ, G., Sterling, S., van Dam-Mieras, R., Adomßent, M., Fischer, D., & Rieckmann, M. (2014). The role of campus, curriculum, and community in higher education for sustainable development–a conference report. *Journal of Cleaner Production, 62*, 134–137.

Pittock, J., Cork, S., & Maynard, S. (2012). The state of the application of ecosystems services in Australia. *Ecosystem Services, 1*(1), 111–120.

United Nations Educational, Scientific and Cultural Organization. (2012). *Shaping the education of tomorrow*. 2012 Report on the UN Decade of Education for Sustainable Development. Paris: UNESCO. http://unesdoc.unesco.org/images/0021/002166/216606e.pdf

United Nations Educational, Scientific and Cultural Organization. (2014). *Shaping the future we want*. UN Decade of Education for Sustainable Development (2005–2014). Final Report. Paris: UNESCO. http://unesdoc.unesco.org/images/0023/002301/230171e.pdf

Vitousek, P. M., Mooney, H. A., Lubchenco, J., & Melillo, J. M. (1997). Human domination of Earth's ecosystems. *Science, 277*(5325), 494–499.

# Part I
# Education Research on Biodiversity

# Conservation of Biological Resources: Why Does It Matter?

Paula Castro, Anabela Marisa Azul and Jorge Paiva

**Abstract**

The world population has grown from 2500 million people in the year 1950 to more than 7300 million people in the year 2015, posing a challenge never faced before in human history. People are less aware about the limitedness of natural resources and the consequences of the present development. Increased population, associated with technological advancement undermines the sustainable development of any nation. Daily, many species are going extinct due to the continuous fragmentation/destruction of habitats, many of which have not been studied or referenced. Presently, we are living in the «Decade of Biodiversity» from the period of year 2011 to the year 2020, which was launched at the end of the year 2011 by the General Secretary of the United Nations, Ban Ki-moon. This poses a great challenge and all societies and nations are saddled with the responsibility of revising their actual models of economic development and increasing their knowledge base, by planning more intelligent and integrative programmes for the conservation of our biological resources and its functions in the ecosystems and human health. This chapter aims to raise awareness on the relevance of biodiversity in people's life. It emphasizes subjects, such as the importance of forests, the unknown biodiversity, and the extinction of species, in order to alert the general public, students, teachers, and other stakeholders to the importance of all biological resources.

P. Castro (✉) · J. Paiva
Department of Life Sciences, Centre for Functional Ecology,
University of Coimbra, Coimbra, Portugal
e-mail: pcastro@ci.uc.pt

A.M. Azul
Center for Neuroscience and Cell Biology, University of Coimbra,
Coimbra, Portugal

 
P. Castro et al. (eds.), *Biodiversity and Education for Sustainable Development*,
World Sustainability Series, DOI 10.1007/978-3-319-32318-3_2

# 1 Introduction

A comprehensive knowledge of the natural systems and of the living beings that inhabit them, together with the relationship they establish between themselves and the environment, is far from being assembled and understood. Despite new technologies and study methodologies, such as the application of Remote Sensing or the use of Geographic Information Systems (GIS) which allowed us to study areas that were inaccessible and observe organisms at scales that were never considered. However, the huge biodiversity of the Earth's ecosystems is yet to reveal all its secrets!

Among the numerous living organisms that inhabit our Planet, *Homo sapiens*, is certainly the species that cause more and serious environmental impacts. The tools we have at our disposal have shown impressive signs of the habitat destruction, particularly, the forest system (Food and Agriculture Organization [FAO] 2015). Given the undeniable facts of environmental destruction (FAO 2015; Vitousek et al. 1997), the traditional human perception of nature's equilibrium is severely affected. In this present phase in which we live, the unbridled consumption of resources (European Environment Agency [EEA] 2015; Giljum et al. 2009; Kovanda and Hak 2011) is evident and brings incalculable costs.

The extreme and rapid expansion of human population (Fig. 1) and its model of economic development that encourages the unbridled consumption of goods and services demean the services provided by the different ecosystems and its resources (EEA 2015). In the mid-2015, the total population reached an incredible number of 7349 million (United Nations [UN] 2015) (Fig. 1a), and it is expected to grow above 50 % in the year 2100 (Fig. 1b).

Each living human, needs basic resources and almost all people seek to utilize significantly more and more resources. These expected demands multiplied by a factor of 7.3 billion (and growing rapidly) compromise the stability of the planet's system. Thus, the people do not only realize how we pollute the "cage" (Earth) in

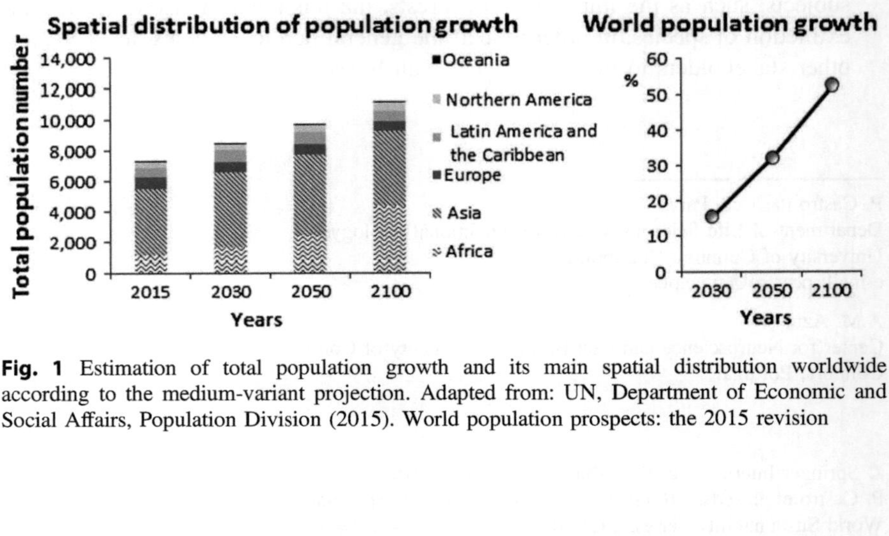


**Fig. 1** Estimation of total population growth and its main spatial distribution worldwide according to the medium-variant projection. Adapted from: UN, Department of Economic and Social Affairs, Population Division (2015). World population prospects: the 2015 revision

which we live, but also destroy the nature, daily eliminating species, many of which, have not been studied.

The present chapter aims to contribute to the knowledge and understanding on the relevance of biodiversity in every one's life, and clarify the reasons to take action towards the conservation of the biological resources. It underlines on issues, such as the role of forest ecosystems, alert to the unknown biodiversity, and to the extinction of species.

## 2 Reasons for Taking Action

The laws of physics (mass conservation, conservation of energy, and entropy laws) are fundamental to the understanding and conservation of ecosystems: no system can create or eliminate matter, it can only be transformed; energy can neither be created nor destroyed (the energy conversion process involves the loss of quality); and everyone constantly needs energy to maintain its low entropy (Fig. 2).

These messages are quite clear: the earth is finite and its ability to absorb wastes and polluted effluents is also finite. The capacity to provide all the resources for human survival is thus finite, and current economic practices which damage the environment, in both developed and underdeveloped nations, cannot be continued.

All human activities make use of the services of the ecosystems and put pressure as well, on the biodiversity that supports these services: (a) Habitat loss, alteration, and fragmentation—for instance, through the land use change for aquaculture, industrial or urban use; the construction of dams and other changes in river systems for irrigation, hydropower or adjustment of current and harmful fishing activities; (b) Overexploitation of populations of wild species—for instance, the harvesting or killing of animals or plants for food, materials or medicine at unsupported natural rates of its reproduction; (c) Pollution—for instance, the excessive use of pesticides in agriculture and aquaculture; urban and industrial effluents and waste mining; (d) Climate change—for instance, the rising levels of greenhouse gases in the

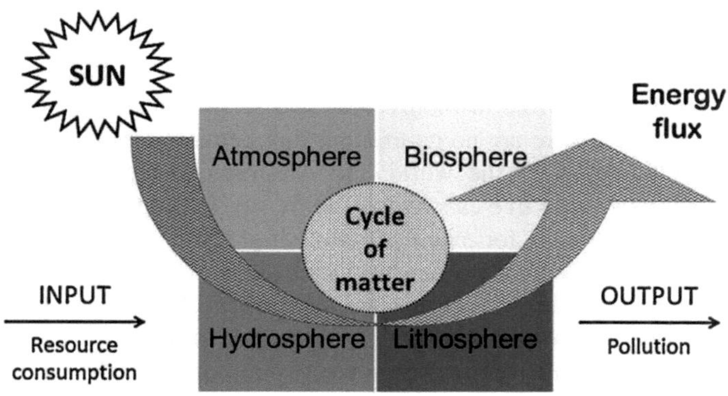


**Fig. 2** Schematic representation of energy and matter fluxes in our planet

atmosphere, mainly caused by burning of fossil fuels, deforestation and industrial processes; and (e) Invasive species—for instance, the introduction (on purpose or inadvertently) of exotic species that become very competitive, parasites or predators of native species (Alberti 2015; Davidson et al. 2014; Doney et al. 2012; EEA 2015).

Very few people realize that, presently, we are living in the "Decade of Biodiversity" launched by the UN in the year 2011. This decade is of extreme importance, so that people may realize that we cannot survive without biodiversity. Taking actions to inform and raise more awareness about the problems caused by the continuous destruction of habitats, loss of biodiversity and the link between biodiversity, ecosystem's services and human well-being is therefore, urgent. The extinction of species cannot be reverted, but it is possible to avoid future extinctions of other species if appropriate programmes are implemented, mainly for those who are at high risk of extinction.

## 3 Forests' Ecosystems and Species Richness

The knowledge of spatial variation in species richness and the diversity along environmental gradients is a central theme in ecology, as they harbour a large part of the terrestrial biodiversity and provide a wide range of ecosystem services and economic growth (EEA 2010). We know that the forests, particularly, the equatorial forest (*pluvisilva*), due to a higher plant biomass production, are unique systems, embracing high levels of biodiversity (Kier et al. 2005; Kraft et al. 2011).

Among plant species, there are huge differences in the amount of biomass produced and on the volume of $CO_2$ consumed. Among the dominating tree's ecosystems, moist tropical forests (*pluvisilva*) are hot spots for plant richness (Barthlott et al. 2007; Gaston 2000), because, by virtue of being on the equatorial zones, they have a constant energy and water source available. Another recent example of these equatorial zones is in the work of Collen et al. (2014), which showed that absolute freshwater diversity was highest in the Amazon Basin.

A study conducted by Kier et al. (2005) proved that tropical and subtropical moist broadleaf forests were the ecosystems with higher plant richness (3161 taxa), followed by Mediterranean forests with 2294 taxa. These two types of forest systems play an important role in biodiversity and in the survival of other species. Despite the fact that there are no co-existing link between tree canopy height and diversity in all regions of the world, the Afrotropic realm richness shows an increase in diversity with tree canopy height, mostly for amphibians, a very slight increase in birds and none for mammals (Roll et al. 2015).

It is also known, that living organisms (biodiversity) constitute our food source, provide us with medicinal substances, clothing (practically, everything we wear is of animal or vegetable origin), energy (for example, firewood, petroleum, waxes, resins), construction materials, and furniture (wood), among other goods. A large part of the electrical energy consumption would not be possible without the contribution of other living beings. Other evidence is that *Homo sapiens* appeared in ecosystems which support most of the earth's biodiversity—the African tropical forests. In addition, it is constantly discovered new uses of plants, animals, and other organisms.

Despite this knowledge, the cutting down of forests continues and land is drastically reclaimed for other uses (Barthlott et al. 2007; Chen et al. 2013). It can be observed from the report on forest global assessment (2015) by FAO of the UN that the loss of forest systems is still happening. In the year 1990, the global forest area was estimated at 4,128,269 ha and in the year 2015, 3,999,134 ha was the total area measured by FAO (FAO 2015). For example, in Brazil, which is among the five countries with the largest area of forest, recent losses (from the year 2010 to 2015) reached 984,000 ha. Subtropical and tropical forests are the most affected type of forest, when compared with boreal and temperate systems (FAO 2015). In the Mediterranean region, the forested area is estimated to be 85 Mha (2 % of the world's forest area), however, this is unevenly distributed between countries. Of this, approximately 1.67 million ha is a primary forest (FAO 2010).

Despite these worrying values, forest management is quite different when comparing between Nordic countries, with an annual net forest gained since the year 1990, with those countries from the South Pole (FAO 2015). Apparently, Mediterranean forests witnessed a spatial expansion of about 1 %, but in contrast, native forest and biodiversity is declining (FAO 2012). Many other examples of native forest/biodiversity loss worldwide may be assessed in current literature (Abood et al. 2015; Baltzer et al. 2014; Butchart et al. 2010; Ferreira et al. 2015; Miranda et al. 2015).

## 3.1 The Easter Island as an Example of Human Unsustainable Practices

The Easter Island situated in the Pacific Ocean (Eastern Polynesia) was a subtropical forest covered by palm trees, before the arrival of Polynesian people, approximately, in the 4th century (Kirch and Ellison 1994). Other evidences suggested a later arrival (Hunt and Lipo 2006). This forest was completely devastated by the Rapa Nui, and together with the concomitant erosion of primeval soils, practically caused its extinction.

If we continue to destroy forests at this rate, it is estimated that before the end of this century, the planet, will virtually have no forests. It will be transformed into an "island" without forests, as what happened in Easter Island.

John Dransfield discovered that the most abundant palm tree that existed in the Easter Island was extremely similar (or perhaps the same species) to the palm tree of Chile (*Jubaea chilensis*), that once had a vast spatial distribution and currently, only occurs in a strictly central area of Chile (between 32 and 35°S) (González 1998). Its fruit is highly appreciated for the purpose of eating and for extracting oil. From its elaborated sap, they produce an alcoholic drink, which is also very valued by the Chilean people. The removal of the sap produced by incision is made on top of the stipe, which causes it to stop producing viable fruit and, most of the time, the tree dies. According to the International Union for Conservation of Nature [IUCN], it is classified as a vulnerable species (González 1998).

Despite already extinct in the subtropical forest of this island this native palm tree was classified as a new species to science: *Paschalococos disperta* (Zizka 1991) which probably became extinct due to the overexploitation of these palm populations. Not only was the respective fruit edible, it was used for many other purposes as well (they eat the heart of palm, used the wood for boats, probably produced an alcohol drink, and used the leaves to cover their houses). These and other unsustainable uses of the species and the services provided to the population, almost led to the extinction of the local population.

This example is a model of human-induced environmental degradation and illustrates very well what may happen to our planet if we continue to foster deforestation practices and devalue forest biodiversity and its services. Forests are the largest producers of biomass, with an extraordinary capacity to depollute (through the amount of carbon dioxide ($CO_2$) consumed) and behave as enormous natural factories of oxygen ($O_2$). Continuing with the present models of development, human population will not survive and the land will be a universal "island", deforested and uninhabited.

## 4 The Relevance of Biological Resources: From Basic Needs to Economic Development

Everyone knows that he/she needs to eat in order to live and grow, and that the food consists of biological materials (plants, animals, and other organisms). It is also known that, for any engine to work, it needs a fuel that, through exothermic chemical reactions (combustion) releases enough heat (energy) for the engine to operate. The fuels (for example, gasoline, diesel, alcohol, gas) are organic compounds with carbon (C), hydrogen ($H_2$) and $O_2$. When a chemical reaction occurs, $CO_2$ is expelled into the atmosphere.

Making an analogy with this example, we may look at our body as a group of several "engines". If the heart, lungs, brain, for example, stops working, the body as a whole also stops. These biological engines also need "fuel" to work. This fuel (food/nutrients) comes from plant products, livestock, and other living sources (yeasts, for example) which are then transformed into energy (heat), through exothermic reactions (digestion), similar to the combustion referred above. Food is the source of the combustible substances, C, $H_2$, $O_2$, and other elements crucial for our survival [for example Nitrogen (N)].

All living beings need nutrients to survive (consumers). The plants (producers), however, are able to synthesize their own food by taking sunlight, to generate endothermic chemical reactions (photosynthesis) with the help of $CO_2$ and water, present in the atmosphere. The plants therefore, produce biomass. Humans, like any other animal, need to consume plants and other consumers in order to produce their own energy.

In addition to these basic services, other services provided by forests and biological resources are well known, clearly described in the Millennium Ecosystem Assessment [MEA] report (2003) and other work studies (Barbeta et al. 2015; Baró et al. 2014; Thompson et al. 2014). Despite the direct link, biodiversity and ecosystem services need to be better studied and understood (Balvanera et al. 2014).

Forests may be used to effectively generate other services, such as income, and employment. They are important systems for socioeconomic and political development. These systems play important roles in the society such as providing land for agriculture; timber and non-timber products, environmental services (for example, to regulate local, regional and global climate, store carbon, and purify air and fresh water), and employment (contributing to poverty alleviation) (Azul et al. 2009, 2014; Sunderlin et al. 2005; Verkerk et al. 2015). Indirect services may also be of benefit to the various stakeholders who depend on these systems (Azul et al. 2014; Duarte et al. 2013).

A new paradigm in forest exploitation is growing and, in addition to having more knowledge about the ecological functions (Barbeta et al. 2015), investment in exploring endogenous resources (non-timber products) and research on bioactive compounds are examples of new key ways that contribute to local, regional, and national socioeconomic activities (Azul et al. 2014). Integrating people in intelligent research and management of native resources, forests, and biodiversity are, thus, essential in forest exploitation and conservation of biological resources. The interest in buying green, natural, and native products is increasing. Environmental concern is pulling the investment in biotechnology and bio-industry as an emerging economy that may reverse the trends in the loss of forests and biodiversity as well as ensure the ecosystem's resilience (Azul et al. 2014; Pizarro-Tobías et al. 2015; Kingston 2010).

## 5 Biodiversity Unknown

From all our heritages (material, cultural, and biological), the only one essential to our survival, is the biological heritage (biodiversity), which has received less attention. In addition, the majority of our biological diversity is not yet known. From the almost 4 million species listed, including oceans, a large part is not sufficiently studied. At the end of the last century, the American biologist, Erwin (1982) after several studies in tropical rain forests (*pluvisilva*) of Central and South America, has calculated it could be as many as 30 million Arthropod species worldwide, and not 1.5 million as estimated at the time. It may be assumed that not even 10 % of the global biological diversity is known. The kingdom Fungi represent another example in which estimations increased from 0.5 to 10 million (Bass and Richards 2011; Blackwell 2011) over the last two decades. Every year, several new species are descried all over the world (see some recent examples on Table 1).

**Table 1** Examples of macro-species discovered in the 20th and 21st centuries

| Organism | Species | Year of description | Common name | Local | Reference |
|---|---|---|---|---|---|
| Animals | *Muntiacus vuquangensis* | 1994 | Giant Muntjac | Vietnam | Tuoc et al. (1994) |
| | *Muntiacus puhoatensis* | 1997 | Puhoat Muntjac | Vietnam | Chau (1997) |
| | *Muntiacus truongsonensis* | 1998 | Truong Son Muntjac | Vietnam | Giao et al. (1998) |
| | *Muntiacus putaoensis* | 1999 | Leaf Muntjac | Myanmar | Amato et al. (1999) |
| | *Callicebus bernhardi* | 2002 | Prince Bernhard's Titi Monkey | Brazil | Van Roosmalen et al. (2002) |
| | *Callicebus stephennashi* | 2002 | Stephen Nash's Titi Monkey | Brazil | |
| | *Lophocebus kipunji* | 2005 | Kipunji | Tanzania | Jones et al. (2005) |
| | *Diopatra micrura* | 2010 | – | Portugal | Pires et al. (2010) |
| | *Rhinopithecus strykeri* | 2011 | Myanmar Snub-nosed Monkey | Myanmar | Geissmann et al. (2011) |
| | *Squamatinia algharbica* | 2012 | – | Portugal | Reboleira et al. (2012) |
| | *Nactus kunan* | 2012 | – | Papua New Guinea | Zug and Fisher (2012) |
| | *Crocidura fingui* | 2015 | Shrew-Fingui | Island of Principe | Ceríaco et al. (2015) |
| Plants | *Wollemia nobilis* | 1995 | Wollemi Pine | Australia | Jones et al. (1995) |
| | *Labramia mayottensis* | 1997 | – | Comoro Islands | Labat et al. (1997) |
| | *Arabis beirana* | 2001 | – | Portugal | Silveira et al. (2001) |
| | *Zygodon catarinoi* | 2006 | – | Portugal | Garcia et al. (2006) |
| | *Narcissus* x *caramulensis* | 2007 | – | Portugal | Ribeiro et al. (2007) |
| | *Tahina spectabilis* | 2008 | Tahina Palm | Madagascar | Dransfield et al. (2008) |
| | *Dendroceros paivae* | 2012 | – | São Tomé e Príncipe Island | Garcia et al. (2012) |
| | *Stachys caroliniana* | 2014 | Hedge-nettle | USA | Nelson and Rayner (2014) |
| Fungi | *Psilocybe germanica* | 2015 | – | Germany | Gartz and Wiedemann (2015) |

(continued)

**Table 1** (continued)

| Organism | Species | Year of description | Common name | Local | Reference |
|---|---|---|---|---|---|
| | *Phallus drewesii* | 2015 | – | São Tomé Island | Desjardin and Perry (2015) |
| | *Inocybe praetervisoides* | 2015 | – | Mediterranean region | Esteve-Raventós et al. (2015) |
| | *Mutinus albotruncatus* | 2015 | – | Brazil | da Silva et al. (2015) |
| Macroalgae | *Fucus guiryi* | 2011 | Seaweed | Portugal | Zardi et al. (2011) |
| | *Phymatolithon lusitanicum* | 2015 | – | Portugal, Spain | Peña et al. (2015) |

Fungi, habitat soil, water, and organisms, are major drivers of ecosystems life cycles.

Another example of an extremely high biodiversity of insects of *pluvisilva* was the work conducted by Wilson (1987). This myrmecologist collected in one Fabaceae tree in the forest of Peru, 43 species of ants, which was approximately, equal to the ant diversity throughout the United Kingdom. Not to mention the enormous group of fungi (Blackwell 2011), and microscopic beings, constantly being discovered by science, as bacteria (Albuquerque et al. 2014) or archaea (Albuquerque et al. 2012), invisible to the naked eye.

Generally, when people think about unknown diversity they almost associate it to organisms of small dimension. Nonetheless, this is not always the case. There is probably more unknown micro biodiversity, but new macrofauna and macroflora are also constantly being discovered. For animals, it may be cited, for example, the discovery in the year 1994 in the *pluvisilva* of Laos, the species *Muntiacus vuquangensis*, the Giant Muntjac, an antelope larger than a goat (Table 1). Since then, 3 more *Muntiacus* species have been discovered (Table 1). In total, 1/3 of all the known Muntjacs (12), were discovered at the end of the 20th century.

Already in the 21st century, were described, in the year 2002, in the Brazilian Amazon, two new species of apes (*Callicebus bernhardi* and *Callicebus stephennashi*) (Table 1). In the year 2005, a new species was seen in Africa (*Lophocebus kipunji*) (Table 1), which was already in danger of extinction in the mountains of Southern Tanzania. More recently, a new species of a shrew mouse (*Crocidura fingui*) endemic, was observed in the Island of Principe. Many other examples are presented in Table 1.

New techniques are available for researchers, as the use of DNA barcodes (Kress et al. 2015), are revolutionizing the methods of identification and increasingly new species are discovered each year.

## 6 The Extinction of Species

Why should we be concerned about the loss of biodiversity? For the first time, one single species (*Homo sapiens*) may cause mass extinction, triggering its own demise and the first cause of the loss of biodiversity is habitat loss, which is due to human activities. The majority of people believe that the only species that are vital for us are those that we currently use (for example, for cooking) and that other species do not present any significant value.

One of the most obvious examples of this indifference is what is happening with the rhino. The 5 species of rhinoceros [2 African: the white rhino (*Ceratotherium simum*) and the black rhino (*Diceros bicornis*), and 3 Asian: the Indian rhinoceros (*Rhinoceros unicornis*), the rhino of Java (*Rhinoceros sondaicus*) and the rhino of Sumatra (*Dicerorhinus sumatrensis*)] are endangered, mainly because of (prohibited) hunt practices to remove their front "horns", which supposedly have medicinal attributes (cancer and sexual impotence). They are also used as adornment pieces, similarly to what happens to elephants, particularly in Africa (*Loxodonta africana*). Despite the strict prohibition of the hunting of the rhino (and other species), even in Natural Parks, created for conservation purposes, illegal/legal hunt creates new ways to bend the established rules of conservation. Who does not remember the killing of the Cecil lion, a major attraction of the Hwange National Park in Matabele land North, Zimbabwe? This shocking case broke out in the media and at least, served to draw the attention for this kind of practices that lead to the extinction of these type of animals.

This "folklore" of aphrodisiac attributes also occurs for other species. The coconut of Seychelles (*Lodoicea maldivica*), due to its anatomical form (Fig. 3), make people believe on its powerful aphrodisiac ability. Presently, it only exists in two islands and collecting the fruit is strictly prohibited. Another example of this stupid aphrodisiac panacea is the "Pau-de-Cabinda", family Rosaceae, *Prunus African* (*Pygeum africanum*), whose bark has chemical products (alkaloids) with some effect in the treatment of prostatic hyperplasia and contractile dysfunction. It is not, in fact, a good "aphrodisiac" and may cause death.

Living beings of greater volume (greater biomass)—plants—are also threatened by human practices. Some examples are the Californian (USA) sequoias (*Sequoia sempervirens*) (ca. 120 m height and 9 m in diameter), the Sierra Redwood (*Sequoiadendron giganteum*) (ca. 100 m height and 12 m in diameter and 2000 tonnes of biomass), and the American Poplar (*Populus tremuloides*) (ca. 6000 tonnes of biomass). The animal with the highest biomass is the blue-whale (*Balaenoptera musculus*). Larger animals of this species (35 m in length and 210 tonnes), were annihilated in the 20th century.

Plants are authentic factories of biomass and oxygen production and many authors consider the preservation of plant diversity as a prerequisite, not only for the maintenance of animals, but also for their evolution.

**Fig. 3** Image of a coconut of *Lodoicea maldivica*

There are still many examples which oblige us to act in order to preserve all species without distinction, because, as it has already been mentioned, not all species are sufficiently studied.

In the Plant Kingdom, a good example is the species *Taxus baccata*, a rare species believed not to have any usefulness. It was a relatively common tree in the Mediterranean forests, which grows at very slow rates, It is an extremely poisonous plant, because it produces a mixture of alkaloids (taxine), lethal for all animals, and is used by populations since remote times (wood, bows and arrows, ornamental gardens, churches and cemeteries, as abortive, killing many times the foetus and also the mother, and even for suicide purposes). However, in the year 1993 (Guenard et al. 1993), it was proved to be of inestimable value. From the American *Taxus brevifolia*, it was isolated the taxol. This compound is an inhibitor of mitosis, by increasing the polymerization of tubulin, with the consequent stabilization of microtubules which prevents nuclear and cellular divisions. Unfortunately, a centenary *Taxus* tree provides only 300 mg of taxol, being necessary the bark of 6 centenarian trees to produce enough taxol to treat one patient. Fortunately, in the month of February 1994, the semi-lab synthesis of the substance was announced (Holton et al. 1994). Therefore, if *Taxus* had been extinct, this substance would never have been found.

In the animal kingdom, we present the case of the lizard *Heloderma suspectum*, the Gila Monster, native of south west of the United States and north of Mexico, that pastors killed whenever they visualized an individual, because it killed their animals (causing hypoglycaemia). The saliva of the lizard contains a protein (exendin-4) which stimulates the pancreas to produce insulin. Since the year 2009, a medicinal product is authorised for diabetes type 2, that is, the synthetic version of exendin-4 (exenatide). Currently, it is prohibited to collect this species of lizard.

## 7 Conclusion

The well-being of mankind is directly connected to the way we treat the biological resources of our planet. Strategies to prevent deforestation, the extinction of species and habitats, pollution, and loss of biological diversity embody a major paradigm to societies, including the scientific community. Our consumerist society must take into consideration, the choice of more environmentally friendly goods, services, and economic activities. Together with public authorities, managers, scientists, land owners, and other stakeholders, new holistic management actions must be developed.

The conservation of our biological resources is crucial, due to the known services provided for human survival, but also encloses other services that we still cannot diagnose. New species are always being discovered and who knows what we may find. Without the biological heritage there is no food, medicinal drugs, energy, and other services. Therefore, we must assume the commitment to change our behaviour towards the sustainability of the ecosystems, because without biodiversity we will endanger the survival of our very own species.

## References

Abood, S. A., Lee, J. S. H., Burivalova, Z., Garcia-Ulloa, J., & Koh, L. P. (2015). Relative contributions of the logging, fiber, oil palm, and mining industries to forest loss in Indonesia. *Conservation Letters, 8*(1), 58–67.

Alberti, M. (2015). Eco-evolutionary dynamics in an urbanizing planet. *Trends in Ecology & Evolution, 30*(2), 114–126.

Albuquerque, L., Johnson, M. M., Schumann, P., Rainey, F. A., & da Costa, M. S. (2014). Description of two new thermophilic species of the genus Rubrobacter, Rubrobacter calidifluminis sp. nov. and Rubrobacter naiadicus sp. nov., and emended description of the genus Rubrobacter and the species Rubrobacter bracarensis. *Systematic and Applied Microbiology, 37*(4), 235–243.

Albuquerque, L., Taborda, M., La Cono, V., Yakimov, M., & da Costa, M. S. (2012). Natrinema salaciae sp. nov., a halophilic archaeon isolated from the deep, hypersaline anoxic Lake Medee in the Eastern Mediterranean Sea. *Systematic and Applied Microbiology, 35*(6), 368–373.

Amato, G., Egan, M. G., & Rabinowitz, A. (1999). A new species of muntjac, *Muntiacus putaoensis* (Artiodactyla: Cervidae) from northern Myanmar. *Animal Conservation, 2*(01), 1–7.

Azul, A. M., Castro, P., Sousa, J. P., & Freitas, H. (2009). Diversity and fruiting patterns of ectomycorrhizal and saprobic fungi as indicators of land-use severity in managed woodlands dominated by *Quercus suber*-a case study from southern Portugal. *Canadian Journal of Forest Research, 39*(12), 2404–2417.

Azul, A. M., Nunes, J., Ferreira, I., Coelho, A. S., Veríssimo, P., Trovão, J., et al. (2014). Valuing native ectomycorrhizal fungi as a Mediterranean forestry component for sustainable and innovative solutions 1. *Botany, 92*(2), 161–171.

Baltzer, J. L., Veness, T., Chasmer, L. E., Sniderhan, A. E., & Quinton, W. L. (2014). Forests on thawing permafrost: fragmentation, edge effects, and net forest loss. *Global Change Biology, 20*(3), 824–834.

Balvanera, P., Siddique, I., Dee, L., Paquette, A., Isbell, F., Gonzalez, A., et al. (2014). Linking biodiversity and ecosystem services: current uncertainties and the necessary next steps. *BioScience, 64*(1), 49–57.

Barbeta, A., Mejía-Chang, M., Ogaya, R., Voltas, J., Dawson, T. E., & Peñuelas, J. (2015). The combined effects of a long-term experimental drought and an extreme drought on the use of plant-water sources in a Mediterranean forest. *Global Change Biology, 21*(3), 1213–1225.

Baró, F., Chaparro, L., Gómez-Baggethun, E., Langemeyer, J., Nowak, D. J., & Terradas, J. (2014). Contribution of ecosystem services to air quality and climate change mitigation policies: The case of urban forests in Barcelona, Spain. *Ambio, 43*(4), 466–479.

Barthlott, W., Hostert, A., Kier, G., Küper, W., Kreft, H., Mutke, J., et al. (2007). Geographic patterns of vascular plant diversity at continental to global scales (Geographische Muster der Gefäßpflanzenvielfalt im kontinentalen und globalen Maßstab). *Erdkunde, 61*(4), 305–315.

Bass, D., & Richards, T. A. (2011). Three reasons to re-evaluate fungal diversity "on earth and in the ocean". *Fungal Biology Reviews, 25*, 159–164.

Blackwell, M. (2011). The fungi: 1, 2, 3, … 5.1 million species? *American Journal of Botany, 98*(3), 426–438.

Butchart, S. H., Walpole, M., Collen, B., Van Strien, A., Scharlemann, J. P., Almond, R. E., et al. (2010). Global biodiversity: Indicators of recent declines. *Science, 328*(5982), 1164–1168.

Ceríaco, L. M., Marques, M. P., Jacquet, F., Nicolas, V., Colyn, M., Denys, C., et al. (2015). Description of a new endemic species of shrew (Mammalia, Soricomorpha) from Príncipe Island (Gulf of Guinea). *Mammalia, 79*(3), 325–341.

Chau, B. (1997). Another new discovery in Vietnam. *Vietnam Economic News, 47*, 46–47.

Chen, Y., Li, X., Liu, X., & Ai, B. (2013). Analyzing land-cover change and corresponding impacts on carbon budget in a fast developing sub-tropical region by integrating MODIS and Landsat TM/ETM + images. *Applied Geography, 45*, 10–21.

Collen, B., Whitton, F., Dyer, E. E., Baillie, J. E., Cumberlidge, N., Darwall, W. R., et al. (2014). Global patterns of freshwater species diversity, threat and endemism. *Global Ecology and Biogeography, 23*(1), 40–51.

da Silva, B. D. B., Sousa, C. T., Martín, M. P., Marinho, P., Calonge, F. D., & Baseia, I. G. (2015). Mutinus albotruncatus (Phallales, Agaricomycetes), a new phalloid from the Brazilian semiarid, and a key to the world species. *Phytotaxa, 236*(3), 237.

Davidson, D. J., Andrews, J., & Pauly, D. (2014). The effort factor: Evaluating the increasing marginal impact of resource extraction over time. *Global Environmental Change, 25*, 63–68.

Desjardin, D. E., & Perry, B. A. (2015). A new species of Phallus from São Tomé, Africa. *Mycologia, 101*(4), 545–547.

Doney, S. C., Ruckelshaus, M., Duffy, J. E., Barry, J. P., Chan, F., English, C. A., et al. (2012). Climate change impacts on marine ecosystems. *Marine Science, 4*.

Dransfield, J., Rakotoarinivo, M., Baker, W. J., Bayton, R. P., Fisher, J. B., Horn, J. W., et al. (2008). A new Coryphoid palm genus from Madagascar. *Botanical Journal of the Linnean Society, 156*(1), 79–91.

Duarte, G. V., Moura, A. I., Moreira, R., Nunes, J., Figueiredo, M. M., & Carvalho, M. G. (2013). Evaluation of several forest residues as potential raw material for bioethanol production in Portugal. *Journal of Bioprocess Engineering and Biorefinery, 2*(1), 73–78.

Erwin, T. L. (1982). Tropical forests: their richness in Coleoptera and other arthropod species. *Coleopterists Bulletin, 36*(1), 74–75.

Esteve-Raventós, F., Horcajada, G. M, Alvarado, P., & Ibarguren, I. O. (2015). Unraveling the *Inocybe praetervisa* group through type studies and ITS data: *I. praetervisoides* sp. nov. from the Mediterranean region. *Mycologia.*

European Environment Agency. (2010). *Assessing biodiversity in Europe (5/2010 report).* Copenhagen: European Environment Agency.

European Environment Agency. (2015). *The European environment—State and outlook 2015: Synthesis report.* Copenhagen: European Environment Agency.

Ferreira, M. P., Alves, D. S., & Shimabukuro, Y. E. (2015). Forest dynamics and land-use transitions in the Brazilian Atlantic Forest: The case of sugarcane expansion. *Regional Environmental Change, 15*(2), 365–377.

Food and Agricultural Organization. (2010). *Global forest resources assessment, 2010. Main report.* FAO Forestry Paper 163. Rome, Italy: UN Food and Agriculture Organization.

Food and Agricultural Organization. (2012). *The state of the world's forests 2012.* Rome, Italy: UN Food and Agriculture Organization.

Food and Agriculture Organization. (2015). *Global forest resources assessment 2015.* Rome, Italy: UN Food and Agriculture Organization.

Garcia, C., Lara, F., Sérgio, C., Sim-Sim, M., Garilleti, R., & Mazimpaka, V. (2006). *Zygodon catarinoi* (Orthotrichaceae, Bryopsida), a new epiphytic species from the Western Mediterranean Basin. *Nova Hedwigia, 82*(1–2), 247–256.

Garcia, C., Sérgio, C., Villarreal, J. C., Sim-Sim, M., & Lara, F. (2012). The hornworts Dendroceros Nees and Megaceros Campb. in São Tomé e Príncipe (Africa, Gulf of Guinea) with the description of Dendroceros paivae sp. nov. *Cryptogamie, Bryologie, 33*(1), 3–21.

Gartz, J., & Wiedemann, G. (2015). Discovery of a new caerulescent Psilocybe mushroom in Germany: *Psilocybe germanica* sp.nov. *Drug Testing and Analysis, 7*(9), 853–857.

Gaston, K. J. (2000). Global patterns in biodiversity. *Nature, 405*(6783), 220–227.

Geissmann, T., Lwin, N., Aung, S. S., Aung, T. N., Aung, Z. M., Hla, T. H., et al. (2011). A new species of snub-nosed monkey, genus Rhinopithecus Milne-Edwards, 1872 (Primates, Colobinae), from Northern Kachin state, Northeastern Myanmar. *American Journal of Primatology, 73*(1), 96–107.

Giao, P. M., Tuoc, D., Dung, V. V., Wikramanayake, E. D., Amato, G., Arctander, P., et al. (1998). Description of *Muntiacus truongsonensis*, a new species of muntjac (Artiodactyla: Muntiacidae) from central Vietnam, and implications for conservation. *Animal Conservation, 1*(01), 61–68.

Giljum, S., Hinterberger, F., Bruckner, M., Burger, E., Frühmann, J., Lutter, S., et al. (2009). *Overconsumption? Our use of the world's natural resources.* The Sustainable Europe Research Institute (SERI), GLOBAL 2000, Friends of the Earth Europe, Vienna. http://www.foeeurope.org/publications/2009/Overconsumption_Sep09.pdf

González, M. (1998). *Jubaea chilensis. The IUCN Red List of Threatened Species* 1998: e.T38586A10128158. Downloaded December 08, 2015. http://dx.doi.org/10.2305/IUCN.UK.1998.RLTS.T38586A10128158.en

Guenard, D., Gueritte-Voegelein, F., & Potier, P. (1993). Taxol and taxotere: discovery, chemistry, and structure-activity relationships. *Accounts of Chemical Research, 26*(4), 160–167.

Holton, R. A., Somoza, C., Kim, H.-B., Liang, F., Biediger, R. J., Boatman, P. D., et al. (1994). First total synthesis of taxol. 1. Functionalization of the B ring. *Journal of the American Chemical Society, 116*(4), 1597–1598.

Hunt, T. L., & Lipo, C. P. (2006). Late colonization of Easter Island. *Science, 311*(5767), 1603–1606.

Jones, T., Ehardt, C. L., Butynski, T. M., Davenport, T. R., Mpunga, N. E., Machaga, S. J., et al. (2005). The highland mangabey *Lophocebus kipunji*: A new species of African monkey. *Science, 308*(5725), 1161–1164.

Jones, W. G., Hill, K. D., & Allen, J. M. (1995). *Wollemia nobilis*, a new living Australian genus and species in the Araucariaceae. *Telopea, 6*(2–3), 173–176.

Kier, G., Mutke, J., Dinerstein, E., Ricketts, T. H., Küper, W., Kreft, H., et al. (2005). Global patterns of plant diversity and floristic knowledge. *Journal of Biogeography, 32*(7), 1107–1116.

Kingston, D. G. (2010). Modern natural products drug discovery and its relevance to biodiversity conservation. *Journal of Natural Products, 74*(3), 496–511.

Kirch, P. V., & Ellison, J. (1994). Palaeoenvironmental evidence for human colonization of remote oceanic islands. *Antiquity, 68*(259), 310–321.

Kovanda, J., & Hak, T. (2011). Historical perspectives of material use in Czechoslovakia in 1855–2007. *Ecological Indicators, 11*, 1375–1384.

Kraft, N. J., Comita, L. S., Chase, J. M., Sanders, N. J., Swenson, N. G., Crist, T. O., et al. (2011). Disentangling the drivers of β diversity along latitudinal and elevational gradients. *Science, 333* (6050), 1755–1758.

Kress, W. J., García-Robledo, C., Uriarte, M., & Erickson, D. L. (2015). DNA barcodes for ecology, evolution, and conservation. *Trends in Ecology & Evolution, 30*(1), 25–35.

Labat, J. N., Pignal, M., & Pascal, O. (1997). Une nouvelle espèce de *Labramia* (Sapotaceae) de l'Ile de Mayotte dans l'Archipel des Comores. *Adansonia, 19*(2), 213–216.

Millennium Ecosystem Assessment. (2003). *Ecosystems and human well-being: A framework for assessment*. Washington, DC: World Resources Institute.

Miranda, A., Altamirano, A., Cayuela, L., Pincheira, F., & Lara, A. (2015). Different times, same story: Native forest loss and landscape homogenization in three physiographical areas of south-central of Chile. *Applied Geography, 60*, 20–28.

Nelson, J. B., & Rayner, D. A. (2014). New hedge-nettle (Stachys: Lamiaceae) from South Carolina, USA. *Journal of the Botanical Research Institute of Texas, 8*(2), 431–440.

Peña, V., Pardo, C., López, L., Carro, B., Hernandez-Kantun, J., Adey, W. H., et al. (2015). *Phymatolithon lusitanicum* sp. nov. (Hapalidiales, Rhodophyta): the third most abundant maerl-forming species in the Atlantic Iberian Peninsula. *Cryptogamie, Algologie, 36*(4), 1–31.

Pires, A., Paxton, H., Quintino, V., & Rodrigues, A. M. (2010). Diopatra (Annelida: Onuphidae) diversity in European waters with the description of *Diopatra micrura*, new species. *Zootaxa, 2395*(1), 17–33.

Pizarro-Tobías, P., Fernández, M., Niqui, J. L., Solano, J., Duque, E., Ramos, J. L., et al. (2015). Restoration of a Mediterranean forest after a fire: Bioremediation and rhizoremediation field-scale trial. *Microbial Biotechnology, 8*(1), 77–92.

Reboleira, A. S. P. S., Goncalves, F., Oromi, P., & Mendes, L. F. (2012). *Squamatinia algharbica* gen. n. sp. n., a remarkable new Coletiniinae silverfish (Zygentoma: Nicoletiidae) from caves in Southern Portugal. *Zootaxa, 3260*, 33–46.

Ribeiro, P. M., Paiva, J. A., & Freitas, H. (2007). *Narcissus* x *caramulensis* (Amaryllidaceae), a new hybrid for the portuguese flora. *Anales del Jardín Botánico de Madrid, 64*(1), 43–46.

Roll, U., Geffen, E., & Yom-Tov, Y. (2015). Linking vertebrate species richness to tree canopy height on a global scale. *Global Ecology and Biogeography,*. doi:10.1111/geb.12325.

Silveira, P., Paiva, J., & Samaniego, N. (2001). A new endemic species of Arabis L. (Cruciferae) from Central Portugal. *Botanical Journal of the Linnean Society, 135*(3), 299–303.

Sunderlin, W. D., Angelsen, A., Belcher, B., Burgers, P., Nasi, R., Santoso, L., et al. (2005). Livelihoods, forests, and conservation in developing countries: an overview. *World Development, 33*(9), 1383–1402.

Thompson, I. D., Okabe, K., Parrotta, J. A., Brockerhoff, E., Jactel, H., Forrester, D. I., et al. (2014). Biodiversity and ecosystem services: Lessons from nature to improve management of planted forests for REDD-plus. *Biodiversity and Conservation, 23*(10), 2613–2635.

Tuoc, D., Dung, V. V., Dawson, S., Arctander, P., & MacKinnon, J. (1994). Introduction of a new large mammal species in Vietnam. *Science and Technology News, 3*, 4–12.

United Nations, Department of Economic and Social Affairs, Population Division. (2015). *World population prospects: The 2015 revision, key findings and advance tables*. Working Paper N. ° ESA/P/WP.241. New York, United Nations.

Van Roosmalen, M. G., Van Roosmalen, T., & Mittermeier, R. A. (2002). A taxonomic review of the titi monkeys, genus Callicebus Thomas, 1903, with the description of two new species, *Callicebus bernhardi* and *Callicebus stephennashi*, from Brazilian Amazonia. *Neotropical Primates, 10*(suppl).

Verkerk, P. J., Levers, C., Kuemmerle, T., Lindner, M., Valbuena, R., Verburg, P. H., et al. (2015). Mapping wood production in European forests. *Forest Ecology and Management, 357*, 228–238.

Vitousek, P. M., Mooney, H. A., Lubchenco, J., & Melillo, J. M. (1997). Human domination of earth's ecosystems. *Science, 277*(5325), 494–499.

Wilson, E. O. (1987). The arboreal ant fauna of Peruvian Amazon forests: A first assessment. *Biotropica*, 245–251.

Zardi, G. I., Nicastro, K. R., Canovas, F., Ferreira Costa, J., Serrão, E. A., & Pearson, G. A. (2011). Adaptive traits are maintained on steep selective gradients despite gene flow and hybridization in the intertidal zone. *PloS one, 6*(6), e19402–e19402.

Zizka, G. (1991). Flowering plants of Easter Island. *Palmarum Hortus Francofurtensis, 3*, 1–108.

Zug, G. R., & Fisher, R. N. (2012). A preliminary assessment of the *Nactus pelagicus* species group (Squamata: Gekkonidae) in New Guinea and a new species from the Admiralty Islands. *Zootaxa, 3257*, 22–37.

# Urban Biodiversity and Cities' Sustainable Development

Lurdes Barrico and Paula Castro

**Abstract**

Urban growth and human competition for land have led to deep structural changes in the composition and dynamics of the landscape, significantly affecting the fragile rural/urban equilibrium. Migration from rural to urban areas has forced cities to expand into the surrounding environments which created substantial environmental impacts on the functioning of natural ecosystems, affecting land use of the rural-urban interface. Understanding the human causes and consequences of land use change, particularly in urban environments, has presently become a major challenge worldwide. Strategic actions and integrated responses involving several stakeholders, including residents, policy-makers, scientists or managers will surely originate the necessary pathways and implement a more adequate planning framework to create sustainable and resilient cities. With a correct model of urban development we may preserve native biodiversity, ecosystem services, and diminish natural hazards in urban environment. We highlight the importance of greening cities and engaging residents and other stakeholders in the planning process and decision-making, as well as the importance of teaching, training or raising awareness as key actions to achieve these goals.

L. Barrico (✉) · P. Castro
Centre for Functional Ecology, Department of Life Sciences,
University of Coimbra, Coimbra, Portugal
e-mail: lurdesbarrico@gmail.com



P. Castro et al. (eds.), *Biodiversity and Education for Sustainable Development*,
World Sustainability Series, DOI 10.1007/978-3-319-32318-3_3

# 1 Introduction

Interest in the science of ecosystem and landscape functions and services has grown especially since the release of the Millennium Ecosystem Assessment report in 2003, where the value of ecosystems' services to human activities and well-being was globally recognized. The conservation of the landscape structure and ecosystems' processes are crucial to the maintenance of their functions which in turn provide the services to human practices and quality of life (de Groot et al. 2002; Millennium Ecosystem Assessment [MEA] 2003; Schuhmann and Mahon 2015). Therefore, an effective assessment of the trade-offs between the human benefits and the future costs of environmental damage or ecosystems' capacity to provide goods and services for future generations need to be further investigated.

Biodiversity, "the diversity of life on Earth", is thus essential for the functioning of ecosystems that underpin the provisioning of ecosystem services that ultimately affect humans (European Environment Agency [EEA] 2015; MEA 2005). Different species play specific functions and changes in the species composition, richness, and communities' structures directly disturb the efficiency in which resources are processed within an ecosystem (Alberti 2005; Leung 2015). Consequently, increases in biodiversity may lead to increases in plant community's productivity, greater nutrient retention in ecosystems, and superior ecosystem stability (DeClerck et al. 2011; McCann 2000).

Habitat destruction, fragmentation, and its degradation caused by land use change, over-exploitation of natural resources, unsustainable human practices, invasive species, ocean acidification, pollution or climate change are jeopardising the provision of several key ecosystem services, threatening biodiversity (EC 2010a, 2014).

# 2 Urban Ecosystems

The typology of ecosystems used in the EEA report (2010a) was discussed and further refined by the Working Group on Mapping and Assessment of Ecosystems and their Services (MAES WG) (EC 2013a). Presently, they recommend distinguishing ecosystems into 12 types considered more adequate for European biodiversity assessments. Those include the typology of an urban ecosystem.

Urban areas constitute a particular type of ecosystem, that contain organisms, physical conditions and entities, and the interactions between them (Pickett et al. 2013; Pickett and Grove 2009). Urban ecosystems balance between the artificial and the natural ecological systems. They are open and very dynamic systems which behave like any other ecosystem consuming/releasing and transforming materials and energy, and interact with other ecosystems. These ecosystems are highly artificial areas, dominated by humans, who reclaimed natural land to build their settlements and to implement their numerous activities (EEA 2010b; Guidotti 2010). They can only survive and deliver quality of life by using the basic services provided by nature, their complex structure and growth affects natural ecosystem's functions (EEA 2010b).

Nowadays, we are entering a new era in which the ecology of the planet is increasingly influenced by human activities, with cities as "hotspots" of demand for ecosystem services and sources of environmental degradation (Hodson and Marvin 2010; Solecki et al. 2013). One of the most salient features that characterize human civilization during the past century is the accelerating urban development (Wu 2008a; Wu et al. 2014). According to the United Nations [UN] (2014), the year 2007 was a historic moment in human civilization, where for the first time more than half of the world's population (slightly over 50 %) was living in urban areas. Nowadays, 54 % of the global population lives in cities and in 2050 it is expected to reach a mean value of 66 %. In developed countries, in 2015, about 73 % of the population lives in cities and in 2050 this is expected to increase to ca. 85 %.

While the urbanized areas occupy a surprisingly tiny fraction (roughly 3 %) of the Earth's surface, their impact has been global (Grimm et al. 2008; Schneider et al. 2010). Although urban development often corresponds to higher levels of economic and social development at national and regional scales, these citizens usually have substantially greater and more diverse demands for natural resources than those from rural areas. Therefore, this urban development, most of the times, adversely affects natural ecosystems and landscapes at local and regional scales (Alberti 2015; Wu et al. 2014).

## 3 Urban Development, Land Use Change, and Environmental Impacts

The growth of the human population and its migration from rural to urban areas has forced cities to find solutions for its expansion. The most alarming phenomenon of urban growth and simultaneously a modern model of development is called urban sprawl. The EEA (2006) has described urban sprawl as "the physical pattern of low-density expansion of large urban areas, under market conditions, mainly into the surrounding agricultural areas". This is an unaffordable growth pattern in the long-term due to a higher consumption of resources and energy. The loss of biodiversity is an urgent issue that cities need to deal with owing to the continuous fragmentation and habitat loss as a result of this model of development (Convention on Biological Diversity [CBD] 2007).

Urban sprawl emerges as an artificial growth to serve society, in opposition to the traditional neighbourhoods characterized by a mixed composition, who evolved as a function of the needs of these societies. Urban sprawl illustrates the complexity of interactions and feedback mechanisms between human decisions and ecological processes in urban ecosystems (European Union [EU] 2011). It fragments forests and croplands, alters biogeochemical cycles, contributes to climate change, degrades hydro-systems, decreases native biodiversity, and reclaims land for infrastructure, altering the structure and functioning of natural ecosystems (Alberti 2008; EU 2011; Vitousek et al. 1997).

Urban sprawl has become a very remarkable characteristic of European urban development (Arribas-Bel et al. 2011; Kasanko et al. 2006). Many examples of this trend can be found in the literature, for example, in Milan (Camagni et al. 2002), Madrid (López de Lucio 2003), Porto (EEA 2006), Barcelona (Catalán et al. 2008), Rome (Frondoni et al. 2011), or Coimbra (Barrico 2015). The area of the city of Coimbra, considered the main urban centre of the municipality by the National Institute of Statistics [INE] (www.ine.pt), expanded from 5873 ha in 2001 to 8318 ha in 2011 (mean growth value of 245 ha/year) (Fig. 1). In this process, land was mostly taken at the expense of the surrounding cropland areas. This expansion, however, was not followed by the corresponding population's growth population growth rate; instead, the population density in the city decreased about 29 % during the same period. This physical pattern of low-density expansion in built-up areas contributed to the evolution of a less compact city, clearly an indicator of urban sprawl (Barrico 2015).

Changes in land use to yield goods and services represent the most substantial human alteration of the Earth system and the urbanization process (the spatial expansion of the built human-constructed elements, such as buildings, roads or runways) is a major driver of this land conversion (Alberti 2010; Vitousek et al. 1997).

Land use is determined by the interaction in space and time of biophysical factors, such as soils, climate, topography, and human factors like population, technology or economic conditions. Consequently, although the land use change is

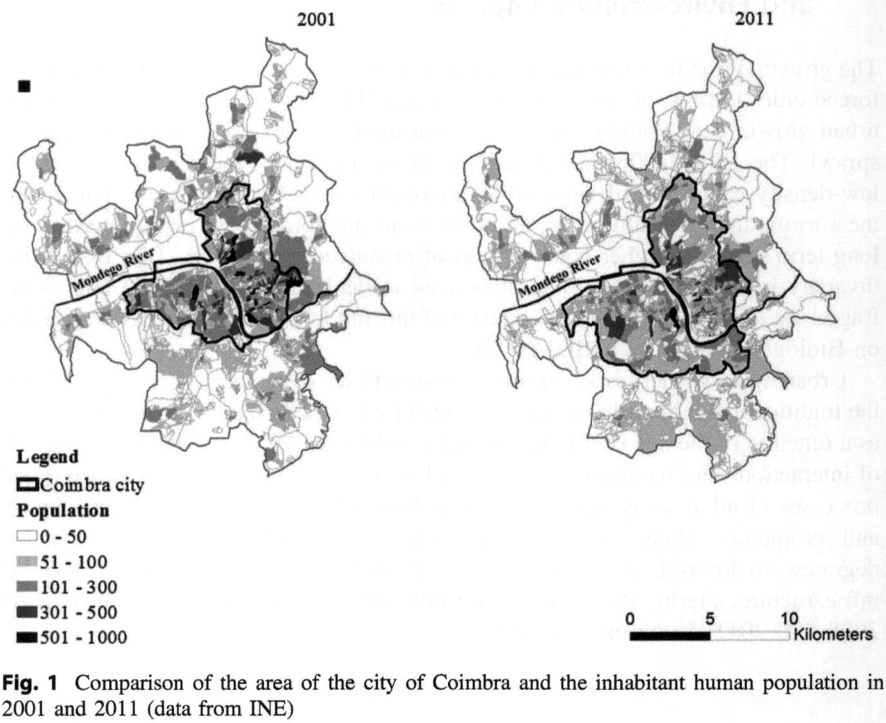


**Fig. 1** Comparison of the area of the city of Coimbra and the inhabitant human population in 2001 and 2011 (data from INE)

necessary and essential for economic and social progress, it also drags negative socio-economic and environmental impacts (Wu 2008b).

Urban areas' expansion implies an increase in impermeable land which affects both geomorphological and hydrological processes, thus causing changes in water and sediment fluxes (Grimm et al. 2008; Wu 2008a). These impervious surfaces and the generation of heat from various combustion processes in urban areas modify its microclimate and air quality (Alberti 2010, 2015). A best-known example of inadvertent climate modifications is the urban heat island effect once the urban areas tend to have higher air and surface temperatures than their surrounding suburban and rural areas (Arnfield 2003; Rosenthal et al. 2008). This effect occurs due to the greater heat retention of buildings and artificial surfaces, compared to the lesser heat retention and cooling properties of vegetation, which is more abundant in the countryside (Rosenthal et al. 2008; Winguth and Kelp 2013). The urbanized areas usually have fewer trees and other vegetation to shade buildings and cool off the air by evapotranspiration and thus, they tend to retain less surface water from precipitation (Rosenthal et al. 2008).

Urbanization also affects biogeochemical processes by modifying the mechanisms that control the spatial and temporal variability of nutrient sources and sinks (Grimm et al. 2008; Kaye et al. 2006). Humans modify the ways in which nutrients are transported across the landscape and their cycles, for example, when nutrients are released from municipal wastewater and from combined sewer-storm water overflow systems in urban surface waters (Alberti 2015).

The expansion of urban areas also drastically affects water resources due to the increased *per capita* use of freshwater and contamination of water bodies by sewage and wastes (Wu 2008a). In addition, the urban centres, especially those in the developed world, are the major producers of greenhouse gases and other air pollutants that cause health problems for humans and the environment (Grimm et al. 2008; Wu 2008a).

## 4 Recognizing the Importance of Urban Biodiversity

Whilst cities pose major challenges for the protection of biodiversity, they have received little consideration in the existing global debate. The impact of urbanization on biodiversity and other natural resources was considered by the CBD in 1992, but a major step towards recognizing the potential of cities for increasing biodiversity was made in Curitiba in 2006 (COP 8) (CBD 2006). Here was initiated a global partnership to promote the discussion on "Cities and Biodiversity" with the objective of encouraging local authorities to contribute to the Convention's 2010 target of significantly reducing the biodiversity loss rate (CBD 2006). The "Curitiba Declaration", adopted at the meeting reaffirmed the urgency to achieve the CBD objectives in urban areas and to engage local authorities for the "Battle of life on Earth", where according to the words of the Executive Secretary "The battle for life on Earth will be won or lost in urban areas". Particular emphasis was placed on raising public awareness and educating future generations as well as on disseminating best practices and lessons learned through cooperation between cities (CBD 2007). At the 9th COP, in Bonn, a decision to promote the engagement of cities and local authorities (Decision IX/28) in national biodiversity strategies and action plans was adopted for the fist time (CBD 2008).

The UN declared 2010 the International Year of Biodiversity (https://www.cbd.int/2010/welcome/) to celebrate life on earth and the value of biodiversity for our lives. The world was invited to take action in 2010 to safeguard biodiversity. In a series of assessments under the title "10 messages for 2010", five key messages were specifically addressed to urban ecosystems (EEA 2010b) (Fig. 2).

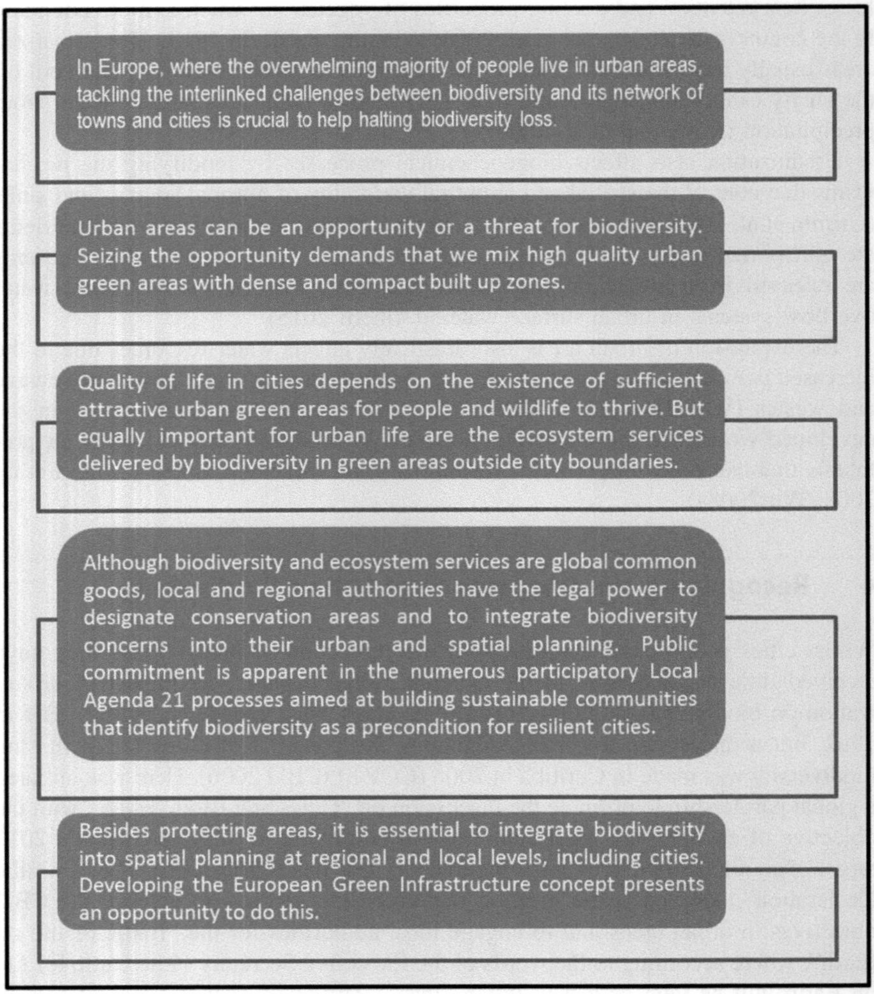


**Fig. 2** Key messages relating to urban ecosystems (EEA 2010b)

Presently we are living the "Decade on Biodiversity" (2011–2020), as declared by the United Nations General Assembly at its 65th session (UN 2011), (https://www.cbd.int/2011-2020/) aiming to contribute to the implementation of the Strategic Plan for Biodiversity for the period 2011–2020 (https://www.cbd.int/sp/) elaborated by tenth meeting of the Conference of the Parties (CBD 2010).

Recently, the COP 12, held in Pyeongchang, Republic of Korea (CBD 2014) reiterated the need for the post-2015 development agenda and sustainable development goals to support the conservation and sustainable use of biodiversity. An important goal was placed on making cities and human settlements inclusive, safe, resilient and sustainable.

## 5 Urban Sustainable Development

According to the United Nations Educational, Scientific and Cultural Organization [UNESCO] (2000), one of the greatest challenges facing the 21st century will be the attainment of sustainable development. This concept has encouraged policy-makers to formulate new strategies to achieve a balanced economic and technological pathway to safeguard the environment now and into the future (EC 2014; Nijkamp and Vreeker 2000). With the unprecedented growth of urbanization it is mandatory to look upon the urban ecosystem, considering the economic, social, and environmental impacts of cities on cities themselves and on other ecosystems (Childers et al. 2015).

The explosive growth of cities and resident population aspiration for a better quality of life needs to be carefully addressed and managed. Therefore, improving the ability of policy-makers to better plan their cities and achieve sustainable development goals are pressing needs of this century (Childers et al. 2015; EC 2014).

The concept of sustainability has become an important paradigm in urban planning, as cities play a key role in our society. Cities are important generators of wealth, employment, and productivity, and often serve as the engines of their national economies (EC 2014; EEA 1995). Consequently, strategies to make cities more sustainable have been addressed more prudently by governments and institutions all over the world, focusing on the protection of their environmental resources (e.g., air quality, biodiversity) as well as the social, cultural, and economic resources (e.g., liveability, prosperity) (EC 2014; EEA 1995; EU 2011; European Federation of Metropolitan and Periurban Natural and Rural Areas [FEDENATUR] 2004).

Urban sprawl is generally believed to result from an uncontrolled and inefficient urban dispersion accompanied by low building and population density, over rural or semi-rural areas (Altieri et al. 2014; Zhang 2001). As a consequence, this model of urban expansion leads to negative effects on the environment. Sustainable planning is an opportunity for cities to address in a more innovative and effective way the challenges they are facing, as well as to create a vision for their future considering all socio-economic and ecological aspects (Sustainable Cities International [SCI] 2012). Hence, the knowledge of the driving forces behind land-use change and urban sprawl, particularly about the natural characteristics of the landscape, the importance of land use history and its relation to the planning framework, are of great importance (Tavares et al. 2012). Pato et al. (2015) found that the importance of physical variables on the planning process on small-scale hydrological basin in the city of Coimbra, Portugal, decreased during the past decades, and their importance for the planning framework was very small when compared to political actors and planning managers.

In Europe, the first step towards sustainability of towns and cities was signed in 1994 with the first European Conference on Sustainable Cities and Towns held in Aalborg, where participants undertook the compromise to develop and implement their own local development agenda—the Aalborg Charter (more information in http://www.sustainablecities.eu/). Since then, five more conferences on this topic have been held. The latest was organized in Geneva in 2013 focusing on "A green and socially responsible economy: a solution in times of crisis?" (www.sustainablegeneva2013.org) and five key outcome messages were produced, highlighting the importance of cities' functions and strengthening communication between stakeholders.

## 5.1 Greening Cities

In 2010 the Environment Council agreed in a new vision and target for biodiversity beyond 2010, explicitly calling for the development and investment in "Green Infrastructures" to support biodiversity and ecosystems' processes. It particularly emphasized the restoration of natural ecosystems as to improve cities' resilience, to sustain the services provided, and also to reduce cities' vulnerability to climate change (EC 2011). More recently, the EC launched a strategy untitled "Green Infrastructure—Enhancing Europe's Natural Capital" based on the principle that protecting and enhancing nature and natural processes, and the many benefits human society gets from nature, are consciously integrated into spatial planning and territorial development (EC 2013b).

Building a green infrastructure can reconnect fragmented natural areas and improve their functional connectivity (Andersson et al. 2014). It can promote quality of life and human well-being, for example, by establishing recreational areas, help to better adapt to climate change through natural flood management, and soil protection or enhance water quality by wetland restoration. Specifically, the second target of the European Biodiversity Strategy to 2020 focuses on maintaining and enhancing ecosystems and their services by establishing green infrastructure and restoring degraded ecosystems across the Europe, in line with the global goal set in 2010 to restore degraded ecosystems at least on 15 % (EC 2011).

Green Infrastructure is a strategically planned network of natural and semi-natural areas with other environmental features designed and managed to deliver a wide range of ecosystem services. It incorporates green spaces (or blue if aquatic ecosystems are concerned) and other physical features in terrestrial (including coastal) and marine areas (EC 2013b).

Greening cities though the creation or restoration of green spaces is thus of much importance to cities' sustainability. Urban green spaces are public or private open spaces in urban areas, mostly covered by vegetation, which are directly (e.g. active or passive recreation) or indirectly (e.g. positive influence in urban functions) available for their users (Baycan-Leven et al. 2002). Public green space includes parks and reserves, sporting fields, riparian areas, greenways and trails, community gardens, street trees, and nature conservation areas. Conventional spaces such as green walls, green roof tops, green alleyways, and cemeteries (Roy et al. 2012) may also be classified as green spaces. Private green spaces includes private backyards, communal grounds of apartment buildings, and corporate campuses (Wolch et al. 2014).

There is a broad consensus about the importance and value of urban green spaces in cities towards protecting and maintaining biodiversity (Haq 2011). Although urban ecosystems tend to have less biodiversity than natural forest habitats, efforts at mitigating global biodiversity must be also a goal in urban areas, especially if they maintain or restore forest fragments (e.g. remaining forests) (Alvey 2006; Goddard et al. 2009).

Remaining forests within urban areas provide important refuges for endangered species and species of high conservation value, and are more frequently occupied by native rather than exotic species (Barrico et al. 2012; LaPaix and Freedman 2010). For example, the study carried out by Barrico (2015) in Coimbra city, comparing public gardens and remaining forest areas, showed that despite the similar values found for the plant species richness and diversity indices on both areas, important differences were obtained regarding the native *taxa* and *taxa* with higher ecological and conservation value, which mostly occurred in the native forests.

## 5.2 Public Participation

Individuals around the world recognize that current economic development trends are not sustainable and that public awareness, education, and training are important in pointing society towards sustainability (EC 2015; Krasny et al. 2014; Tilbury et al. 2002). Therefore, more effective strategies are needed to raise awareness and improve communication and education efforts on the importance of biodiversity and ecosystem functioning in urban environments (CBD 2014). It is also necessary to identify trends and assess the effectiveness of current and future policies to improve individuals' participation and involvement in actions regarding biodiversity conservation, urban ecology, and sustainability (CBD 2014; Krasny et al. 2014).

Individuals are generally aware of their local environmental problems, but the level of public involvement is usually low. To solve problems we need to act locally, so the importance of local knowledge and approaches on continued education and coordination between local/regional stakeholders are needed to achieve of sustainable urban public policies (Table 1).

**Table 1** Proposals for biodiversity conservation and sustainability of urban areas at local level of governance

| |
|---|
| *Measurement and mapping of biodiversity* |
| Engaging society (e.g. public, NGOs, private sector, and other actors in the planning process) |
| Reinforce the communication and people involvement on impact assessments and other decision-processes so to make well-informed choices (e.g. environmental impact assessment and strategic environmental assessment or the Local Agenda 21 processes) |
| Environmental education through teaching, training or raising awareness actions for the importance of biodiversity tailoring different stakeholders |
| Allocation of more resources (e.g. funds, green infrastructure, and personnel) |
| A more clear understanding of the systematic representation of the different driving forces affecting land-use change |
| Improve the network between the scientific, political, and the private sectors |
| Control of alien invasive species and other threats to native biodiversity |
| Greening cities |

The provision, design, management and protection of urban green spaces are the main purposes of the plan of sustainability and liveability of modern cities (Baycan-Leven et al. 2002; Haq 2011). This requires large economic efforts and future commitment to their conservation by the government and local authorities. This should be followed by actions to promote environmental citizen awareness, so that the urban environment can be protected and preserved (Gomes and Panagopoulos 2008).

## 6 Final Remarks

It is not easy to understand the true nature of cities. They are key systems of social, cultural, and economic growth. However, the strategic model of development and its spatial expansion is a crucial challenge to manage. Changes in land use associated with urban sprawl/urbanization drastically affect biodiversity, ecosystem functioning, and environmental quality as well as human behaviour, community structure, and social organization. Both the loss and fragmentation of natural habitats due to urbanization also have direct and indirect impacts on the diversity, structure, and distribution of vegetation leading to important consequences in the distribution, movement, and survival of species (Alberti 2015).

The Europe 2020 strategy sets out a vision of Europe's social market economy for the 21st century, putting forward three mutually reinforcing priorities: smart, sustainable, and inclusive growth (EC 2010b) and under this work programme specific funding calls are addressed to cities' development. Promoting a more resource efficient, greener, and more competitive economy are objectives of sustainable growth which includes preventing biodiversity loss, sustainable use of resources, improving green procurement, promoting civic participation, and the adoption of biodiversity conservation practices.

## References

Alberti, M. (2005). The effects of urban patterns on ecosystem function. *International Regional Science Review, 28*(2), 168–192.

Alberti, M. (2008). *Advances in urban ecology: Integrating humans and ecological processes in urban ecosystems*. New York: Springer.

Alberti, M. (2010). Maintaining ecological integrity and sustaining ecosystem function in urban areas. *Current Opinion in Environmental Sustainability, 2*(3), 178–184.

Alberti, M. (2015). Eco-evolutionary dynamics in an urbanizing planet. *Trends in Ecology and Evolution, 30*(2), 114–126.

Altieri, L., Cocchi, D., Pezzi, G., Scott, E. M., & Ventrucci, M. (2014). Urban sprawl scatterplots for urban morphological zones data. *Ecological Indicators, 36*, 315–323.

Alvey, A. A. (2006). Promoting and preserving biodiversity in the urban forest. *Urban Forestry and Urban Greening, 5*(4), 195–201.

Andersson, E., Barthel, S., Borgström, S., Colding, J., Elmqvist, T., Folke, C., et al. (2014). Reconnecting cities to the biosphere: Stewardship of green infrastructure and urban ecosystem services. *Ambio, 43*(4), 445–453.

Arnfield, A. J. (2003). Two decades of urban climate research: A review of turbulence, exchanges of energy and water, and the urban heat island. *International Journal of Climatology, 23*(1), 1–26.

Arribas-Bel, D., Nijkamp, P., & Scholten, H. (2011). Multidimensional urban sprawl in Europe: A self-organizing map approach. *Computers, Environment and Urban Systems, 35*(4), 263–275.

Barrico, L. (2015). *Ecological processes related to urban development and land use change in the municipality of Coimbra*. Doctoral dissertation. Retrieved from http://hdl.handle.net/10316/26494

Barrico, L., Azul, A. M., Morais, M. C., Coutinho, A. P., Freitas, H., & Castro, P. (2012). Biodiversity in urban ecosystems: Plants and macromycetes as indicators for conservation planning in the city of Coimbra (Portugal). *Landscape and Urban Planning, 106*(1), 88–102.

Baycan-Leven, T., van Leeuwen, E., Rodenburg, C., & Nijkamp, P. (2002). Development and management of green spaces in European cities: A comparative analysis. In *38th International Planning Congress on "The Pulsar Effect" Planning with Peaks, Athens, Greece*, September 21–26, 2002.

Camagni, R., Gibelli, M. C., & Rigamonti, P. (2002). Urban mobility and urban form: The social and environmental costs of different patterns of urban expansion. *Ecological Economics, 40*(2), 199–216.

Catalán, B., Saurí, D., & Serra, P. (2008). Urban sprawl in the Mediterranean? Patterns of growth and change in the Barcelona Metropolitan Region 1993–2000. *Landscape and Urban Planning, 85*(3–4), 174–184.

Childers, D. L., Cadenasso, M. L., Grove, J. M., Marshall, V., McGrath, B., & Pickett, S. T. A. (2015). An ecology for cities: A transformational nexus of design and ecology to advance climate change resilience and urban sustainability. *Sustainability, 7*(4), 3774–3791.

Convention on Biological Diversity. (2006). *Report of the eighth meeting of the parties to the convention on biological diversity* (UNEP/CBD/COP/8/31).

Convention on Biological Diversity. (2007). *Report of the cities and biodiversity: Achieving the 2010 biodiversity target* (UNEP/CBD/Cities/1/3).

Convention on Biological Diversity. (2008). *Report of the conference of the parties to the convention on biological diversity on the work of its ninth meeting* (UNEP/CBD/COP/9/29).
Convention on Biological Diversity. (2010). *Report of the tenth meeting of the conference of the parties to the convention on biological diversity* (UNEP/CBD/COP/10/27).
Convention on Biological Diversity. (2014). *Report of the twelfth meeting of the conference of the parties to the convention on biological diversity* (UNEP/CBD/COP/12/29).
DeClerck, F. A. J., Fanzo, J., Palm, C., & Remans, R. (2011). Ecological approaches to human nutrition. *Food and Nutrition Bulletin, 32*(1), S41–S50.
de Groot, R. S., Wilson, M. A., & Boumans, R. M. J. (2002). A typology for the classification, description and valuation of ecosystem functions, goods and services. *Ecological Economics, 41*(3), 393–408.
European Commission. (2010a). *Options for an EU vision and target for biodiversity beyond 2010 (COM (2010) 4 final)*. Brussels: European Commission.
European Commission. (2010b). *A strategy for smart, sustainable and inclusive growth (COM (2010) 2020)*. Brussels: European Commission.
European Commission. (2011). *Our life insurance, our natural capital: An EU biodiversity strategy to 2020 (COM (2011) 244 final)*. Brussels: European Commission.
European Commission. (2013a). *Mapping and Assessment of Ecosystems and their Services (MAES): An analytical framework for ecosystem assessments under action 5 of the EU biodiversity strategy to 2020*. Brussels: European Commission.
European Commission. (2013b). *Green Infrastructure (GI)—enhancing Europe's Natural Capital (COM (2013) 249 final)*. Brussels: European Commission.
European Commission. (2014). *General Union environment action programme to 2020: Living well, within the limits of our planet*. Brussels: European Commission.
European Commission. (2015). *Horizon 2020 work. Work programme 2016–2017* (European Commission Decision C (2015) 6776).
European Environment Agency. (1995). *Environment in the European Union 1995: Report for the review of the fifth environmental action programme*. Copenhagen: European Environment Agency.
European Environment Agency. (2006). *Urban sprawl in Europe—the ignored challenge* (Report No. 10). Copenhagen: European Environment Agency.
European Environment Agency. (2010a). *EU 2010 biodiversity baseline* (EEA Technical report No. 12). Copenhagen: European Environment Agency.
European Environment Agency. (2010b). *10 messages for 2010: Urban ecosystems*. Copenhagen: European Environment Agency.
European Environment Agency. (2015). *The European environment—state and outlook 2015: Synthesis report*. Copenhagen: European Environment Agency.
European Federation of Metropolitan and Periurban Natural and Rural Areas. (2004). *The place of periurban natural spaces for a sustainable city*. Brussels: European Commission, Direction General Environment.
European Union. (2011). *Cities of tomorrow: Challenges, visions, ways forward*. Brussels: European Commission—Directorate General for Regional Policy.
Frondoni, R., Mollo, B., & Capotorti, G. (2011). A landscape analysis of land cover change in the municipality of Rome (Italy): Spatio-temporal characteristics and ecological implications of land cover transitions from 1954 to 2001. *Landscape and Urban Planning, 100*(1–2), 117–128.
Goddard, M. A., Dougill, A. J., & Benton, T. G. (2009). Scaling up from gardens: Biodiversity conservation in urban environments. *Trends in Ecology and Evolution, 25*(2), 90–98.
Gomes, P. D. P. M., & Panagopoulos, T. (2008). Opportunities for sustainable design in renovation of the Stanaway Park, Dublin, Ireland. In *1st WSEAS International Conference on Landscape Architecture, Algarve, Portugal*.
Grimm, N. B., Faeth, S. H., Golubiewski, N. E., Redman, C. L., Wu, J., Bai, X., et al. (2008). Global change and the ecology of cities. *Science, 319*(5864), 756–760.

Guidotti, T. L. (2010). Health and urban ecosystems. *Archives of Environmental and Occupational Health, 65*(1), 54–55.
Haq, S Md A. (2011). Urban green spaces and an integrative approach to sustainable environment. *Journal of Environmental Protection, 2*(5), 601–608.
Hodson, M., & Marvin, S. (2010). Urbanism in the anthropocene: Ecological urbanism or premium ecological enclaves? *City, 14*(3), 298–313.
Kasanko, M., Barredo, J. I., Lavalle, C., McCormick, N., Demicheli, L., Sagris, V., et al. (2006). Are European cities becoming dispersed? A comparative analysis of 15 European urban areas. *Landscape and Urban Planning, 77*(1–2), 111–130.
Kaye, J. P., Groffman, P. M., Grimm, N. B., Baker, L. A., & Pouyat, R. V. (2006). A distinct urban biogeochemistry? *Trends in Ecology and Evolution, 21*(4), 192–199.
Krasny, M. E., Russ, A., Tidball, K. G., & Elmqvist, T. (2014). Civic ecology practices: Participatory approaches to generating and measuring ecosystem services in cities. *Ecosystem Services, 7*, 177–186.
LaPaix, R., & Freedman, B. (2010). Vegetation structure and composition within urban parks of Halifax regional municipality, Nova Scotia, Canada. *Landscape and Urban Planning, 98*(2), 124–135.
Leung, J. Y. S. (2015). Habitat heterogeneity affects ecological functions of macrobenthic communities in a mangrove: Implication for the impact of restoration and afforestation. *Global Ecology and Conservation, 4*, 423–433.
López de Lucio, M. (2003). Transformaciones territoriales recientes en la región urbana de Madrid. *Urban, 8*, 124–161. [in Spanish].
McCann, K. S. (2000). The diversity—stability debate. *Nature, 405*, 228–233.
Millennium Ecosystem Assessment. (2003). *Ecosystems and human well-being: A framework for assessment*. Washington, DC: World Resources Institute.
Millennium Ecosystem Assessment. (2005). *Ecosystems and human well-being: Biodiversity synthesis*. Washington, DC: World Resources Institute.
Nijkamp, P., & Vreeker, R. (2000). Sustainability assessment of development scenarios: Methodology and application to Thailand. *Ecological Economics, 33*(1), 7–27.
Pato, R. L., Castro, P., & Tavares, A. O. (2015). The relevance of physical forces on land-use change and planning process. *Journal of Environmental Planning and Management*. doi:10.1080/09640568.2015.1035773
Pickett, S. T. A., & Grove, J. M. (2009). Urban ecosystems: What would Tansley do? *Urban Ecosystems, 12*(1), 1–8.
Pickett, S. T. A., Cadenasso, M. L., & McGrath, B. (Eds.). (2013). *Resilience in ecology and urban design. Linking theory and practice for sustainable cities*. Dordrecht: Springer Science +Business Media.
Rosenthal, J. K., Crauderueff, R., & Carter, M. (2008). *Urban heat island mitigation can improve New York city's environment: Research on the impacts of mitigation strategies on the urban environment*. Sustainable South Bronx Working Paper.
Roy, S., Byrne, J., & Pickering, C. (2012). A systematic quantitative review of urban tree benefits, costs, and assessment methods across cities in different climatic zones. *Urban Forestry and Urban Greening, 11*(4), 351–363.
Schneider, A., Friedl, M. A., & Potere, D. (2010). Mapping global urban areas using MODIS 500-m data: New methods and datasets based on 'urban ecoregions'. *Remote Sensing of Environment, 114*(8), 1733–1746.
Schuhmann, P. W., & Mahon, R. (2015). The valuation of marine ecosystem goods and services in the Caribbean: A literature review and framework for future valuation efforts. *Ecosystem Services, 11*(1), 56–66.
Solecki, W., Seto, K. C., & Marcotullio, P. J. (2013). It's time for an urbanization science. *Environment: Science and Policy for Sustainable Development, 55*(1), 12–17.
Sustainable Cities International. (2012). *Indicators for sustainability: How cities are monitoring and evaluating their success*. Vancouver: Canadian International Development Agency.

Tavares, A. O., Pato, R. L., & Magalhães, M. C. (2012). Spatial and temporal land use change and occupation over the last half century in a peri-urban area. *Applied Geography, 34*, 432–444.

Tilbury, D., Stevenson, R. B., Fien, J., & Schreuder, D. (Eds.). (2002). *Education and sustainability: Responding to the global challenge*. Gland, Switzerland and Cambridge, UK: Commission on Education and Communication, IUCN.

United Nations. (2011). *Resolution adopted by the general assembly: Convention on biological diversity* (A/RES/65/161).

United Nations. (2014). *Population division. World urbanization prospects: The 2014 revision*. Retrieved from http://esa.un.org/unpd/wup/

United Nations Educational, Scientific and Cultural Organization. (2000). *World conference on science. Science for the twenty-first century: A new commitment*. Paris: UNESCO.

Vitousek, P. M., Mooney, H. A., Lubchenco, J., & Melillo, J. M. (1997). Human domination of Earth's ecosystems. *Science, 277*(5325), 494–499.

Winguth, A. M. E., & Kelp, B. (2013). The urban heat island of the north-central Texas region and its relation to the 2011 severe Texas drought. *Journal of Applied Meteorology and Climatology, 52*(11), 2418–2433.

Wolch, J. R., Byrne, J., & Newell, J. P. (2014). Urban green space, public health, and environmental justice: The challenge of making cities "just green enough". *Landscape and Urban Planning, 125*, 234–244.

Wu, J. (2008a). Making the case for landscape ecology: An effective approach to urban sustainability. *Landscape Journal, 27*(1), 41–50.

Wu, J. (2008b). Land use changes: Economic, social, and environmental impacts. *Agricultural and Applied Economics Association, 23*(4), 6–10.

Wu, J., Xiang, W.-N., & Zhao, J. (2014). Urban ecology in China: historical developments and future directions. *Landscape and Urban Planning, 125*, 222–233.

Zhang, T. (2001). Community features and urban sprawl: The case of the Chicago metropolitan region. *Land Use Policy, 18*(3), 221–232.

# Education on Biodiversity in the Polar Regions

José C. Xavier, Gerlis Fugmann, Inga Beck, Louise Huffman and Eric Jensen

**Abstract**

The polar regions are famously associated with extreme temperatures, ice, snow, legendary explorers, indigenous people, polar bears, penguins and other impressive fauna and flora. The past decades have witnessed a revolution in the amount of data collected in the polar regions, with considerable advances in the knowledge of numerous areas, including in polar biodiversity. Educationally, the polar regions can be perfect vehicles to transfer educational concepts related to biodiversity, but unfortunately, the evaluation of the impact of educational activities related to polar biodiversity is scarce. This chapter provides a general

J.C. Xavier (✉)
Marine and Environmental Research Centre (MARE), Department of Life Sciences, University of Coimbra, 3001-401 Coimbra, Portugal
e-mail: jxavier@zoo.uc.pt

J.C. Xavier
British Antarctic Survey, Natural Environment Research Council, High Cross, Madingley Road, CB3 0ET Cambridge, UK

G. Fugmann
Association of Polar Early Career Scientists (APECS), UiT, the Arctic University of Norway, Tromsø, Norway

I. Beck
Heidelberg University, Institute of Geography, Berliner Straße 48, 69120 Heidelberg, Germany

L. Huffman
Polar Educators International (PEI), ARCUS, 3535 College Rd., Suite 101, Fairbanks, AK 99709-3710, USA

E. Jensen
Department of Sociology, University of Warwick, Coventry CV4 7AL, UK



P. Castro et al. (eds.), *Biodiversity and Education for Sustainable Development*, World Sustainability Series, DOI 10.1007/978-3-319-32318-3_4

review of the importance of the polar regions, the increasing status of polar education in the last decade, examples of polar educational activities on biodiversity, and a resource to stakeholders interested in polar science and education. With pivotal inputs to polar education during the International Polar Year, 2007–2008, three of the most important challenges to be addressed in the future are to assess the needs of polar educators, strengthen the network of information sharing of educational materials (e.g. in a validated, multi-lingual and easily accessible online mode) and to develop robust evaluation of the educational activities. Recent improvements in internet technologies may provide a major source of ideas and the ability to effortlessly spread polar information relevant to biodiversity education.

## 1 Importance of Polar Regions

The polar regions are the cornerstones of the global ecosystem, barometers of the health of the planet, and the places where global processes connect across the Earth (Kennicutt et al. 2014; Krupnik et al. 2011; Smetacek and Nicol 2005). Indeed, polar environments are changing faster than any other region on Earth (with regional and global implications for societies, economies and ecosystems), processes in polar regions have profound influence on the global environment (e.g. climate, sea level, ocean systems), the polar regions are home (particularly for the Arctic) to more than four million people who face changes in their natural environment faster than elsewhere and, finally, within the polar regions lie important scientific and technological challenges yet to be investigated (Allison et al. 2007, 2009; IPCC 2007; Kennicutt et al. 2014; Krupnik et al. 2011; Sarmiento et al. 2004). Furthermore, human impacts on polar regions, such as pollution, invasive species and development of non-renewable resources may contribute to extend these changes (Bennett et al. 2015; Convey et al. 2012). Indeed, the intensity and governance of human activities also vary greatly between the Arctic and the Antarctic; the Arctic has been continuously inhabited for millennia (and most Arctic land masses belong to sovereign states) while much of the Antarctic has been claimed by various states but governed by the international Antarctic Treaty System which sets issues of sovereignty (Bennett et al. 2015). At a time when the world's population is exerting an increasing influence on Earth and its environments, and the human living conditions are rapidly affected by global changes, the polar regions are critical to any vision of humanity's overall prospects on the "over-stressed" earth. It has become clear that the polar regions provide a litmus test and the insight to help society as a whole recognize the planetary limits of our behavior (Allison et al. 2007; Bennett et al. 2015; Rockstrom et al. 2009).

## 2 Importance of Polar Biodiversity

Organisms that live permanently in the polar regions face extreme environmental conditions including chronic low temperatures, high winds and solar radiation, freezing and/or desiccation stress, environmental variability on both short and long time-scales, and extreme and acute events (Convey et al. 2012; Meltofte 2013; Meltofte et al. 2013; Peck et al. 2006; Thomas et al. 2008; Xavier and Peck 2015). For example, the terrestrial habitats of the higher latitudes of the Arctic and Antarctic are characterized by the combination of long winters, short-productive cool summers with short growing seasons, and extremely climatic variability with low air temperatures of up to −40 to −80 °C or even lower (although such extremes are not reached at lower latitudes or in habitats protected by winter snow) and large regional differences (Convey et al. 2012; Meltofte 2013).

There are contrasting patterns of biodiversity in the Arctic and Antarctic in terms of species diversity, both on land and in the sea (Aronson et al. 2011; Convey et al. 2012) (Fig. 1).

Excluding endo-parasites and microbes, the Arctic is home to 21,000 species of animals, plants and fungi (Meltofte et al. 2013). Although large and globally significant populations of seabirds and marine mammals breed on the fringes of the Antarctic continent, macroscopic terrestrial biodiversity is low and comprised almost entirely of invertebrates and cryptogams (Convey 2007). There are no native terrestrial vertebrates in the Antarctic, contrasting with large populations of birds and mammals (including land predators) in the Arctic (e.g. Polar bears, narwhals, caribou, muskoxen, walruses) (Convey et al. 2012; Meltofte 2013).

Arctic terrestrial ecosystems are characterized by high numbers of migratory vertebrates, providing an important connection with global biodiversity (Convey et al. 2012; Thomas et al. 2008). In contrast, there is no analogous migratory component within Antarctic or sub-Antarctic terrestrial ecosystems.

Arctic and Antarctic marine environment are amongst the most thermally stable on Earth (Convey et al. 2012), although in some regions significant changes have been noticed in these marine environments (Gutt et al. 2015; Walczowski and Piechura 2006). The extreme seasonality in light regime at high latitudes leads to intense pulses of biological production in the short and cold polar summers (i.e. June–Sept. in the Arctic and Dec.–March in the Antarctic), this being particularly apparent in the short phytoplankton blooms in spring that form the basis of the Arctic and Antarctic marine food webs (Knox 2007; Thomas et al. 2008). In terms of plants, 2218 species and subspecies are regarded as part of the regular Arctic flora whereas the Antarctic vascular plant flora includes only two native species on the continent (Convey 2007; Elven et al. 2011). The reverse pattern is apparent at sea; today, Antarctic benthic marine diversity is second only to that of coral reefs. The Arctic Ocean has lower diversity in some groups than typifies the Antarctic with its lesser sea ice extent (Verde et al. 2012). However, the Arctic also has a much more complex oceanographic system than does the Antarctic, and in other biological groups, such as fish, there is higher diversity as well as representation of

**Fig. 1** Examples of **a** polar animals (Black-browed albatross *Diomedea melanophrys*, Gentoo penguins *Pygoscelis papua*, and the squid *Galiteuthis glacialis*, **b** Polar bear *Ursus maritimus*, caribou *Rangifer tarandus*, walrus *Odobenus rosmarus* and polar cod *Boreogadus saida*, **c** educators (including indigenous of the Arctic) working together on biodiversity activities. Photos by José C. Xavier, Inga Beck, Hauke Flores and Louise Huffman

more than one higher taxonomic level (Convey et al. 2012). Both regions support large breeding populations of marine mammals and birds during their respective summers (Thomas et al. 2008).

A special situation can be found in the Arctic due to the connection of humans and biodiversity. The harvesting of Arctic flora and fauna (mammals, birds, fish and plants) has for millennia been not only the basis for human existence and survival in this harsh climate, but also an integral part in the culture, identity and spiritual life of Arctic peoples and communities (Meltofte et al. 2013).

As some of the most rapidly changing ecosystems on our planet are located in the polar regions, polar ecosystems and the biodiversity they hold are already responding to these changes and it is expected that even more profound impacts will occur this century (Convey et al. 2012). In the Arctic today, climate change is seen as the greatest threat to its biodiversity (CAFF 2015). Compounding the risk to polar biodiversity is the fact that many polar ecosystems have limited functional redundancy; in the event of the loss of a single keystone species, they may potentially be exposed to cascading effects and complete ecosystem restructuring (Post et al. 2009). In several parts of the Arctic, shifts are already witnessed including changes in the ecosystem and the northward expansion of many species (Meltofte et al. 2013; Richter-Menge and Overland 2010).

## 3 Importance of Linking Education and Biodiversity Issues Related to the Polar Regions

Biodiversity provides numerous essential services to society, yet biodiversity continues to decline, even though worldwide conservation efforts are increasing (Barnosky et al. 2012; Rands et al. 2010; Rockstrom et al. 2009). Filling gaps in our knowledge and building on success, through scaling up and further long-term investment in conservation that works and bringing this knowledge to society are all critical steps that must be taken. The UN declared 2010 the International Year of Biodiversity with goals to adopt a strategic plan containing a vision for 2050 and new biodiversity targets to be achieved in 2020 (Hill et al. 2015; Rands et al. 2010; Velasco et al. 2015). The Arctic Council published the Arctic Biodiversity Assessment 2013 which included 17 key recommendations to be implemented through an implementation plan by 2021 (CAFF 2015; Meltofte 2013). Biodiversity must be made an integral element of social, economic and political decision-making, as is happening with carbon and climate change. Members in a civil society (along with governments and businesses) all have crucial roles in education related to biodiversity.

The polar regions continue to spark our curiosity and imagination, coupled with a sense of adventure and fear of the unknown (May et al. 2014; Kaiser et al. 2010), providing perfect ingredients for both education and public outreach related to biodiversity. The polar regions do benefit from public enthusiasm for images and exploration and public interest in ice and ice-dependent biodiversity such as the charismatic mammals. Substantial national investments in polar formal education and outreach activities, including films, exhibitions, expeditions, educational materials, and books, will undoubtedly increase public awareness (Carlson 2009). Polar examples can be an excellent way to transmit basic concepts about a wide range of STEM (science, technology, engineering and mathematics) disciplines. Indeed, numerous countries conducted their education and outreach programs in the last few years (particularly during the International Polar Year, in 2007–08) focusing on a wide range of disciplines, including biodiversity, while sharing knowledge about the importance of the polar regions (Kaiser et al. 2010; Zicus et al. 2011). An assessment of the hundreds of education, outreach and communication activities during the 4th International Polar Year (IPY) 2007–2008 concluded that they touched more than 14 million people in 70 countries (Provencher et al. 2011).

Globally, habitat loss and degradation directly caused by local human activities appear to be the greatest threats to biodiversity (IPCC 2013; Vié et al. 2009). In polar environments, the primary threats are largely a result of human activities in distant lower latitudes, particularly as a result of $CO_2$ emissions and other pollutants, and increased pressure to exploit polar resources as those elsewhere become depleted; consequently anthropogenic climate change is the predominant threat to the polar regions, directly precipitating massive ecological change and interacting synergistically with other threats (e.g. fisheries overexploitation, pollution, invasive species) (Bennett et al. 2015). Such issues related to biodiversity must be addressed under a concerted plan to inform the broader educational system worldwide, as well as policy makers (e.g. at the Arctic Council and at the Antarctic Treaty Consultative Meetings) and the general public.

## 4 Examples of Polar Activities Related to Biodiversity

Educational initiatives on biodiversity are manifold and include various levels of audiences. For example, these can be schools (through student outreach programs and other classroom activities, teaching resource projects, student and teacher expeditions and science fairs); universities (through field courses, expeditions and professional development programs) and the general public (through multilevel initiatives presentations, exhibitions, science events, music, books, films and contests) (Kaiser et al. 2010). There were hundreds of examples of projects from small, local initiatives to larger international projects during the IPY related to polar biodiversity. Among others, the Global POP: International School Education—future scientists at work project (http://sustain.no/projects/globalpop) focused on environmental contaminants in fish and "included hands-on school activities and scientific research" (Kaiser et al. 2010). Seasons and Biomes of the GLOBE Program (http://www.globe.gov/do-globe/measurement-campaigns/past-projects/earth-as-a-system-projects/seasons-and-biomes) focused on monitoring seasons of the taiga/tundra forest and tundra biomes by engaging school teachers and students (Kaiser et al. 2010). The book, Polar Science and Global Climate: An International

Resource for Education and Outreach, developed during the IPY contained among others a series of classroom materials on a variety of biodiversity-related topics (e.g. "Penguin Family Reunion", "Polar Feasts" including Antarctic and Arctic Food Web Cards, and "Optimal Foraging") (Kaiser et al. 2010). Student expeditions, organized by Students on Ice (http://studentsonice.com/) and Students on Board (http://www.arcticnet.ulaval.ca/sb/index.php), among others, venture into both the Arctic and Antarctic and include strong components on polar biodiversity education. They are successfully continuing today (Kaiser et al. 2010). During the IPY, also several larger multilevel-initiatives and projects were organized. LATITUDE60! (http://www.portalpolar.pt/latitude60—o-que-foi.html) was the polar educational program in Portugal for the IPY, and was recognized as a success story internationally. It reached all corners of the entire country, from children to adults, from kindergarten to University level, and the general public through more than 40 educational activities (e.g. national contests, field courses, theatre plays, exhibitions, films, polar calendar). The activities were organized by polar scientists and educators around the country while addressing issues such as biodiversity, climate change and polar research (Kaiser et al. 2010; Schiermeier 2009; Xavier et al. 2013; Zicus et al. 2011). The EALÁT-Network Study (http://icr.arcticportal.org/about-ealat) on reindeer husbandry and climate change focused on the adaptability of reindeer herding communities to environmental change and the knowledge of reindeer herders, and had strong project components on education and outreach (Kaiser et al. 2010). Polar Weeks initiated during the IPY with the theme "What happens at the poles affects us all" provided an opportunity for focused events worldwide on polar regions and a variety of issues (Zicus et al. 2011).

When funding ended at the end of the IPY, many of the IPY initiatives stopped or evolved into new projects. The Association of Polar Early Career Scientists (APECS) and Polar Educators International (PEI) both had their roots in the International Polar Year and are working to continue the momentum and interest in the polar regions, by continuing many initiatives with e.g. the contribution of early career researchers and teachers around the world (e.g. during Polar Weeks). PEI is a vital international network of educators and researchers that aims to provide a deeper understanding of current polar science and to inspire appreciation and knowledge of the polar regions, their connectedness to Earth's systems and biodiversity, and importance to all humans across latitudes and cultures. These PEI goals are met through continuing professional development activities such as webinars and international workshops (Huffman et al., in press; Walton et al. 2013). Another recent example are the implementation actions of the recommendations provided as part of the Arctic Biodiversity Assessment 2013 published by the Conservation of Arctic Flora and Fauna (CAFF) Working Group of the Arctic Council, which also emphasizes the importance of promoting public training and education on Arctic biodiversity as well as the development of communication and outreach tools (CAFF 2015). An "Arctic Biodiversity Through the Lens" Photography Contest was organized in connection with the Arctic Biodiversity Congress 2014, and other tools (e.g. publications, films, social media campaigns) and educational kits (e.g. on Arctic ecology) are among the resources developed (CAFF 2015).

## 5 Tackling the Challenging Issues on Polar Education Related to Biodiversity: Step 1. Evaluation of Impact

While many types of polar education *could* make a difference in addressing the human dimensions of biodiversity conservation, it is important to identify which types of initiatives contribute and how. While impact evaluation can be difficult in many contexts, the way it operates is straightforward conceptually: You try to isolate the effect of an intervention (for example, by measuring a child's thinking about a polar concept before and after an intervention). The social sciences have invested decades of effort into developing the tools to validly research phenomena relating to attitudes, interests, motivations, thought processes and emotions, outcomes which are the subject of most informal learning and engagement goals. Indeed, few educational objectives are impervious to robust measurement using social scientific methods of one kind or another. For example, if we know how the same person understands the concept of polar biodiversity before encountering an educational programme and after, then we can identify the 'effect' or 'impact' of the educational programme on this outcome (Moss et al. 2015). Qualitative data or multiple data collection time points within the informal learning or engagement experience can further close the inferential gap to be able to robustly attribute impact (whether positive or negative) to an experience (Wagoner and Jensen 2015).

Turning to children's attitudes or learning, accurate impact evaluation of polar education programmes requires gathering data directly from the children (not a proxy such as a teacher or parent) using appropriate language and good survey design techniques. For example, after pilot testing different impact evaluation question options (Wagoner and Jensen 2015) for evaluating children's learning at London Zoo, the following question was used: 'Please draw your favourite wildlife habitat and all the plants and animals that live there (put names and labels on everything)'. This item yielded annotated drawings from questionnaire data gathered from pupils by their teachers before and after their visit to London Zoo. These annotated drawings were then subjected to rigorous content analysis to quantify the impact patterns (and variables that predicted impact), as well as a conventional qualitative analysis to elaborate understanding of the details of what was happening with children's understanding of wildlife habitats over the course of a zoo visit (Jensen 2014a). In the following example, a UK pupil visiting the zoo shows a small improvement in the accuracy of the representation, with a shift from having a polar bear and penguin side by side to showing only a penguin (female, age 13; Fig. 2). To elicit evaluation data, polar educators could ask children to draw a polar landscape and all the plants and animals that live there both before and after an educational intervention.

When conducting an impact evaluation of polar education with limited resources, the limitations in available budget, staff and methodological expertise are real problems for many polar education organisations (Jensen 2014b), undermining their ability to use robust evaluation methods. Indeed, those working in polar education

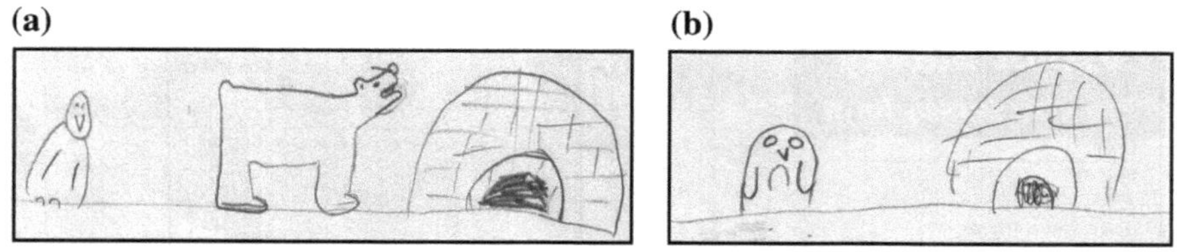

**Fig. 2** Pre- **a** and post-visit **b** drawings for same child from impact evaluation at London Zoo

organisations are busy people. Most of polar educators have not been trained in the social scientific research methods required to produce valid evaluations (example in Fig. 3) or to be a critical consumer of impact evaluations conducted by others such as external evaluation consultants. Below it is discussed two main options for overcoming these challenges: Improving methodological knowledge within polar education institutions and using technology-enhanced methods of evaluation.

Some methodological understanding would certainly be beneficial for polar education staff who will encounter evaluation evidence over the course of their careers. It is important to be a savvy consumer of polar education evaluation research, able to identify and avoid common limitations. Today however, there is less of a need for polar education practitioners to develop the practical knowledge and skills to be able to implement high quality evaluations for themselves. This is because recent improvements in open source technology bring good quality evaluation within easy reach of many more polar education professionals and organisations.

For many, if not most, polar education organisations, technology-enhanced evaluation could be a real solution for embedding robust evidence within the fabric of polar education practice. Automated evaluation tools enable answers to questions such as, what proportion of polar education participants are satisfied with their experiences? And, what factors are affecting the quantity and type of impact on polar education participants? Automated methods of evaluation can eliminate the need for on-going costs and expensive external consultants in order to gain evaluation evidence. New technologies enable the design of evaluation systems that can be fully automated after an initial customization and set-up. Using these technologies, a one-time infusion of expertise can create a system used by polar education practitioners without any skills in social scientific analysis. Recently developed options using open source technology include:

- A web-based system for gathering impact data before, during and after a polar education programme using automated evaluation technologies for data collection and analysis (www.qualiaanalytics.org);
- A system of evaluation that is integrated into a polar education institution's visitor smartphone app to automatically gather, analyse and display for the institution evaluation results such as visitor timing and tracking information and micro-survey feedback through the app. (www.qualia.org.uk);

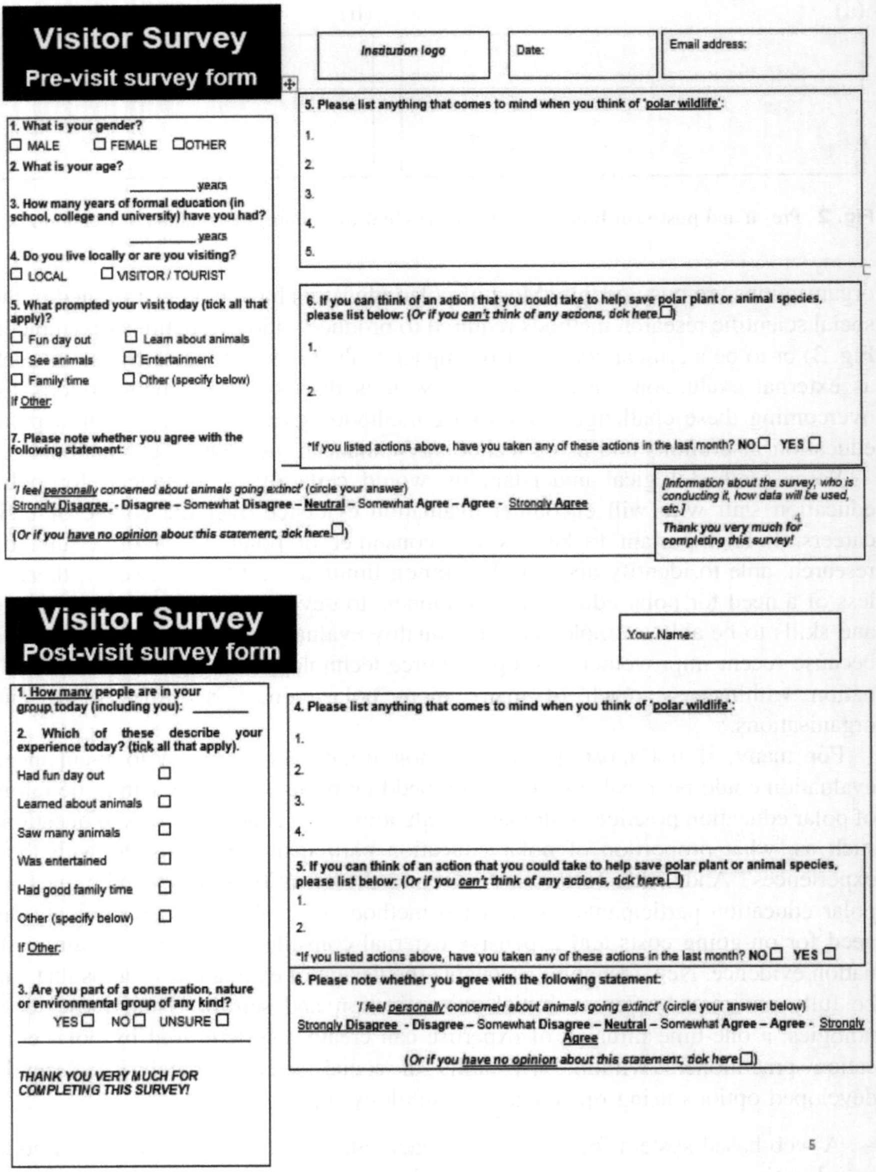
Visitor Survey
Pre-visit survey form

Institution logo | Date: | Email address:

1. What is your gender?
☐ MALE ☐ FEMALE ☐OTHER
2. What is your age?
________ years
3. How many years of formal education (in school, college and university) have you had?
________ years
4. Do you live locally or are you visiting?
☐ LOCAL ☐ VISITOR / TOURIST
5. What prompted your visit today (tick all that apply)?
☐ Fun day out ☐ Learn about animals
☐ See animals ☐ Entertainment
☐ Family time ☐ Other (specify below)
If Other:
7. Please note whether you agree with the following statement:
*'I feel personally concerned about animals going extinct'* (circle your answer)
Strongly Disagree - Disagree – Somewhat Disagree – Neutral – Somewhat Agree – Agree - Strongly Agree
*(Or if you have no opinion about this statement, tick here* ☐*)*

5. Please list anything that comes to mind when you think of 'polar wildlife':
1.
2.
3.
4.
5.

6. If you can think of an action that you could take to help save polar plant or animal species, please list below: *(Or if you can't think of any actions, tick here* ☐*)*
1.
2.
*If you listed actions above, have you taken any of these actions in the last month? NO ☐ YES ☐

*[Information about the survey, who is conducting it, how data will be used, etc.]*
*Thank you very much for completing this survey!*
4

Visitor Survey
Post-visit survey form

Your Name:

1. How many people are in your group today (including you): ________
2. Which of these describe your experience today? (tick all that apply).
Had fun day out ☐
Learned about animals ☐
Saw many animals ☐
Was entertained ☐
Had good family time ☐
Other (specify below) ☐
If Other:
3. Are you part of a conservation, nature or environmental group of any kind?
YES ☐ NO ☐ UNSURE ☐
*THANK YOU VERY MUCH FOR COMPLETING THIS SURVEY!*

4. Please list anything that comes to mind when you think of 'polar wildlife':
1.
2.
3.
4.
5. If you can think of an action that you could take to help save polar plant or animal species, please list below: *(Or if you can't think of any actions, tick here* ☐*)*
1.
2.
*If you listed actions above, have you taken any of these actions in the last month? NO ☐ YES ☐
6. Please note whether you agree with the following statement:
*'I feel personally concerned about animals going extinct'* (circle your answer below)
Strongly Disagree - Disagree – Somewhat Disagree – Neutral – Somewhat Agree – Agree - Strongly Agree
*(Or if you have no opinion about this statement, tick here* ☐*)*

5

**Fig. 3** Pre-visit survey and pos-visit survey to evaluate the impact of educational activities on polar biodiversity

– An automated system that identifies indicators of 'quality of experience' in social media messages posted by visitors, using categories developed through online ethnographic research with people tweeting about public engagement experiences (www.culturesmile.org).

In each of these cases, data collection and analysis are fully automated, with visualisations of the results displayed for institutions using the systems in real time. This means that institutions can immediately act on incoming evaluation results, rather than waiting for cumbersome processes of data collection, data entry and analysis to take place (also see www.qualiaanalytics.org).

Of course, any polar education evaluation approach has strengths and limitations. Automated evaluation methods are not suitable for every evaluation challenge. However, greater adoption of these technologies could raise the minimum standard of evaluation in the polar education sector, and provide institutions with a finger on the pulse of their audiences. Because these systems have been built using robust open source software development, the systems can be easily adopted by polar education organisations and interface with other commonly used software such as Eventbrite. The result is greater availability of higher quality evaluation evidence, while organisations conserve resources by replacing existing consultancy costs or diverted staff time. Ideally, if the resource burden of on-going polar education evaluation and market research could be removed, the sector would then be able to focus on strategic investment in in-depth rigorous research on aspects of engagement that require particular attention across the globe's polar regions (for example, reaching new audiences for polar education (Dawson and Jensen 2011; Jensen et al. 2011)).

## 6 Final Considerations

In this book chapter, we reviewed the importance of the polar regions to our planet, assessed how education on the polar regions has been developed in the last decade, provided key educational activities related to the polar regions on biodiversity, and provided a resource to stakeholders (such as scientists, educators, teachers and policy makers) interested in polar science and education.

Various steps have been taken to have an efficient international network of polar educators, particularly with the recent establishment of Polar Educators International. Several themes, related directly or indirectly to polar biodiversity, must be considered for a sustainable growth of polar educational issues: (1) Funding is needed to build a more reliable and vibrant online presence and platform for connecting polar educators and researchers internationally, (2) There is a need to continue to provide face-to-face opportunities for collaborations: educator/educator and educator/researcher and educators with other stakeholders, (3) The need to build an online "matching" service for educators to find researchers and researchers to find help with their broader impact and communication needs, (4) Seek creative solutions for making resources and dialogues easily shared across multiple languages, (5) Clearly design goals for polar education and communication in such a way that they can be quantified and evaluated, and (6) Continue to leverage what individuals within the polar community are doing and build on lessons learned during education outreach projects from the International Polar Year.

Within other future challenges to polar education, using robust evaluation to ensure the steady improvement of polar education initiatives should be viewed as essential. Developing appropriate skills and knowledge to use impact evaluation to inform the practice of polar education and engagement practice should not be an insurmountable barrier. Moreover, recent improvements in evaluation technology have opened up new options for implementing systems that provide on-going evaluation insights on an automated basis with a one-time infusion of expertise at the survey design stage (e.g. see www.qualiaanalytics.org, www.qualia.org.uk or artory.co.uk). Given the logistical challenges involved in developing high quality, practical evaluation methods training for practitioners whose primary responsibilities lie elsewhere, such technology may be a big part of the answer to the challenge of implementing robust evaluation in informal learning and engagement institutions. On-going evaluation systems would allow polar education organisations to be much better attuned to their audiences' needs, less reliant on speculation about which interventions are effective and therefore more likely to deliver positive impacts.

**Acknowledgements** This work is a culmination of numerous discussions with members of the Association of Polar Early Career Scientists and Polar Educators International, and is a product of the educational legacies from the International Polar Year. JX is supported by the Investigator FCT program (IF/00616/2013) and is part of the SCAR Ant-ERA, SCAR EGBAMM, PROPOLAR and ICED programs.

## References

Allison, I., Béland, M., Alverson, K., Bell, K., Bell, R.R., Carlson, D., et al. (2007). *The scope of science for the International Polar Year 2007–2008*. WMO/TD-No. 1364, Geneva.

Allison, I., Béland, M., Alverson, K., Bell, R., Carlson, D., Cutler, P., et al. (2009). *The state of polar research*. Geneva: World Meteorological Organization.

Aronson, R. B., Thatje, S., McClintock, J. B., & Hughes, K. A. (2011). Anthropogenic impacts on marine ecosystems in Antarctica. *Annals of the New York Academy of Sciences, 1223*, 82–107.

Barnosky, A. D., Hadly, E. A., Bascompte, J., Berlow, E. L., Brown, J. H., Fortelius, M., et al. (2012). Approaching a state shift in Earth's biosphere. *Nature, 486*, 52–58.

Bennett, J. R., Shaw, J. D., Terauds, A., Smol, J. P., Aerts, R., Bergstrom, D. M., et al. (2015). Polar lessons learned: Long-term management based on shared threats in Arctic and Antarctic environments. *Frontiers in Ecology and the Environment, 13*, 316–324.

CAFF. (2015). *Actions for Arctic biodiversity, 2013–2021: Implementing the recommendations of the Arctic biodiversity assessment*. Akureyri, Iceland: Conservation of Arctic Flora and Fauna.

Carlson, D. (2009). International viewpoint and news. *Environmental Geology, 56*, 1735–1738.

Convey, P. (2007). Antarctic ecosystems. In S. A. Levin (Ed.), *Encyclopedia of biodiversity*. San Diego: Elsevier.

Convey, P., Aitken, S., di Prisco, G., Gill, M. J., Coulson, S. J., Barry, T., et al. (2012). The impacts of climate change on circumpolar biodiversity. *Biodiversity, 13*, 134–143.

Dawson, E., & Jensen, E. (2011). Towards a 'contextual turn' in visitor research: Evaluating visitor segmentation and identity-related motivations. *Visitor Studies, 14*, 127–140.
Elven, R., Murray, D. F., Razzhivin, V., & Yutsev, B. A. (2011). *Checklist of the pan-Arctic flora (PAF)*. Oslo: Natural History Museum, University of Oslo.
Gutt, J., Bertler, N., Bracegirdle, T. J., Buschmann, A., Hosie, G., Isla, E., et al. (2015). The Southern Ocean ecosystem under multiple climate change stresses—an integrated circumpolar assessment. *Global Change Biology, 21*, 1434–1453.
Hill, R., Dyer, G., Lozada-Ellison, L.-M., Gimona, A., Martin-Ortega, J., Munoz-Rojas, J., et al. (2015). A social–ecological systems analysis of impediments to delivery of the Aichi 2020 targets and potentially more effective pathways to the conservation of biodiversity. *Global Environmental Change, 34*, 22–34.
Huffman, L., Xavier, J.C., Beck, I. & Lehman, R. (in press). Bringing polar research into the classroom. *Polar Record, 5*.
IPCC. (2007). *Climate change 2007: Synthesis report*. Geneva, Switzerland: IPCC.
IPCC. (2013). Summary for policymakers. In T. F. Stocker, D. Qin, G.-K. Plattner, M. Tignor, S. K. Allen, J. Boschung, et al. (Eds.), *Climate change 2013: The physical science basis. Contribution of working group I to the fifth assessment report of the Intergovernmental Panel on Climate Change* (p. 27). Cambridge, UK: Cambridge University Press.
Jensen, E. (2014a). Evaluating children's conservation biology learning at the zoo. *Conservation Biology, 28*, 1004–1011.
Jensen, E. (2014b). The problems with science communication evaluation. *Journal of Science Communication, 1*, C04.
Jensen, E., Dawson, E., & Falk, J. (2011). Dialogue and synthesis: Developing consensus in visitor research methodology. *Visitor Studies, 14*, 158–161.
Kaiser, B., Zicus, S., & Allen, B. (2010). *Polar science and global climate: An international resource for education and outreach*. London: Pearson Custom Publishing.
Kennicutt, M. C, I. I., Chown, S. L., Cassano, J. J., Liggett, D., Massom, R., Peck, L. S., et al. (2014). Six priorities for Antarctic science. *Nature, 512*, 23–25.
Knox, G. A. (2007). *The biology of the Southern Ocean*. CRC Press and Taylor & Francis: USA.
Krupnik, I., Allison, I., Bell, R., Cutler, P., Hik, D., López-Martinez, J., et al. (2011). *Understanding Earth's polar challenges: International Polar Year 2007–2008*. Rovaniemi, Finland: University of the Arctic and CCI Press.
May, I., Huffman, L. T., Xavier, J. C., & Walton, D. W. H. (2014). Education and polar research: Bringing polar science into the classroom. *Journal of Geological Resource and Engineering, 4*, 217–221.
Meltofte, H. (2013). *Arctic biodiversity assessment: Status and trends in Arctic biodiversity*. Akureyri, Iceland: Conservation of Arctic Flora and Fauna.
Meltofte, H., Barry, T., Berteaux, D., Bültmann, H., Christiansen, J. S., Cook, J. A., et al. (2013). *Arctic biodiversity assessment: Synthesis*. Akureyri, Iceland: Conservation of Arctic Flora and Fauna (CAFF).
Moss, A., Jensen, E., & Gusset, M. (2015). Evaluating the contribution of zoos and aquariums to Aichi biodiversity target 1. *Conservation Biology, 29*, 537–544.
Peck, L., Convey, P., & Barnes, D. K. A. (2006). Environmental constraints on life histories in Antarctic ecosystems: Tempos, timings and predictability. *Biological Reviews, 81*, 75–109.
Post, E., Forchhammer, M. C., Bret-Harte, M. S., Callaghan, T. V., Christensen, T. R., Elberling, B., et al. (2009). Ecological dynamics across the Arctic associated with recent climate change. *Science, 325*, 1355–1358.
Provencher, J., Baeseman, J., Carlson, D., Badhe, R., Bellman, J., Hik, D., et al. (2011). *Polar research education, outreach and communication during the fourth IPY: How the 2007–2008 International Polar Year has contributed to the future of education, outreach and communication* (p. 48). Paris: International Council for Science (ICSU).
Rands, M. R., Adams, W. M., Bennun, L., Butchart, S. H., Clements, A., Coomes, D., et al. (2010). Biodiversity conservation: Challenges beyond 2010. *Science, 329*, 1298–1303.

Richter-Menge, J., & Overland, J.E. (2010). *Arctic report card 2010*. http://www.arctic.noaa.gov/reportcard
Rockstrom, J., Steffen, W., Noone, K., Persson, A., Chapin, F. S., Lambin, E. F., et al. (2009). A safe operating space for humanity. *Nature, 461*, 472–475.
Sarmiento, J. L., Gruber, N., Brzezinski, M. A., & Dunne, J. P. (2004). High-latitude controls of thermocline nutrients and low latitude biological productivity. *Nature, 427*, 56–60.
Schiermeier, Q. (2009). In from the cold. *Nature, 457*, 1072–1077.
Smetacek, V., & Nicol, S. (2005). Polar ocean ecosystems in a changing world. *Nature, 437*, 362–368.
Thomas, D. N., Fogg, G. E., Convey, P., Fritsen, C. H., Gili, J.-M., Gradinger, R., et al. (2008). *The biology of the polar regions*. Oxford: Oxford University Press.
Velasco, D., García-Llorente, M., Alonso, B., Dolera, A., Palomo, I., Iniesta-Arandia, I., et al. (2015). Biodiversity conservation research challenges in the 21st century: A review of publishing trends in 2000 and 2011. *Environmental Science & Policy, 54*, 90–96.
Verde, C., Giordano, D., di Prisco, G., & Andersen, O. (2012). The haemoglobins of polar fish: Evolutionary and physiological significance of multiplicity in Arctic fish. *Biodiversity, 13*, 228–233.
Vié, J.-C., Hilton-Taylor, C., & Stuart, S. N. (2009). *Wildlife in a changing world—an analysis of the 2008 IUCN red list of threatened species*. Gland, Switzerland: IUCN.
Wagoner, B., & Jensen, E. (2015). Microgenetic evaluation: Studying learning in motion. In G. Marsico, R. Ruggieri, & S. Salvatore (Eds.), *The yearbook of idiographic science: Reflexivity and change in psychology*. Charlotte, USA: Information Age Publishing.
Walczowski, W., & Piechura, J. (2006). New evidence of warming propagating toward the Arctic ocean. *Geophysical Research Letters, 33*, L12601.
Walton, D., Xavier, J. C., May, I., & Huffman, L. (2013). Polar educators international—a new initiative for schools. *Antarctic Science, 25*, 473.
Xavier, J. C., Barbosa, A., Agusti, S., Alonso-Sáez, L., Alvito, P., Ameneiro, J., et al. (2013). Polar marine biology science in Portugal and Spain: Recent advances and future perspectives. *Journal of Sea Research, 83*, 9–29.
Xavier, J. C., & Peck, L. S. (2015). Life beyond the ice. In D. Liggett, B. Storey, Y. Cook, & V. Meduna (Eds.), *Exploring the last continent* (pp. 229–252). Switzerland: Springer International Publishing.
Zicus, S., Almeida, M., Edwards, K., Hik, D., Huffman, L., Kaiser, B., et al. (2011). IPY education activities. In I. Krupnik, I. Allison, R. Bell, P. Cutler, D. Hik, J. López-Martinez, et al. (Eds.), *Understanding Earth's polar challenges: International Polar Year 2007–2008* (pp. 481–496). Rovaniemi, Finland: University of the Arctic and CCI Press.

# Teaching Mycology Worldwide

María P. Martín and Roy Watling

**Abstract**

Taxonomy is the discipline responsible for the classification of organisms and involves sampling, discovery, and description of species. Fungi are a large group of eukaryotic organisms very diverse in morphology, physiology, and ecology. It is very difficult to delimit a species concept which is common to all fungal groups. In general, taxonomists continue using morphological characters; however, DNA sequence analysis (barcoding) is now essential to discover the true identity of new fungal species. In this chapter some aspects of teaching mycology will be summarized, including websides where fungi from different countries are well-documented, as well as scientific databases available for various groups of fungi and the application of new technologies. Some examples when training taxonomists all around the World are included, such as collecting in the rain forest with Malaysian and Thai students, teaching general and master courses in Australia, Brazil, Ecuador and India, and on-line-tools needed for the effective training of students and teachers when they are particularly when separated by long distances.

M.P. Martín (✉)
Departamento de Micología, Real Jardín Botánico, RJB-CSIC,
Plaza de Murillo 2, Madrid, Spain
e-mail: maripaz@rjb.csic.es

R. Watling
Caledonian Mycological Enterprises, Crelach, 26 Blinkbonny Avenue,
Edinburgh EH4 3HU, UK



P. Castro et al. (eds.), *Biodiversity and Education for Sustainable Development*,
World Sustainability Series, DOI 10.1007/978-3-319-32318-3_5

## 1 Introduction

Prof. Frederick DeForest Heald (1872–1954), pioneer in teaching plant pathology wrote: "the minimum time for a course in mycology according to the plan outlined is six hours of laboratory work per week throughout one school year" (Heald 1922). Although the number of hours is important, even more is how to teach. The plan outlined by Prof. Heald was teaching to construct diagrammatic keys to the various groups of fungi through the observation of specimens and discussion among students, instead of teaching directly how to use artificial keys to determine the species (to put the specimens in their respective species pigeon-holes, as he said). Prof. Heald recommends his method because of: "(1) The creation of greater interest on the part of the student in his work; (2) The development of students' ability to reason and weigh evidences; (3) The cultivation of the scientific imagination; (4) A better understanding of evolution and what it means; and (5) The possibility of emphasizing natural descent of the various groups and bringing out the fact that classification is in reality but a means to an end—an expression of relationships." Today, when studying fungi, students not only have the figures in books to interpret the features, as in the time of Prof. Heald, but now there is also a wide range of new tools.

Fungi are a large group of very diverse eukaryotic organisms and it is very difficult to delimit a species concept to apply to all the different fungus groups (Webster and Weber 2007). However, today in order to distinguish and to describe new species morphological, ecological or physiological characters are not just relied upon but DNA sequence analyses are available. In particular, the sequences of nuclear ribosomal DNA are very useful in the identification of fungi at different taxonomic levels (Bruns et al. 1991). The results from the last technique demonstrate that even if it is a well understood area the mycodiversity is surprisingly more diverse than previously thought. Documentation in tropical areas is even more difficult but the indication is that even the morphological approach uncovers many more species and some genera not previously recognise. Although recent molecular methods have moved onto another plane which still confirms this high fungal diversity morphological methods are still important where difficulties are experienced in communication, availability of relevant materials, etc. After applying good documented collecting techniques the dried material will always be available for future molecular study. It is now possible for some students learn how to obtain and interpret other characters, such as the small portion of DNA that has been selected by Schoch et al. (2012) as the first bar-coding option for fungi, although more recently there have been other regions suggested (Stielow et al. 2015). What, however, is most important is for the student to appreciate that what is imbued in the morphological approach is still relevant and ideally new and classical methods should be married. Present-day students have a great advantage in this digital era in which they are comfortable using the new methodology over those of earlier years but it is everyone's loss if the classical approach is totally ignored!

In the next pages, how mycology has developed from just utilizing books to the use now of free tools available through the new technologies is presented. Such tools include movies, webpages, applications to mobile devices, etc. and it is shown how these facilities have developed further with the advent of molecular analysis.

However, there is no doubt that learning mycology in any country and knowing its mycodiversity, at least at the scholar and high school level, should start with a class in a near-by forest, in a Botanical Garden, or a place were teachers can attract attention and demonstrate the diversity of fungi. If for any reason this is not possible, teachers could call upon some of the excellent movies found free on the internet, where curiosity in fungi can be ignited in young minds. Some examples of these kind of movies: Natural History, https://www.youtube.com/watch?v=70LA0mijzCM; Kingdom of the Forest-Fungi, National Geographic, https://www.youtube.com/watch?v=zb4y40kFhL4; Fungi, BBC's The Private Life of Plants documentary series, such as https://www.youtube.com/watch?v=vZ5Me4N_XXE, and https://www.youtube.com/watch?v=ETRX1-3fqRo). Good DVDs by Taylor F. Lockwood covering various aspects of larger fungi are available from www.kingdomoffungi.com.

Teachers should utilize the wide range of fungi available to offer selected examples of fungal diversity and not just rely on the fruit-bodies of the well-known macromycetes, available in stores such as those from the genera *Agaricus* or *Boletus*, although nowdays the range of species available is increasing. Lichenized fungi (Lichens), mouldy food-stuffs, plant parasites, and soil samples should all be included. Then under both binocular and monocular microscopes students can discover other important dimensions of fungi by observing hyphae or mycorrhizas, features which set fungi apart.

Although this chapter focuses on teaching mycology to secondary students, the advice is that in the university, the mycological course should start also with fieldwork and develop the subject around careful observation and documentation. If this is not possible, students should locate good internet pages about biodiversity of fungi, share and discuss them in an internet forum or in their university intranet, if available, and their tutors (http://www.nifg.org.ik/other_websites.htm), allow access to British Isles fungus groups, and also to many other fungus sites, some of them dedicated to special groups, such as one devoted to waxcaps (http://www.aber.ac.uk/waxcap/). Moreover, both students and tutors should be aware of the information available in the well-developed fungus databases of Index Fungorum (http://www.indexfungorum.org) and Mycobank (http://www.mycobank.org), where one can find not only the nomenclatural novelties, such as new names and combinations, but also find the first description of a fungus, and in the case of Mycobank the DNA sequences associated with them. Also, the website Global Biodiversity Information Facility (GBIF 2015; http://www.gbif.org offers); this page offers the opportunity to explore and contribute to the global body of evidence documenting the huge diversity of life on our planet; actually there are more that 579,000,000 records shared freely by hundreds of institutions worldwide, and 10,103,232 correspond to fungi, which around 7,000,000 are included with coordinates (Fig. 1).

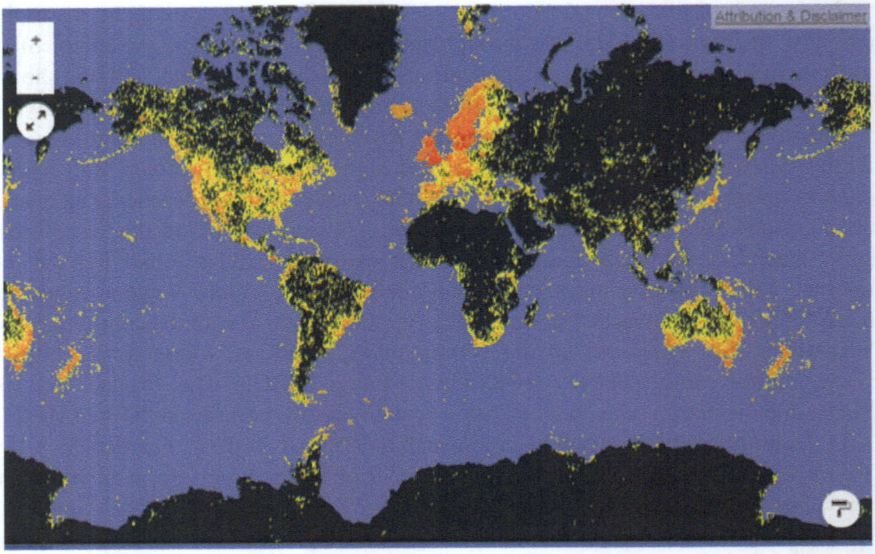


**Fig. 1** Map of the world showing around 7,000,000 fungi occurrences (GBIF.org)

## 2 Collecting

As mentioned before the fungi are such a diverse group of organisms that even the collecting methods to be adopted are very different from fungus to fungus. Macrofungi even the tiniest specimens can be preserved for later examination; if the fieldwork is for more than one day, a different approach depending of the country should be adopted (Fig. 2).

When collecting fungi make sure that all the fungus and not just the top has been collected, as sometimes they may be buried deeply in the soil, so carefully excavate the specimens. In order to protect the specimens an array of small disused tins and/or plastic (not glass) containers can be used to place the specimens in. This keeps them separate and avoids accidents by not using glass containers. The important aspect of collecting is to keep the specimens in good condition once picked and so several professional mycologists wrap their material loosely in twists of wax-proofed paper or even in metal 'bacon' foil. Both these can be obtained from a local store being used extensively for culinary activities. Do not put specimens in polythene bags as they will sweat in transit and start to decompose rapidly, destroying many important characters. Specimens can be maintained in the cool part of a household refrigerator over-night if they cannot be examined immediately. Working with macrofungi in the tropics is sometimes difficult when a source of electricity is not available, so that the use of a European fruit drier cannot be set-up at a base camp. Failing the use of such a drier, larger specimens should be placed on

**Fig. 2** Field work in Thailand. *Author* Roy Watling

a drying rack set up with an encircling cardboard chimney and small source of heat; smaller specimens can be placed directly into indicator silica gel (this can be used also for bulkier specimens, in which case they must be sliced appropriately), changing the silica gel next day, or at least twice for leathery and tougher specimens. If using a drier or the drying action of a fire, then place the specimens in small open mesh sacks e.g. old nylon stockings or similar are appropriate. This keeps collections separate, as once dry, specimens which looked very different in the field may look like each other. Keep a record of the specimen, and always make sure the collection number accompanies the specimen at all times. The old silica gel can be reused by drying it near to a source of heat, on the rack, etc., and leaving for 24 h. Delicate specimens can be protected with moss, within the silica gel phial, and the whole retained for shipping back to the laboratory. Also, it is very important to choose the correct silica gel, since fine silica gel is damaging to cover-slips when examining on return; round silica gel pellets tend to grind the specimens when dry; so use medium coarse calibre silica gel. Collecting and Examination of larger fungi is described in Henderson et al. (1969), illustrating important features to observe when the specimens are in the fresh condition.

Once collected, in many groups of fungi, the spore print is important for successful identification and should be obtained, by using a second specimen from the collection or part of any larger specimen. Label the specimen taken for the spore-print with the same number as the original collection so that they can be reunited at a later date. Moreover, it may be rewarding to check close to the place of collection, since mushrooms make the spore-print in nature; if the mushroom is mature one can notice coloured dust covering the surrounding ground, leaves or branches, and this can be carefully collected. Some videos are available on the Internet, of how to obtain spore prints (such as https://www.youtube.com/watch?v=5XztAeKjKHY or https://www.youtube.com/watch?v=-fNHTmNy5to). Always make a note of where the specimen was found and with what species it might have been associated. It might not be possible to identify individual trees or plants when in a woodland but some broad description of the habitat is better than nothing and could assist in any future identification. Keep the same numerals for the notes as on the specimens. With small specimens and when collecting microscopic specimens a piece of the substrate is best dried with the actual fruiting body.

Students should be encouraged to make illustrations of their finds by either photography or sketches, preferably coloured. Digital cameras which young people are so familiar with make photography much easier but in the latter case crayons whose colour is released when dampened allows a whole range of colours to be taken in the wettest rain forest. As many fungi are not recorded or even described from an area, colour illustrations play an important role in documenting that specific area. If the specimens cannot be identified, then with the collected material and an illustration, an expert might be able to identify them in the future. Make sure the number for the collections is kept also with the illustrations. Even without a name, the illustrations add to the knowledge of the fungal diversity as there are still many fungi to be formally recognised.

## 3 Morphological Analyses

Exploring the fungal world will continue in school, since students can prepare a series of exhibitions. The exhibitions can be set on the walls of the classroom covering the different aspects of fungi. Such an exhibition could include edible (and poisonous—if known) mushrooms, wood-rotters including bracket fungi, crust-fungi, rotting fruit and vegetables, and even pharmaceutical products and domestic utilities such as biological washing powders. The exhibition should be prepared with fresh specimens, but if this is not possible, with images obtained from different sources: books, colour images or internet pages. Different countries or even areas of countries have produced attractive illustrative books which will show the range of colours and textures and sometimes the bizarre nature of some fruiting bodies. Excellent collections of images (including Spanish and Catalan description) have been published by the Societat Catalana de Micologia under the abbreviate name Bolets de Catalunya (http://www.micocat.org; 34 collections already published with 50 planches each). Generally, edible fungi and rotten examples of vegetable and fruit can be obtained fairly easily, the latter two from the discards of green-grocers!

The focus of the exhibition should first be related to macroscopic morphology. Students can arrange their specimens or images according to the different morphologies they observe. Later, the students can be distributed in teams and choose a group of fungi to focus upon: team A, stomach (gasteroid) fungi; team B, fungi with a stem (or stipe); team C, fungi lacking a stem; team D: lichens, and so on. After discussing the forms that they observed, each team, will look for information on the internet (some examples included below) or use any mobile device application (some exampled below), in order to prepare a short talk in which they should describe the group they have studied using the appropriate names for the different macroscopic morphological characters observed. Tutors will decide which team will be the first to give the talk as it depends on the specimens and whether they are resilient and less likely to decay. It is very important for students also to prepare spore-prints (https://www.youtube.com/watch?v=UFgg8ZBLruY), because they can use them as a source of spores which they can observe directly under the microscope after registering their colour in deposit but later also act as source by which to grow the fungus in culture in the following days. This activity will demonstrate the diversity of fungi and their importance in everyday life as visual impact is very important and relates to young people.

There is a series of very good webpages that students can look through in order to find information. Here are some examples, such as the North American Mycological Association, http://www.namyco.org/join.php, to general information; http://www.mycokey.com/newMycoKeySite/MycoKeyIdentQuick.html, with interactive keys; some Wikipedia pages, such as the page of gasteroid fungi (in English), https://en.wikipedia.org/wiki/Gasteroid_fungi; or some specific pages, such as the page about crustose lichens, http://centexnaturalist.com/, maintained by Jerry Evans.

The application to mobile devices that students could use is dependent on the countries and their languages. Some popular fungus applications are: Roger's Mushrooms (Pro), by Roger's Plants Ltd, this App is based upon the works of Roger Phillips, author of 'Mushrooms', the worldwide best-selling mushroom and fungus book; and others, such as the Great Encyclopedia of Fungi by AppGrade, or Audubon Mushrooms, a field guide to North American Mushrooms by National Audubon Society, and Mushrooms PRO, NATURE MOBILE—For Safe Enjoyment! by NATURE MOBILE G.m.b.H; and, iMushroom Hunter by iCartel s.r.o. Also, the App FunKey: Key to Agarics of Australia, by LucidMobile. The first author, together with a Spanish entreprise (Wake App!), has developed an application called FungiNote, with simple ways to observe the mascroscopic features, in order to try to find which fungus is in front of you (more about this application, and experience gained is included in the next section of this chapter).

Returning to the student exhibitions, if it was done with fresh collections, the next class could focus on microscopic characters of fungi. As a minimum exercise, students will need to observe the spore morphology, basidia, asci and cystidia, and structure of the cap (pileus) preparing their own microscopic slides. In this second activity they can discuss how the new information they have discovered can change their first grouping (classification). A second interpretation should be prepared based on microscopic observations, including their own microscopic drawings. Micrographs of these fungal structures can be viewed in the series of manuals 'How to Identify Fungi' published by Mad River Press (California, USA) (Largent 1977).. This will allow comparisons to be made to establish between what the students observe down the microscope with what is seen in the different images. Also, they can look at the New Zealand's virtual Mycota webpage (http://virtualmycota.landcareresearch.co.nz/webforms/vM_Mushroom.aspx?PK=0), from where fungi are distributed according to the spore colour and other morphological features.

To study fungi from a very different angle, cultures can be prepared. Students can make soil dilution plates or rub soil along the base of a Petri dish before flooding it with luke-warm agar. Equally a small amount of a spore print can be dispersed in water and added to luke-warm agar before pouring the agar out. After incubation, the students can compare what they have obtained with colour illustrations of soil fungi and discuss their role in the soil. If for any reason, poor or negative cultures result, tutors should have some extra cultures, prepared earlier, to show to the students what ideally they should have obtained. These cultures can be the source also of material, which will later under-go molecular study. There are some good books on cultures, such as Watanabe (2010) or St Germain and Summerbell (1995), and part of the series mentioned above produced by Mad River Press; as well as the classic book on soil fungi by Barron (1968) and Domsch et al. (1980). Also, students can look through different papers published on line in order to observe the different cultures obtained from a variety of sources, such as those isolates from butternut (https://www.purdue.edu/htiKrc/newsletter/2012/May.html), culture fungal endophytes from Australian rainforests (http://eatlas.org.au/media/736), or fungi present in indoor environments (http://scialert.net/fulltext/?doi=ajbs.2012.304.313).

After some days, observing and discussing about fungi, students should be able to answer a series of general questions related to these incredible organisms, such as those included in the files http://www.namyco.org/docs/Answers_to_Fun_in_Fungi.pdf and http://www.namyco.org/docs/Answers_to_More_Questions_About_Fungi.pdf (English/Spanish). If students need help, they can go through other books and webpages, such as the Fungi Kingdom of the Kew Botanical Garden, UK (http://www.kew.org/science-conservation/plants-fungi/fungi).

## 4 Scientists and Company Collaboration to Education

Although scientists and private companies collaborate in many ways, such as in the production of new medicines or innovations in different fields of engineering; the collaboration is less habitual when we speak about education. Two years ago, the first author had the opportunity to collaborate with the Spanish company Wake App! a company that produce and edits applications for mobile devices. As young people are now very familiar with the use of apps in an ever-expanding field of digital communication it was seen that a Mycological Application should be considered. As explained to the different media, at the beginning the company wanted an app. focused just on mushrooms or toadstools (fungi with a stem and a domed cap), but after discussion it was agreed to produce an app. suitable for macroscopic fungi in general (http://www.dicat.csic.es/rdcsic/index.php/en/recursos-naturales-2/96-historiasde-exito/236-cientificos-y-empresa-colaboran-en-la-creacion-de-una-app).

The company Wake App! which had previous experience with medical apps., focused on the design and usability whilst an independent developer carried out the programming with the scientists at the Real Jardín Botánico (CSIC) providing the scientific advice and guidance for the content.

Thus with FungiNote students can compare morphological aspects of the fungi with clear drawings, and progressively filter the characters until arriving at the most probable species (Fig. 3).

Moreover, the app can filter by habitat and by season of the year when fungi are found, both in Northern and Southern Hemispheres. Also, the app. allows the creation of a personalized field notebook where the user can annotate the findings and share them with other students or people interested in fungi using the same application (Fig. 4).

This app will be improved, since at present it is only in Spanish and for devices with the IOS operating system; however, it was well noted in some educational webpages (e.g. www.thebiologistapprentice.com/blog/categoru/funginote).

**Fig. 3** FungiNote screens showing some morphological aspects of the fungi with clear drawings, and progressively filters until arriving at the most probable species. The app. can filter by habitat and by season of the year when fungi are found

**Fig. 4** FungiNote allows creation of a personalized field notebook where the user can annotate the findings and share them with other students or people interested in fungi using the same application

To develop this app. was a very rich experience, since there is the chance to explore a totally new sector in which to transmit contributions to the knowledge of fungi as well as the ability to encourage curiosity for the fungi. To us, there is not doubt that "the combination of scientists and commercial companys makes knowledge transfer to society possible. This allows knowledge to be passed onto society to the point of putting it in the hands of thousands of fans, teachers and citizens, in an innovative scientific outreach activity".

## 5 Conclusions

Current knowledge of fungal diversity in many areas of the World is poorly documented, even though fungi are vital for terrestrial function of ecosystems. One of the biggest problems in the description and monitoring of the mycodiversity of a site is undoubtedly the decline of the number of taxonomists in the world. It is vital that new generations are introduced in the study of fungi and thankfully this is being made easier with the widespread availability of digital systems. Even with the stimulus of molecular data, the study of mycology will wane if no new taxonomists are forthcoming. Thus teachers and tutors should use all the tools available, from books to the use of new technologies. Potential mycologists should be versed in not just molecular work and digital communication but also in the long tested classical methods. The marriage of all these techniques ensures an exciting future for young people. They will not only be able to expand our present knowledge of the mycodiversity but by digital means link with information sourced from other organisms and ecosystems.

**Acknowledgments** We would like to acknowledge to all the students and colleagues for helpful discussions when preparing our mycological courses. M.P.M. would like to thank to those involved in FungiNote: Wake App! (production and design), Ricardo Sánchez Sotres (technological development and programming), colleagues who collaborate on the first 150 scientific contents (F.D. Calonge, L. Suz, R. Pino-Bodas, M. Tabarés and D. Sierra), and to the CSIC. To F. Pando (www.gbif.es) for helping with Fig. 1.

## References

Barron, G. (1968). *Genera of hyphomycetes from soil*. New York, USA: Kreiger Publishing Co.

Bruns, T. D., White, T. J., & Taylor, J. W. (1991). Fungal molecular systematics. *Annual Reviews of Ecology and Systematics, 22*, 525–564.

Domsch, K. H., Gams, W., & Anderson, T. H. (1980). *Compendium of soil fungi* (p. 865). London, England: Academic Press.

Global Biodiversity Information Facility. (GBIF.org; 2015, November 10). GBIF occurrence, download http://doi.org/10.15468/dl.hzimk7

Heald, F. D. (1922). Some suggestions for teaching mycology. *Transactions of the American Mycological Society, 41*, 175–178.

Henderson, D. M., Orton, P. D., & Watling, R. (1969). *British fungus flora: Agarics and Boleti—Introduction*. Edinburgh, UK: HMSO.

Largent, D. (Ed.). (1977). How to identify fungi. California, US: Mad River Press.

Schoch, C., Seifert, K. A., Huhndorf, S., Robert, V., Spouge, J., Levesque, A., et al. (2012). Nuclear ribosomal internal transcribed spacer (ITS) region as a universal barcode marker for fungi. *PNAS, 109*(16), 5907–6354.

St Germain, G., & Summerbell, R. (1995). Identifying filamentous fungi: A clinical laboratory handbook. Star Publishing Company.

Stielow, J. B., Lévesque, C. A., Seifert, K. A., Meyer, W., Irinyi, L., Smits, D., et al. (2015). One fungus, which gene? Development and assessment of universal primers for potential secondary fungal DNA barcodes. *Persoonia, 35*, 242–263.

Watanabe, T. (2010). *Pictorial atlas of soil and seed fungi: Morphologies of cultured fungi and key to species* (3rd ed.). CRC Press.

Webster, J., & Weber, R. (2007). *Introduction to mycology* (3rd ed.). UK: Cambridge University Press.

# Bugs and Society I: Raising Awareness About Endemic Biodiversity

Ana Moura Arroz, Rosalina Gabriel, Isabel R. Amorim, Rita São Marcos and Paulo A.V. Borges

**Abstract**

The negative impact of biodiversity loss on ecosystem functioning and services, and ultimately on human well-being, has been unequivocally established; however, despite all efforts, biodiversity is still declining worldwide. It is widely accepted that biodiversity awareness is crucial for its conservation. Nevertheless, after many initiatives to alert society about the consequences of losing biodiversity, biodiversity loss is still perceived as a minor environmental risk compared to others such as climate change. Thus far, most communication strategies have involved conventional venues, targeting people who are already "environmentally-aware", and have not incorporated societal idiosyncrasies and cultural backgrounds. The wicked problem of loss of biodiversity and the existing strategies to promote people's engagement are discussed under the risk communication framework. The *risk perspective* focuses science communication on the mitigation of risk and/or the minimisation of its consequences. Further, it helps to keep the target in mind, to establish activities and strategies that are useful for reaching the proposed goals, and to regulate the processes based on desired outcomes. After presenting the role of communication in risk governance, the principles and strategic options of the Azorean intervention,

A.M. Arroz (✉) · R. São Marcos
Departamento de Ciências da Educação, cE3c—Centre for Ecology, Evolution and Environmental Changes/Azorean Biodiversity Group and Universidade dos Açores, 9700-042 Angra do Heroísmo, Açores, Portugal
e-mail: ana.mm.arroz@uac.pt

R. Gabriel (✉) · I.R. Amorim · P.A.V. Borges
Departamento de Ciências Agrárias, cE3c—Centre for Ecology, Evolution and Environmental Changes/Azorean Biodiversity Group and Universidade dos Açores, 9700-042 Angra do Heroísmo, Açores, Portugal
e-mail: rosalina.ma.gabriel@uac.pt

 
P. Castro et al. (eds.), *Biodiversity and Education for Sustainable Development*, World Sustainability Series, DOI 10.1007/978-3-319-32318-3_6

*Bugs & Society*, are presented. This communicational programme will be analysed and evaluated in another chapter through the presentation of two activities, which are provided as examples for further discussion.

# 1 Framing Biodiversity Loss Within Risk Governance Towards Nature Conservation

## 1.1 Social (Ir)relevance of Biodiversity Loss: A Minor Global Problem

Biodiversity[1]—the remarkable variety of life (CED 1992), appears to be unique to Earth (Cardinale et al. 2012). It is not only a major component of a region's natural heritage, but is also essential to our survival. Ecosystem services including provision (e.g., food, fuel, fresh water) and regulation (e.g., climate, floods, disease) (MEA 2005) rely on species, even on those that are considered too small and/or superfluous.

### 1.1.1 Biodiversity Loss as an Environmental Risk

An increasing consciousness of the importance of biodiversity and its conservation is evident in several established global targets, such as the 2010 target (Adenle 2012) and the "Strategic Plan for Biodiversity 2011–2020" (SPB 2010), and in local, national, regional and international legal regulations concerning species and habitat conservation (e.g., Bern Convention, Habitats Directive). In the Azores, the Regional Legislative Decree 15/2012/A, recently approved the legal regime for the conservation of nature and biodiversity in Azores. However, despite the mounting evidence regarding the negative impact of biodiversity loss on human well-being and health, the efficacy of those initiatives has failed to meet the expectations since: (i) biodiversity is still declining due to human activities (MEA 2005; Butchart et al. 2010; Cardinale et al. 2012); (ii) deadlines to achieve conservation goals keep being pushed forward (e.g. SPB 2010; Butchart et al. 2010; Adenle 2012); and (iii) biodiversity loss is still not presented as a major risk in international agreements on disaster risk management (e.g., Sendai Framework for Disaster Risk Reduction 2015–2030; UNISDR 2015). Despite the broad consensus among biologists on the magnitude of biodiversity loss (Cardinale et al. 2012), the number of studies that demonstrate that the loss of biodiversity is already beyond safe operating planetary boundaries (e.g., Rockström et al. 2009; Liu et al. 2015), and the fact that halting biodiversity loss has become a key sustainability challenge for the 21st century, non-expert citizens tend to downplay the consequences of biodiversity loss.

In fact, when compared with the visibility attained by other environmental risks, such as climate change, scientists dealing with biodiversity loss have not been able to

[1]For the definition of most technical terms see EEA (2015).

effectively persuade society at large, and politicians in particular, of the seriousness of the risk (Pereira et al. 2010). An innovative strategy towards a better awareness of biodiversity in the policy arena is the recently created Intergovernmental Platform on Biodiversity and Ecosystem Services (IPBES) (see Pereira et al. 2013).

This difficulty in communicating the risk of biodiversity loss can be attributed to several factors, some of which are related to the specific characteristics of the problem, namely, the difficulty of evaluating the impact of biodiversity loss, the uncertainty, and pervasiveness of this risk, and the delayed invisibility of its effects. A clear enunciation of this problem thus requires an understanding of its uncertain and ambiguous nature (WBGU 2000). When considering *biodiversity loss*, uncertainty refers mostly to the type and severity of its consequences, while ambiguity refers to the lack of consensus among stakeholders regarding the prevalence and consequences of biodiversity loss—they all present different attitudes and values, focus on different interests and change their minds at different paces. There is also a time lag between the moment that biodiversity is lost and the moment its impacts start to be noticed; consequently, biodiversity loss is not taken as an imminent threat, even though both its probability of occurrence and damage potential are high. In this sense, biodiversity loss may be considered a Cassandra type risk (WBGU 2000; Renn 2008). Finally, the complexity (multiple causes and consequences), ambiguity (different recognition by various stakeholders), and insidious nature (effects only visible after a long time) of biodiversity loss make it a *wicked problem* (e.g. Sharman and Mlambo 2012): ill-defined, with no "best solution", impossible to address by trial and error, but, at the same time, too important not to solve.

### 1.1.2 Engaging Society with Biodiversity Loss

Once a risk has been identified and the probabilities and consequences of risk events have been assessed, the next obvious step is to plan mitigation strategies. With risks as complex as biodiversity loss, many measures need to be put in place to account for different possibilities (flexibility), including precautionary measures (preparedness), which must be able to anticipate and face many scenarios, regardless of ecosystem resilience (e.g., ecological corridors and micro-reserves, projection analyses). However, to achieve a consensus, or at least a strategic tolerance regarding ambiguous problems (e.g., WBGU 2000), it is necessary to engage people.

In Portugal, among other issues that should be addressed to increase communication efficacy, is the concern regarding the low level of scientific literacy (EC 2005) and the estrangement towards science (EC 2005, 2010). In 2005, only 52 % of Portuguese individuals inquired answered scientific questions correctly, while the European average (EU25) was 66 %. In 2010, Portugal ranked 4th among EU27 countries in not acknowledging any relevance to science in everyday life; besides, about one-third (35 %) of the inquiry respondents were not interested in new scientific findings and technological innovation (compared to the 20 % EU27 average), although more than half (56 %) felt that they were poorly informed regarding those subjects. This suggests that science and technology are not adequately communicated, which not only hinders the objective set in 2000 of turning EU into a knowledge economy centred on an ambitious research and innovation agenda (EU 2000), but also may jeopardise sustainable development in Portugal. Risk communication may become a key player in changing this situation.

Promoting science and scientific literacy, and narrowing the gap between research and "real life" are challenges that must cope with the complexity of science and its typically encrypted language, which often makes communication between different scientific fields difficult, and is even recognised by several authors as a paradigmatic incommensurability (Kuhn 1970).

Bridging the gap between people and science does not mean forcing the former to value science according to the same criteria as experts. The aim, instead, is that by combining different agendas and interests, personal and significant reasons will emerge and allow people to understand and value the positive role of science in dealing with everyday life situations. To achieve such a goal, people's interests and universes of significance must be investigated, and science communication must play a mediating role among agendas of different actors to achieve a higher good, such as nature conservation or biodiversity loss mitigation. Moreover, to widen audiences, science communication must diversify its strategies, languages and discourses according to a variety of citizens with unique cultural backgrounds, sensibilities and values to successfully reach the intended public.

Besides, exploring stakeholder's perspectives on biodiversity will provide important undercover arguments, relevant for delineating successful science communication strategies tailored according to people's interests, as well as the opportunity to deconstruct their assumptions and beliefs. Questions like the following are crucial for gathering pertinent information among locals to assist in communication strategies planning:

- How is biodiversity loss perceived by different stakeholders?
- What knowledge do they have on biodiversity? Where did they get this knowledge?
- How do they estimate the probability of risk occurrence and the severity of its consequences?
- How do emotional factors, such as the delight/disgust in a species, affect people's perceptions of the impact of its loss?
- Who is responsible for what, in terms of nature conservation?

## 1.2 Risk Communication as a Tool for Biodiversity Conservation

### 1.2.1 Risk Communication Within the Governance of Risk

Risk communication is a relatively new field (Box 1) that has been expanding rapidly as current societies face new risks and uncertainties (Beck 1999; Giddens 2002) that must be addressed to improve preparedness and build confidence within communities. Risk communication is understood, according to CAC, *as an interactive exchange of information among individuals, groups, and institutions related to the assessment, characterization, and management of risk* (CAC 1997).

**Box 1. A brief overview of the evolution of risk communication**
In the last decades, a paradigm shift has occurred in how risk is communicated, from mainly informative, unidirectional and persuasive strategies, to more dialogical ones (Fischhoff 1995; McComas 2006; Saunders et al. 2006). Early on (1975–1984), risk communication mostly consisted of conveying technical information, and did not consider the perspectives of the target groups. In the following decade (1985–1994), risk communication was dominated by persuasive approaches aimed to make people believe in certain ideas (or raise acceptance levels) and/or to produce changes in behaviour. Around 1995, it began to rely more on participatory approaches, namely by including citizens as privileged partners in negotiations towards collective understanding and actions. These methods bring together citizens and experts in a social learning process intended to create mutual trust (Renn 2008), respect for each other's backgrounds and agendas (Rowe and Frewer 2005), and empower community members (Burgess and Clark 2006) to deal with the issues in private and/or public (legal, institutional) ways (McComas 2006). One thing that is obvious from the history of risk communication is the growing recognition of risk as a social, cultural and psychological construct (Fischhoff 1995; McComas 2006). This shift to dialogical approaches started a new era in risk communication, where the ability to engage various stakeholders, to stimulate dialogue and to foster tolerance plays a central role, i.e., recognising that considering many different viewpoints allows for the building of a strategic consensus, which is a prerequisite for action to take place. Risk communication is seen as a bridge combining the ideas of sound science and democratic participation.

In integrated risk management, risk communication must take the central role in all processes concerning risk, from risk identification and assessment to decision-making and implementation and regulation of risk interventions (Renn 2008; IRGC 2008). Communication is meant to facilitate the understanding of risk specificities by all of society and to promote discussion on everyone's role and responsibilities in the process of governance. The intentional interactive nature of risk communication is meant to "give a voice" to all stakeholders in order to reach agreements that maximise everyone's interests and, consequently, to achieve the expected outcomes of augmenting public audibility and commitment, and communication effectiveness. Partnerships, where civil society as a whole becomes a stakeholder and emphasis is put on the local context where these dynamics take place, allow for the inclusion of all actors as collaborative partners (Fig. 1).

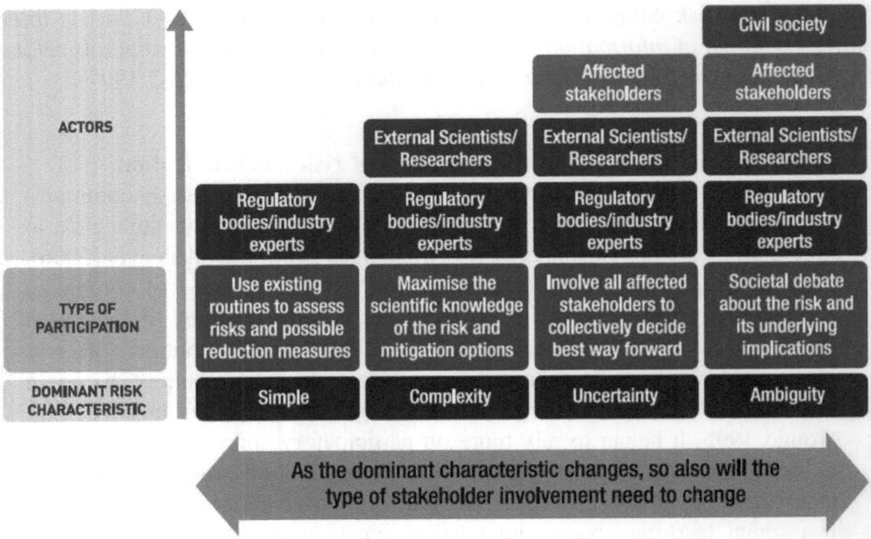


**Fig. 1** Stakeholders involvement in the IRGC risk governance framework (reproduced from IRGC 2008, p. 18)

Once these conditions are met, risk communication becomes, at the same time, an empowerment tool that allows informed decision-making and increases the power to act and, thus, autonomy. It also becomes a tool to promote public trust in the risk management system.

### 1.2.2 Engaging Society Through Risk Communication

It is noteworthy to underline the fact that decision-making in risk management depends on the social, institutional, political and economic characteristics of the place where a specific risk exists. Those characteristics dictate the processes of governance to be implemented, which may facilitate, or compromise, engagement of different actors in the management process itself. Inclusive management is based on the assumption *that all stakeholders have something to contribute to the process of risk governance and that their inclusion improves the final decisions rather than impedes the decision-making process or compromises the quality of scientific input* (IRGC 2008, p. 18). However, this does not mean that stakeholders and civil society should have the same levels of involvement and commitment for all risks. Following the IRGC approach, the appropriate type of involvement by different actors should be according to risk profile and risk knowledge. Given that biodiversity loss falls within the category of high-ambiguity risk, measures should focus on promoting risk visibility and analysis of risk implications. To address these types of risks, society at large must be engaged (Fig. 1), even though some may not be aware or may underestimate all the negative consequences of this particular risk.

A more inclusive philosophy that seeks to obtain and to be influenced by more diverse opinions is believed to increase participation and bring innumerable social benefits (e.g., Fischhoff 1995; Rowe and Frewer 2005), such as the facilitation of policy implementation and the promotion of their efficacy, conflict reduction, consolidation of the democratic processes and increased trust in politicians (Vasconcelos et al. 2009).

Yet, it is possible to recognise different rationales for public engagement practises (Fiorino 1990; Cass 2006): normative, substantive and instrumental (Table 1). Public engagement may be used as a means to reach better practises and policies (substantive), but it can also be a goal in itself, to pursue democratic values towards empowerment and participation (normative), and/or to strategically attain social trust to conquer advantages in the *social game* (instrumental).

Although all involvement strategies aim to increase public participation, typologies of participation may differ, including in the level of engagement pursued. According to the International Association of Public Participation, five levels of engagement may be recognised according to the increasing level of public impact, corresponding to different purposes: inform, consult, involve, collaborate and empower (IAP2 2007). Each goal calls for specific engagement techniques and creates distinct expectations among citizens, but all of them may be useful per se or in combination, depending on the specific situations.

Strategies of public involvement are often organised in a bipolar continuum, ranging from less to more engaging, or from the involvement *of the* public to the involvement *by the* public, according to strategic goals and emphasis on persuasion, mediation or co-production. Regardless of the specific public engagement strategy considered, communication and public participation processes should be guided by ethical values, such as accountability, transparency, respect, inclusion and a true willingness to involve the public (e.g., Rees 2013; OPE 2014).

The importance of public participation in the relationship between science and society, namely in risk communication, but also in policy and public management,

**Table 1** Rationales for public engagement

| | Normative | Substantive | Instrumental |
|---|---|---|---|
| Aim | Goal | Mean | Strategy |
| Key ideas | To pursue democracy | Multiple opinions guarantees quality | Get an advantage in the "social game" |
| Benefits | Increase empowerment and participation | Provide better and more informed policies and practices | Build trust to conquer audibility and legitimate positions |
| Critical points | Democracy and instrumentalization<br>Representation and representativeness | Cost/benefit<br>Believe in a common will to be discovered<br>Support of automatic consensus | Dubious ethics of strategic behaviour<br>Rhetoric<br>Manipulation |

in general, has been well established. However, some authors question the enthusiasm for a so-called *doctrine of engagement* (Walls et al. 2011). Walls and colleagues draw attention to the *rather uncritical acceptance of many of these (assumed) potential benefits* (Walls et al. 2011, p. 243), while Cass (2006) systematises the critiques of participation and deliberation addressing normative legitimacy (democracy and representation), substantive effectiveness of policies, instrumentality (strategic uses of participatory-deliberative public engagement) and incorporation into policy.

One of the acutest problems with establishing the benefits of public engagement initiatives is the difficulty of assessing their efficacy. Firstly, few studies evaluate the outputs and outcomes of interventions. Secondly, even when interventions are critically examined, there is little empirical evidence of the existence of benefits (review in Rowe and Frewer 2004). As Rowe and Frewer (2004) point out, there are theoretical and practical difficulties for the implementation of such an evaluation system, which must be fair to all participants and efficient in getting results.

Despite the criticisms and difficulties presented in evaluating the performance of communication strategies, the design of the public intervention *Bugs & Society* considered an efficacy assessment, using a system to continuously evaluate and regulate intervention outputs and outcomes (see also Amorim et al. 2016; this book).

## 1.3 The Problem: How to Promote Endemic Insects as Azorean Natural Heritage?

### 1.3.1 Endemic Biodiversity in the Azores

The Azores are well known for its breath-taking nature landscapes, dominated by volcanic features and vibrant green vegetation. Moreover, the Azores belong to one of the most diverse biogeographical regions of Europe, the Macaronesian Islands, a group of islands in the North Atlantic Ocean that also includes the archipelagos of Madeira, Canaries and Cape Verde (Derneği 2010; Myers et al. 2000). The Macaronesian Islands host around 23,000 terrestrial and marine species, of which approximately 5600 are endemic species (Izquierdo et al. 2004; Arechavaleta et al. 2005; Borges et al. 2008, 2010), i.e. species that only occur there. The fact that these species can only be found on the Macaronesian Islands may be due to the extinction of all populations that may have existed on continental areas or, alternatively, the endemic species may have originated on the particular archipelagos where they occur, as the result of local evolutionary processes over the course of millions of years.

Focusing on the Azores, roughly 60 % of its terrestrial endemic species (and subspecies) are arthropods (Borges et al. 2010), *those small things that run the world* (Wilson 1987, p. 344), and are crucial for healthy ecosystems. Arthropods include many animal groups such as spiders, mites, springtails, crustaceans, centipedes, millipedes and, most of all, insects. In fact, around three-quarters of the Azorean endemic arthropods are insects, which represent almost half of the terrestrial endemic species of the Azores. Despite the uniqueness of these Azorean endemic species, their important ecosystem functions and growing interest amongst biologists, their small sizes and the occupancy of recondite habitats makes them difficult to observe and, consequently, most, if not all, are unknown to the general public. Azorean endemic arthropods lack, therefore, the charisma of other island endemic species, such as Darwin's finches or the giant tortoises of the Galapagos Islands, which adds to the many challenges that the conservation of Azorean terrestrial biodiversity must face.

### 1.3.2 The Invisibility of Azorean Biodiversity Loss

Almost 600 years after the first settlers arrived in the 15th century, 246,000 people live in the Azores (INE 2012). Most of the territory in the archipelago has been transformed into urban areas, Japanese cedar (*Cryptomeria japonica* (L. fil.) D. Don) plantations and, mainly, into graze lands, with dairy farms and agriculture as the main sources of livelihood (SREA 2011). The native forest has been reduced to less than 3 % of its original extent (e.g., Gaspar et al. 2008), being almost restricted to mid and high elevations, where most of the endemic fauna and flora of the archipelago can be found. Only *circa* 17 % of the territory is protected under the Azorean regional network of Natural Parks (Calado et al. 2009). These land use changes have had an impact on the species that inhabit the Azores, and many endemic species, namely insects and spiders, but also birds and other life forms that once roamed the Azores, are now extinct (Terzopoulou et al. 2015).

Although the need to preserve this unique natural heritage has already been recognised as a priority in Regional and European Development Strategies (RIS3-Açores, H2020), society, in general, is largely unaware, on one hand, of the value of the Azorean biological diversity and the risks associated with its loss, and on the other hand, of the need to preserve it and the responsibility of all citizens in this endeavour. This lack of awareness is echoed in many areas, namely in the growing economic activity of tourism, where the Azores are advertised as a "pristine wilderness" destination. Tourism marketing, including initiatives promoted by local authorities, present the green landscapes of the archipelago as an indicator of unspoiled nature. However, this *green portrait* of the Azores "sold" to visitors and locals is quite misleading, as it fosters a biased and uncorrected view of the biodiversity found in the archipelago (Santos 2014), and overlooks current threats to its conservation. More protagonism is given to the flora compared to the fauna, when, in fact, almost half of all species in the Azores are animals (43.3 vs. 25.8 %), and almost three-quarters (73 %) of all terrestrial species that can only be found in archipelago are also animals (Borges et al. 2010). In addition, having landscapes dominated by *green* is not by any means an environmental quality stamp, since not all green areas may be relevant to the conservation of biodiversity. In the Azores, a *green* patch of native forest and a *green* patch of intensive pasture represent ecosystems with distinct levels of human intervention, hosting different kinds and amounts of biodiversity. The former habitat is the least disturbed—where most of the endemic species can be found; the latter, in spite of being much more common and presenting great economic relevance, is rather poor in species (Borges et al. 2009).

Both graze lands and areas covered by exotic (and even invasive) plant species, such as the Japanese cedar or the kahili ginger (*Hedychium gardnerianum* Sheppard ex Ker-Gawl.), are unfortunately commonly perceived as emblematic of nature in the Azores (Arroz and Gabriel 2011). Also very revealing of this attachment to exotic species is the fact that the most popular and iconic plant of the Azores, hydrangea (*Hydrangea macrophylla* (Thunb.) Ser.), is an exotic species introduced from Asia that, in some of the Azorean islands (e.g., Faial and Flores), has the status of an invasive species (Silva et al. 2008). The inability to distinguish and value species that naturally occur in the Azores (indigenous species), the lack of knowledge regarding species that only exist in the archipelago, and that many of them are threatened are all serious impediments to the conservation of the unique Azorean biodiversity.

A high level of environmental literacy and awareness of the seriousness of the risk of local biodiversity loss, particularly of indigenous species (native and endemic) of a community represents a valuable resource to conservation. Those communities tend to be more involved in the management and mitigation of biodiversity loss, demanding the mobilisation of resources (scientific, technical, financial, social, politics) (Vasconcelos et al. 2009). The first step towards the creation of a public involvement culture is, thus, to promote high-quality public communication, as Rowe and Frewer (2005) proposed in their typology of public involvement considering the flow of information.

### 1.3.3 Challenging the Social Invisibility of Biodiversity Loss with Azorean Insects

Insects are among the least favoured taxonomic groups in most cultures. People have a strong dislike for many insect species; this negative attitude comes from the fact that insects are regarded as useless, dangerous and disease vectors (e.g., malaria) and agriculture pests (Neto 2004; Posey 1983). Besides, insects awake feelings of phobia, dislike, disgust and aversion. Being seen as ugly and dangerous, insects are perceived as a threat, even as a "God's mistake" (Gurung 2003, p. 337). The Azores are no exception, and preliminary studies show similar attitudes towards insects and other arthropods (Silva and Gabriel 2007; Gabriel et al. 2012a, b). The economical and heritage damages caused by a few insects in urban areas (e.g., termites) (Arroz et al. 2012a; Borges and Myles 2007) and agriculture (e.g., Japanese beetle, *Popillia japonica* Newman, 1841) (Lopes et al. 2001) in the Azores certainly contribute to that negative view. Accordingly, 22.8 % of Azorean students from 12–18 years old

(N = 1023) claim that, in the case of a global catastrophe, they would not safe-keep insects in Noah's Ark, basing their attitudes, in order of importance, on the fact that: (i) they "do not like" insects, (ii) insects "are gross", (iii) "are bothersome", (iv) "are scary", (v) "are useless", (vi) "are ugly" and/or (vii) "are irrelevant" (Gabriel et al. 2012a, b; unpublished data).

This project aimed to raise biodiversity awareness, which is a goal clearly in line with European (EU Regulation No 1291/2013 [Horizon 2020]; SPB 2010 [Aichi Targets]) and Regional Development Strategies (EC 2014 [RIS3-Açores]). Specifically, it aimed to:

(i) outline a science communication programme with the specific characteristics of the risk of biodiversity loss, sensitive to the cultural references of the local people;
(ii) develop communication strategies that take into account Azorean biodiversity, focusing on insects—the taxonomic group with the largest number of endemic species; and
(iii) describe the principles and strategies that supported the intervention *Bugs & Society*.

# 2 Planning the *Bugs & Society* Intervention: From Principles to Evaluation

## 2.1 Purposes and Guiding Principles

It is widely accepted that one of the first steps towards conservation is knowledge (e.g., Balmford et al. 2002), and that the primary goal of any conservation initiative must be to acquaint people with the species in need of protection. Given the predominantly negative perspectives of Azorean people towards insects and aiming to promote endemic species as a significant part of the natural heritage of the Azores, the ultimate goals of the intervention *Bugs & Society* were the following:

(1) to make people aware of the existence of these species, particularly of their uniqueness, to strengthen people's identity with their natural heritage;
(2) to foster biodiversity conservation, alerting people to the risks associated with its loss and to their responsibility in the matter;
(3) to develop communication strategies to be implemented and tested inside and outside conventional science venues (schools, science centres, museums, interpretative centres), targeting people that are "environmental unaware", and using multiple discourses and languages; and
(4) to alert researchers to the need for approaching the grammars of the population when communicating science, which involves understanding the way people perceive the issues in question.

To be able to work as part of a multidisciplinary team, there was a need to create a common lexicon among diverse scientific domains, making explicit the principles guiding the science communication intervention on environmental education of the Azorean Biodiversity Group (e.g., Arroz et al. 2006; Cordeiro et al. 2012; Gabriel et al. 2012a). The main guiding principles behind the Azorean intervention *Bugs & Society* are summarised in Box 2.

Ethically, in risk/science communication, as in other public interventions, the golden rule should be to establish relationships between researchers/communicators and citizens that rely on the acknowledgement and respect for all interlocutors and for their diversity of cultural capitals, reasoning and logics.

From an epistemological perspective, the relationship between the researcher/communicator and the subject under analysis should seek to respect the complexity of the problem (biodiversity loss), the different legitimation agencies of *making science* and *communicating science* (Dahlstrom 2014), and of the strategies to address it, adopting a critical posture in order to inform future interventions.

**Box 2. Main guiding principles for the communication intervention *Bugs & Society***

Ethical

*Suitability*: to successfully involve citizens, communication must be tailored according to their cultural references, which implies acknowledging their logic and rationalities, as well as their needs and interests.

*Inclusivity*: science communication must reach all citizens and should particularly target those with less access to scientific culture.

*Autonomy*: a social intervention, based on a collaborative logic, aims to engage, capacitate and empower citizens.

Epistemological

*Multidisciplinary*: addressing the complexity of *wicked* problems, and dealing with the social indifference generally associated with them, demands the engagement of many scientific domains and disciplines.

*Accuracy*: taking into account citizens' logic, interests and perspectives when tailoring communication to specific audiences must not compromise scientific truthfulness.

*Accountability*: the strategic planning of a social intervention must include the systematic and continuous evaluation of its implementation, results and impacts to allow for adjustments during the progress of the intervention and, therefore, ultimately guarantee its efficiency and effectiveness.

Ontological

*Extensiveness*: using a multitude of communication strategies and vehicles promotes message saturation, which facilitates raising awareness among the target group, even on non-focal issues, at the same time that contributes towards reaching a wider audience.

Finally, from an ontological perspective, science and risk communication interventions may be seen as processes of social influence (Weber and Word 2001) by which citizens may or may not allow themselves to be critically persuaded. This will ultimately depend on, for example, the proximity between people's beliefs and the message being conveyed, and the ability to establish a relationship of trust and credibility with communicators.

## 2.2 Logic Model: A Pragmatic Approach for Planning, Implementation and Evaluation

A logic model was used as a pragmatic approach that provides a critical framework: (i) to regulate the entire process of conception, planning, and evaluation of all initiatives within the intervention; and (ii) to operationalise output indicators in order to assess whether planned activities are likely to achieve intended results. This is particularly relevant, given that criteria for measuring public engagement and intervention success are often neglected (Walls et al. 2011).

A logic model is a planning tool to clarify and graphically display what a project intends to do and what it hopes to affect and accomplish (NNLM 2006). Besides summarising key programme elements, such as goals, resources, activities, outputs and outcomes, it also elucidates the rationale behind the activities and facilitates project management. It is a central reference for everyone involved in the intervention, therefore promoting communication and learning (Knowlton and Phillips 2013). However, a logic model differs significantly from an action plan. While an action plan may be seen as a guide for running a project, a logic model tests an explicit hypothesis to the extent that it implies a "theory-based" evaluation. Thus, a logic model potentiates experience-based learning (Schmitz and Parsons 1999), concerning both project implications and knowledge transferability. Equally important for using a logic model framework in the Azorean *Bugs & Society* intervention were the facts that logic models are easily adaptable to specific activities, by deciding on indicators specific to the particular issues being addressed, and that team members have had previous positive experience with these models (Arroz et al. 2012a). Developing a logic model was crucial for structuring the monitoring and regulatory processes of all activities, as it facilitated identifying proper indicators for those activities (Fig. 2).

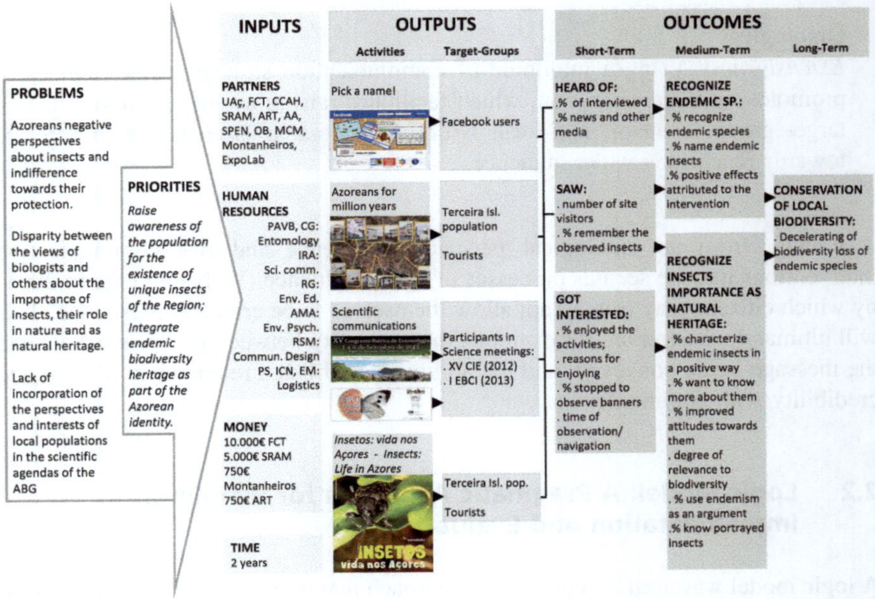


**Fig. 2** Logic model for the Azorean intervention Bugs & Society (initials for the fields partners, human resources and funding refer to research team members, collaborators and institutions—see acknowledgements)

## 2.3 Logic Model: A Pragmatic Approach for Planning, Implementation and Evaluation

Thus, the Azorean intervention *Bugs & Society* comprised: (i) defining action priorities, namely, making local citizens aware of the existence of insect species that can only be found in the Azores, and establishing a link between endemic biodiversity and natural heritage, so that it becomes part of the Azorean identity; (ii) establishing short-, medium- and long-term expected results according to the available time, resources and partnerships negotiated, (iii) planning and implementing specific activities for different target groups; and finally, (iv) comparing actual outputs and outcomes to expected ones.

The initial challenge—to increase awareness of Azorean endemic biodiversity—was briefly characterised according to three axes: (i) negative perspectives of Azorean people concerning insects and their lack of interest concerning insect conservation; (ii) the gap between expert biologists and general public perspectives on the importance of insects and their role in nature, and on the value of natural heritage; and (iii) that the Azorean Biodiversity Group has yet to incorporate in its scientific research agenda the perspectives and interests of Azorean citizens.

Activities that mobilised different resources and languages were planned, at different times, for different target groups. All of which aimed to impact the local conservation of endemic species and, therefore, contribute to halting biodiversity loss (see Box 3 and Amorim et al. 2016).

**Box 3. Activities developed in the communication intervention *Bugs & Society***

*Açorianos há milhões de anos* (Azoreans for millions of years)—an outdoor exhibition of large banners (4 m × 2.75 m) of close-up (focus stacking extreme macro) photographs of Azorean endemic insects, targeting the local population and tourists visiting Terceira Island (see Amorim et al. 2016);

*Insetos: Vida nos Açores* (A bug's life in the Azores)—an indoor exhibition on the diversity of insects (including endemic species) that can be found in the Azores, and their importance in insular ecosystems, developed in collaboration with the Science Centre in Angra do Heroísmo, targeting the local population and tourists visiting Terceira Island;

*Chama-lhe Nomes!* (Pick a name!)—the creative naming of endemic insects—a contest on the social network Facebook, targeting a wider online audience (see Amorim et al. 2016); and

*Communications on scientific meetings* targeting biology and entomology experts to raise awareness of the importance of considering citizens' perspectives about insects on their scientific activities: *XV Congresso Ibérico de Entomologia* 2012 (Arroz et al. 2012b; Gabriel et al. 2012b), *I Encontro sobre Biodiversidade e Conservação de Invertebrados* 2013 (Amorim et al. 2013; Arroz et al. 2013; Gabriel et al. 2013).

The Azorean intervention *Bugs & Society* comprised several stages (Fig. 3).

The logic model was also used to evaluate participation, outputs and outcomes, where indicators and data-collection methods were specific to the particular

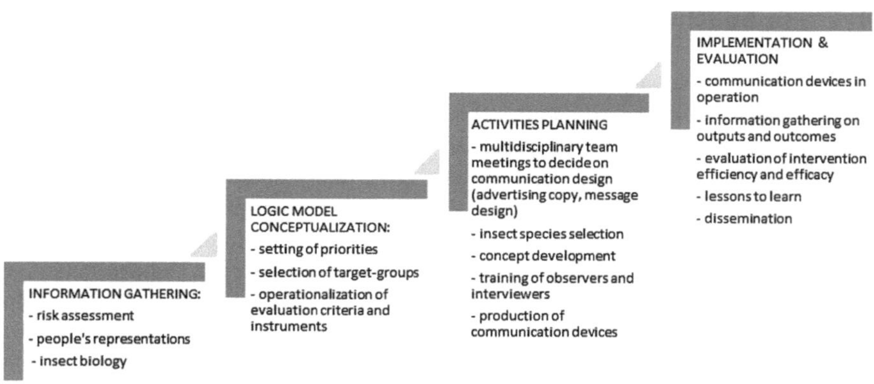


**Fig. 3** Intervention stages and general time schedule

communication activity (see chapter: Amorim et al. 2016): the exhibition *Azoreans for millions of years* and the Facebook contest *Pick a name!*

The evaluation was based upon an existing approach, developed by us and other authors (e.g., Arroz et al. 2012a; Reed 2008; Walls et al. 2011). This approach is:

- *Pragmatically oriented*, focusing on participation as a means to an end, where engagement with the stakeholders aims to promote the quality of environmental decisions and practises;
- *Multi-criteria*, considering various sets of criteria/evaluation perspectives;
- *Multi-method*, comprising a mix of qualitative and quantitative methods; and
- *Reflexive*, identifying underlying implications to improve future interventions, rather than simply evaluating the success of the present intervention.

Regulatory processes involved in the intervention conception and production of communication devices, the characterisation of participants' representations of nature and insects, the characterisation of participants' relationships with nature and insects, and the evaluation of results and impacts were all operationalised (see Amorim et al. 2016). However, prior to that, action indicators concerning results and impacts had to be envisioned:

- results/outputs: indicators mainly concerning regulatory processes of the efficiency and efficacy of devices, interest sparked by the communication devices, and public resonance of the intervention, via both informal channels and social media amplification; and
- impacts/outcomes: indicators concerning the enhancement of basic entomological literacy concerning endemic species, social introjection of endemic insects as valued natural heritage entities, and the willingness of entomologists to incorporate citizens' perspectives on insects into their scientific agendas.

Additionally, unpredicted outcomes and spontaneous activities (e.g., photography contest, other lines of research, education activities) were documented.

## 3 In Synthesis

The rate of biodiversity loss has not decreased despite all efforts, which indicates that the measures taken to address this issue need to be improved, including the strategies to communicate the value of biodiversity and the risk of its loss. The communication strategies adopted to date have likely misjudged the scientific literacy level of the general public and struggled to bridge the gap between science experts and the general public. For instance, the use of highly technical and/or encrypted language (e.g., concepts alien to most people, such as biodiversity—see Lindemann-Matthies et al. 2010), the venues traditionally chosen for science dissemination (e.g., science centres, museums, universities), and/or targeting already environmentally aware audiences (e.g., teachers, students) may have hampered the goal of mainstreaming the message and attaining engagement from all of society.

Additionally, the use of concept terms, such as *endemism*, so dear to conservationists were signalled as having xenophobic resonances (Simberloff et al. 2013) and may be read as morally controversial.

In this chapter, it was presented an example of a science communication intervention established in the integrated risk management framework. The purposes and principles of risk communication were addressed, highlighting the mediation role among stakeholders with different agendas and cultural capital. The *risk perspective* focuses science communication on the mitigation of risk and/or the minimisation of its consequences. These rather concrete purposes help to keep the target in mind, establish activities and strategies useful to achieve goals, and regulate the processes based on desired outcomes.

Therefore, based on the principles and strategies of *Bugs & Society* described above, in the next chapter (Amorim et al. 2016), the description and evaluation of two practical initiatives, *Açorianos há milhões de anos* (Azoreans for millions of years) and *Chama-lhe Nomes!* (Pick a name!) will be presented.

**Acknowledgments** We would like to thank several persons and institutions without which this work would not have been possible. FUNDING: Project PTDC/BIA-BEC/104571/2008 (FCT, Portugal), Project ATLANTISMAR - M2.1.2/I/027/2011 (DRCT, Azores), Grant SFRH/BPD/102804/2014 (IRA); PARTNERS: University of the Azores (UAç), Os Montanheiros, Amigos dos Açores (AA), Sociedade Portuguesa de Entomologia (SPEN), Ordem dos Biólogos (OB), Museu Carlos Machado (MCM), Expo-Lab Lagoa - Açores, Câmara Municipal de Angra do Heroísmo (CCAH), Associação Regional de Turismo (ART), Secretaria Regional do Ambiente e do Mar (SRAM), Ecoteca da Terceira and the owners of private buildings and managers of public institutions that made the outdoor exhibition possible; PHOTOGRAPHERS: J. Torrent (JT), E. Mendonça (EM), PAV Borges (PAVB) and M Gascoigne-Pees (MGP); PARTICIPANTS and COLLEAGUES: all the participants in pilot-studies, surveys and interviews, people involved in logistics and colleagues from the Azorean Biodiversity Group, especially C. Gaspar, for fruitful discussions and general help.

## References

Adenle, A. A. (2012). Failure to achieve 2010 biodiversity's target in developing countries: How can conservation help? *Biodiversity and Conservation, 21*, 2435–2442.

Amorim, I., Arroz, A. M., São Marcos, R., Borges, P. A. V., & Gabriel, R. (2016). Bugs and society II: Testing two communication strategies for public engagement in the Azores. In P. Castro, U. M. Azeiteiro, P. Bacelar Nicolau, W. Leal Filho, & A. M. Azul (Eds.), *Biodiversity and education for sustainable development* (pp. xx–xx). Dordrecht: Springer.

Amorim, I. R., Gabriel, R., Arroz, A. M., São Marcos, R., & Borges, P. A. V. (2013, July). *Chama-lhe nomes! As redes sociais na divulgação científica de informação taxonómica.* I Encontro sobre a Biodiversidade e Conservação de Invertebrados em Portugal. Faculty of Sciences, University of Lisbon, Portugal. At https://repositorio.uac.pt/handle/10400.3/2377

Arechavaleta, M., Zurita, N., Marrero, M. C., & Martín, J. L. (2005). *Lista preliminar de espécies silvestres de Cabo Verde (hongos, plantas y animales terrestres).* Consejería de Medio Ambiente e Ordenación Territorial, Gobierno de Canarias, Santa Cruz de Tenerife.

Arroz, A. M., Amorim, I. R., Gabriel, R., São Marcos, R., Gaspar, C., & Borges, P. A. V. (2012b, September, 1–6). *De marginal a bem patrimonial: Processos de divulgação científica na exposição 'Tesouros de seis patas'*. XV Congresso Ibérico de Entomologia. Angra do Heroísmo, Portugal: University of the Azores.

Arroz, A. M., & Gabriel, R. (2011). Controvérsias em torno da Hortênsia e da Marsília: um contributo para a compreensão do património natural à luz da relação entre ciência e sociedade. In R. Gabriel & R. B. Elias (Eds.), *Livro de resumos do workshop "Prevenção e controlo de espécies invasoras no Arquipélago dos Açores"* (pp. 68). Angra do Heroísmo, Portugal: University of the Azores. At http://hdl.handle.net/10400.3/2407

Arroz, A. M., Gabriel, R., Amorim, I. R., São Marcos, R., Torrent, J., & Borges, P. A. V. (2013, July). *Insectos pela cidade: Uma intervenção urbana em etnoentomologia*. I Encontro sobre a Biodiversidade e Conservação de Invertebrados em Portugal. Faculty of Sciences, University of Lisbon, Portugal. At https://repositorio.uac.pt/handle/10400.3/2383

Arroz, A. M., Gabriel, R., Rodrigues, A. F. F., Palos, A. C. P., & González, P. F. (2006, February). *O educador e a intervenção ambiental: uma experiência de formação transdisciplinar pós-graduada*. V Congreso Internacional de Educación Superior. Havana, Universidad 2006.

Arroz, A. M., São Marcos, R., Neves, I. C., Guerreiro, O., Gabriel, R., & Borges, P. A. V. (2012a). *Relatório final da campanha SOS TERMITAS - unidos na prevenção*. Angra do Heroísmo: University of the Azores. At http://cita.angra.uac.pt/ficheiros/publicacoes/1331251532.pdf

Balmford, A., Clegg, L., Coulson, T., & Taylor, J. (2002). Why conservationists should heed Pokémon. *Science, 295*(5564), 2367b.

Beck, U. (1999). *World risk society*. Cambridge: Polity Press.

Borges, P. A. V., Abreu, C., Aguiar, A. M. F., Carvalho, P., Jardim, R., Melo, I., et al. (Eds.). (2008). *A list of the terrestrial fungi, flora and fauna of Madeira and Selvagens archipelagos*. Funchal and Angra do Heroísmo: DRAM & University of the Azores.

Borges, P. A. V., Azevedo, E. B., Borba, A., Dinis, F. O., Gabriel, R., & Silva, E. (2009). Ilhas Oceânicas. In H. M. Pereira, T. Domingos, & L. Vicente (Eds.), *Ecossistemas e bem-estar humano - Avaliação para Portugal do Millenium Ecosystem Assessment* (pp. 461–508). Escolar Editora: Lisboa.

Borges, P. A. V., Costa, A., Cunha, R., Gabriel, R., Gonçalves, V., Martins, A. F., et al. (Eds.). (2010). *A list of the terrestrial and marine biota from the Azores*. Cascais: Princípia.

Borges, P. A. V., & Myles, T. G. (Eds.). (2007). *Térmitas dos Açores*. Lisboa: Princípia.

Burgess, J., & Clark, J. (2006). Evaluating public and stakeholder engagement strategies in environmental governance. In A. Guimarães Pereira, S. Guedes Vaz, & S. Tognetti (Eds.), *Interfaces between science and society* (pp. 225–252). Sheffield: Greenleaf.

Butchart, S. H. M., Walpole, M., Collen, B., Strien, A. V., Scharlemann, J. P. W., Almond, R. E. A., et al. (2010). Global biodiversity: Indicators of recent declines. *Science, 328*, 1164–1168.

CAC—Codex Alimentarius Commission. (1997). *Report of the twenty-second session of the Codex Alimentarius Commission, Geneva, 23–28 June 7 1997*. At http://www.codexalimentarius.net/web/archives.jsp?year=97

Calado, H., Lopes, C., Porteiro, J., Paramio, L., & Monteiro, P. (2009). Legal and technical framework of Azorean protected areas. *Journal of Coastal Research, SI 56*, 1179–1183.

Cardinale, B. J., Duffy, J. E., Gonzalez, A., Hooper, D. U., Perrings, C., Venail, P., et al. (2012). Biodiversity loss and its impact on humanity. *Nature, 486*, 59–67.

Cass, N. (2006). *Participatory-deliberative engagement: A literature review*. Published by the School of Environment and Development, Manchester University, Manchester M13 9PL, UK. At http://www.manchester.ac.uk/sed/research/beyond_nimbyism

CBD—Convention on Biological Diversity (1992). *Article 2. Use of terms "biological diversity"*. At https://www.cbd.int/convention/articles/default.shtml?a=cbd-02

Cordeiro, L., Arroz, A. M., & Gabriel, R. (2012). O ambiente sob o signo da infância: diversidade de conceitos e expectativas face ao futuro. In R. Gabriel (Ed.), *Abordagens do ambiente em contexto educativo* (pp. 177–289). Cascais: Princípia.

Dahlstrom, M. F. (2014). Using narratives and storytelling to communicate science with nonexpert audiences. *PNAS, 111*(4), 13614–13620.

Derneği, D. (2010). *Ecosystem profile: Mediterranean basin biodiversity hotspot*. CEPF—Critical Ecosystem Partnership Fund.

EC—European Commission. (2005, June). *Science and technology, January–February 2005*. Special Eurobarometer 224/Wave 63.1—TNS Opinion & Social.

EC—European Commission. (2010, June). *Europeans, science and technology, January–February 2010*. Special Eurobarometer 340/Wave 73.1—TNS Opinion & Social.

EC—European Commission. (2014, March). *National/regional innovation strategies for smart specialisation (RIS3). Cohesion Policy 2014–2020*. Brussels: European Commission.

EEA—European Environment Agency. (2015). *EEA glossary*. At http://glossary.eea.europa.eu/

EU—European Union (2000). *The Lisbon special European Council (March 2000): Towards a Europe of innovation and knowledge*. Brussels, EUR-Lex. At http://eur-lex.europa.eu/legal-content/EN/TXT/?uri=uriserv:c10241

EU Regulation No 1291/2013 of the European Parliament and of the Council of 11 December 2013 establishing Horizon 2020—the framework programme for research and innovation (2014–2020). *Official Journal of the European Union* I, no 347/104 of 20.12.2013.

Fiorino, D. J. (1990). Citizen participation and environmental risk: A survey of institutional mechanisms. *Science, Technology and Human Values, 15*(2), 226–243.

Fischhoff, B. (1995). Risk perception and communication unplugged: Twenty years of process. *Risk Analysis, 15*, 137–145.

Gabriel, R., Amorim, I. R., Arroz, A. M., São Marcos, R., & Borges, P. A. V. (2012b, September). *O lugar dos insetos na nova Arca de Noé*. Comunicação apresentada no XV Congresso Ibérico de Entomologia. Angra do Heroísmo, Portugal: University of the Azores.

Gabriel, R., Arroz, A. M., Amorim, I. R., São Marcos, R., & Borges, P. A. V. (2012a). Perspetivas acerca do património Natural dos Açores: Exploração preliminar dos dados obtidos junto de alunos do 3° ciclo e secundário na Região Autónoma dos Açores. In P. A. V. Borges, I. R. Amorim, A. M. Arroz, B. C. Emerson, C. Faria, R. Gabriel, & F. Pereira (Eds.), *Final report "What can the Macaronesian islands teach us about speciation? A case study of* Tarphius *beetles and* Hipparchia *butterflies"* (pp. 89–197). FCT-PTDC/BIA-BEC-104571/2008. Angra do Heroísmo: University of the Azores.

Gabriel, R., Arroz, A. M., Amorim, I. R., São Marcos, R., Torrent, J., & Borges, P. A. V. (2013, July). *Os mal-amados: contributos para a compreensão das atitudes dos jovens face aos insectos e à conservação da natureza. I Encontro sobre a Biodiversidade e Conservação de Invertebrados em Portugal*. Portugal: Faculty of Sciences, University of Lisbon. At https://repositorio.uac.pt/handle/10400.3/2378

Gaspar, C., Borges, P. A. V., & Gaston, K. J. (2008). Diversity and distribution of arthropods in native forests of the Azores archipelago. *Arquipélago. Life and marine Sciences, 25*, 1–30.

Giddens, A. (2002). *Modernidade e identidade*. Rio de Janeiro: Zahar Ed.

Gurung, A. B. (2003). Insects—a mistake in God's creation? Tharu farmers' perception and knowledge of insects: A case study of Gobardiha Village Development Committee, Dans-Deukhuri, Nepal. *Agriculture and Human Values, 20*, 337–370.

IAP2—International Association of Public Participation. (2007). *International association of public participation spectrum of public participation*. At http://c.ymcdn.com/sites/, www.iap2.org/resource/resmgr/imported/IAP2%20Spectrum_vertical.pdf

INE—Instituto Nacional de Estatística. (2012). *Indicadores sociais 2011*. Lisboa: Instituto Nacional de Estatística.

IRGC. (2008). *An introduction to the IRGC risk governance framework—White paper No 1*. International Risk Governance Council: Lausanne.

Izquierdo, I., Martin, J. L., Zurita, N., & Arechavaleta, M. (Eds.). (2004). *Lista de especies silvestres de Canarias (hongos, plantas y animales terrestres)* (2nd ed.). Tenerife: Consejería de Política Territorial y Medio Ambiente del Gobierno de Canarias.

Knowlton, L. W., & Phillips, C. C. (2013). *The logic model guidebook: Better strategies for great results* (2nd ed.). Los Angeles: Sage.

Kuhn, T. S. (1970). *The structure of scientific revolutions*. Chicago: University of Chicago Press.

Lindemann-Matthies, P., Junge, X., & Matthies, D. (2010). Experimental evidence for human preference of biodiversity in grassland ecosystems. *Biological Conservation, 143*, 195–202.
Liu, J., Mooney, H., Hull, V., Davis, S. J., Gaskell, J., Hertel, T., et al. (2015). Systems integration for global sustainability. *Science, 347*, 1258832.
Lopes, D. J. H., Mumford, J. D., Mexia, A. M. M., & Leach, A. (2001). Padrão disseminador do escaravelho japonês (*Popillia japonica* Newman; Coleoptera: Scarabaeidae) na ilha Terceira, Açores. *ZAPATERI Revista Aragonesa de Entomología, 9*, 107–113.
McComas, K. A. (2006). Defining moments in risk communication research: 1996–2005. *Journal of Health Communication, 11*, 75–91.
MEA—Millennium Ecosystem Assessment. (2005). *Ecosystems and human well-being: Synthesis*. Washington, D.C.: Island Press.
Myers, N., Mittermeier, R. A., Mittermeier, C. G., da Fonseca, G. A. B., & Kent, J. (2000). Biodiversity hotspots for conservation priorities. *Nature, 402*, 853–858.
Neto, C. (2004). Estudos etnometodológicos no estado da Bahia, Brasil: uma homenagem aos 50 anos do campo de pesquisa. *Biotemas, 17*(1), 117–149.
NNLM—National Network of Libraries of Medicine. (2006). *Guide 5: Define how a program will work—The logic model*. At https://nnlm.gov/outreach/community/logicmodel.html
OPE. (2014). *Public engagement guide. Office of public engagement*. Newfoundland Labrador: The Government of Newfoundland and Labrador.
Pereira, H. M., Belnap, J., Brummitt, N., Collen, B., Ding, H., Gonzalez-Espinosa, M., et al. (2010). Global biodiversity monitoring. *Frontiers in Ecology and the Environment, 8*, 459–460.
Pereira, H. M., Ferrier, S., Walters, M., Geller, G. N., Jongman, R. H. G., Scholes, R. J., et al. (2013). Essential biodiversity variables. *Science, 339*(6117), 277–278.
Posey, D. A. (1983). Ethnomethodology as an emic guide to cultural systems: The case of the insects and the Kayapó indians of Amazonia. *Revista Brasileira de Zoologia, 1*(3), 135–144.
Reed, M. S. (2008). Stakeholder participation for environmental management: A literature review. *Biological Conservation, 141*, 2417–2431.
Rees, D. (2013). *Engagement strategy for Blaenau Gwent*. Blaenau Gwent: Blaenau Gwent County Borough Council Local Service Board. At http://www.blaenau-gwent.gov.uk/community/14723.asp
Renn, O. (2008). *Risk governance. Coping with uncertainty in a complex world*. London: Earthscan.
Rockström, J., Steffen, W., Noone, K., Persson, Å., Chapin III, F. S, Lambin, E. F., et al. (2009). A safe operating space for humanity. *Nature, 461*, 472–475.
Rowe, G., & Frewer, L. J. (2004). Evaluating public participation exercises: A research agenda. *Science, Technology and Human Values, 29*(4), 512–556.
Rowe, G., & Frewer, L. J. (2005). A typology of public engagement mechanisms. *Science, Technology and Human Values, 30*(2), 251–290.
Santos, I. C. V. (2014). *Objeto de desejo e estratégia de marketing: um estudo do valor de uso da Natureza na publicidade televisiva de dois canais portugueses* (M.Sc., thesis in Gestão e Conservação da Natureza). Angra do Heroísmo, University of the Azores. At http://hdl.handle.net/10400.3/3090
Saunders, C. D., Brook, A. T., & Myers, E. O. (2006). Using psychology to save biodiversity and human well-being. *Conservation Biology, 20*, 702–705.
Schmitz, C. C., & Parsons, B. A. (1999). *Everything you wanted to know about logic models but were afraid to ask*. Battle Creek, MI: W. K. Kellogg Foundation.
Sharman, M., & Mlambo, M. C. (2012). Wicked: The problem of biodiversity loss. *GAIA—Ecological Perspectives for Science and Society, 21*(4), 274–277.
Silva, E., & Gabriel, R. (2007). *As atitudes face ao ambiente em regiões periféricas*. Angra do Heroísmo: FCT & University of the Azores.
Silva, L., Ojeda, E., & Rodríguez, J. L. (Eds.). (2008). *Invasive terrestrial flora & fauna of Macaronesia. TOP 100 in Azores*. Madeira and Canaries. Ponta Delgada: ARENA.

Simberloff, D., Martin, J. L., Genovesi, P., Maris, V., Wardle, D. A., Aronson, J., et al. (2013). Impacts of biological invasions: What's what and the way forward. *Trends in Ecology & Evolution, 28*, 58–66.

SPB. (2010). *Strategic plan for biodiversity 2011–2020: The Aichi biodiversity targets*. Aichi: UNEP.

SREA—Serviço Regional de Estatística dos Açores. (2011). *Censos 2011. Principais resultados definitivos dos censos 1991, 2001 e 2011*. At http://estatistica.azores.gov.pt/conteudos/relatorios/lista_relatorios.aspx?idc=392&idsc=1959&lang_id=1

Terzopoulou, S., Rigal, F., Whittaker, R. J., Borges, P. A. V., & Triantis, K. A. (2015). Drivers of extinction: The case of Azorean beetles. *Biology Letters, 11*, 1–4.

UNISDR—United Nations for Disaster Risk Reduction. (2015). *Sendai framework for disaster risk reduction 2015–2030*. Brussels: United Nations.

Vasconcelos, L., Alho, J. M., & Martins, J. P. (Eds.). (2009). *Cidadãos pelo ambiente, conservação da natureza e biodiversidade em Portugal*. Lisboa: Esfera do Caos.

Walls, J., Rowe, G., & Frewer, L. (2011). Stakeholder engagement in food risk management: Evaluation of an iterated workshop approach. *Public Understanding of Science, 20*(2), 241–260.

WBGU (German Advisory Council on Global Change). (2000). *World in transition: Strategies for managing global environmental risks. Annual report 1998*. Berlin: Springer.

Weber, J. R., & Word, C. S. (2001). The communication process as evaluative context: What do non-scientists hear when scientists hear when scientists speak? *BioScience, 51*(6), 487–495.

Wilson, E. O. (1987). The little things that run the world (the importance and conservation of invertebrates). *Conservation Biology, 1*, 344–346.

# Part II
# Teaching About Biological Invasions and Threats to Biodiversity

# Science and Education at the Centre for Invasion Biology

Sarah J. Davies, G. John Measey, Dorette du Plessis and David M. Richardson

**Abstract**

South Africa has severe problems caused by biological invasions in terrestrial, freshwater and marine ecosystems, and a long history of managing biological invasions. However, appreciation and systematic study of the problems associated with invasive species are relatively recent. In 2004, the Centre for Invasion Biology (CIB) was established as one of the first six national Centres of Excellence funded by the South African government. The aim of the DST-NRF Centres of Excellence is to concentrate existing capacity and resources to enable researchers to collaborate across disciplines on long-term projects that are locally relevant and internationally competitive. Understanding the biological and ecological underpinnings of invasions is crucial, but much emphasis is placed on understanding the 'human dimensions' of invasions, and on seeking solutions for current problems, and techniques for preventing new invasions. Education is a critical component of this knowledge-building process, and the CIB infuses education for sustainable development into all of its activities. The Centre conducts education and outreach at the secondary school, under-graduate, post-graduate and post-doctoral levels, and in the workplace to develop capacity at all levels. The Centre's flagship outreach programme, Iimbovane, aims to increase environmental literacy and inspire secondary school learners to take up scientific careers through facilitating field and laboratory work that is embedded in the life science curriculum; the programme focuses on under-resourced schools. At tertiary level, the under-graduate training course in invasion biology presented at Stellenbosch University provides an introduction to the exciting and important field of invasion science for final-year Bachelor degree students. A distributed network of researchers located in universities and research

S.J. Davies (✉) · G. John Measey · D. du Plessis · D.M. Richardson
Centre for Invasion Biology, Department of Botany and Zoology, Stellenbosch University, Private Bag X1, Matieland 7602, South Africa
e-mail: sdavies@sun.ac.za



P. Castro et al. (eds.), *Biodiversity and Education for Sustainable Development*, World Sustainability Series, DOI 10.1007/978-3-319-32318-3_7

institutions around South Africa also trains 50–60 post-graduate students, and hosts ten to twelve post-doctoral researchers each year, through whom much of the Centre's research is conducted. In the working world, members are involved both in citizen science programmes on invasive species and in collaborative work with partner organisations which implement invasive species management programmes, provide employment opportunities for graduates of the Centre, and form a source of working (part-time) graduate students.

## 1 Introduction

Biological invasions are a substantial component of global change and are widely recognized as a major and growing threat to global biodiversity and the sustained delivery of goods and services from ecosystems (Pyšek and Richardson 2010). Appreciation and systematic study of the problems associated with invasive species are relatively recent, however. The study of invasions originally focussed on biological and ecological issues, and invasions were widely seen as a grand natural experiment in biogeography to gain insights into the determinants of range limits and factors that control membership of communities and ecosystems. Understanding these biological and ecological underpinnings of invasions remains crucial, but increasing emphasis is being placed on understanding the 'human dimensions' of invasions, seeking solutions for current problems and techniques for preventing new invasions. Therefore, in recent decades, most research on invasive species has sought solutions to the many social, ecological and economic problems associated with invasions (Richardson et al. 2011).

Many countries, especially the more developed nations, have major programmes in place to manage biological invasions. These are typically multi-faceted and multi-level initiatives that seek to reduce the extent and impacts of currently invasive species, while simultaneously implementing measures to reduce the chance of new invasions. The initiation and growth of such programmes has contributed to the rapid growth of invasion science, which is currently one of the most popular and vibrant sub-disciplines of ecology and environmental management worldwide. Thousands of scientific papers are published on aspects of biological invasions every year and the field now has a growing number of specialist journals and features prominently in the programmes of academic conferences worldwide (Richardson 2011). As with climate change and other complex environmental problems, there is a crucial need to raise awareness of all aspects of the process, and to highlight options for management among the public.

South Africa has severe problems with biological invasions in its terrestrial, freshwater and marine ecosystems, but is also one of the countries with the longest history of managing biological invasions. Despite many challenges typical of developing countries, South Africa has invested substantially in infrastructure and capacity development for addressing problems associated with biological invasions.

In 2004, the DST-NRF Centre of Excellence for Invasion Biology (the Centre for Invasion Biology; hereafter the CIB or 'the Centre') was launched—one of the country's first six national Centres of Excellence as part of a programme funded by the national government to concentrate existing capacity and resources to enable researchers to collaborate across disciplines on long-term projects that are locally relevant and internationally competitive (van Wilgen et al. 2014).

The CIB is a distributed network of researchers, students and partners managed from a primary hub at Stellenbosch University, with a secondary hub of researchers and support staff at the University of Pretoria. The aim of the Pretoria (or 'northern') hub is to coordinate activities of the northern partners where necessary to complete particular projects, and to support long-term biodiversity research in that area. In addition to the two hubs, CIB researchers are staff members of a range of academic and other research-oriented organisations and the network as a whole covers more than ten academic institutions and all of the country's provinces which have a university or large research organisation. This distributed network broadens the reach of the CIB beyond the higher education institutions, where most research is generated, and the CIB builds external partnerships to inform decision-makers in a range of organisations about biological invasions and their social and ecological impacts. Partners range from conservation agencies (e.g. South African National Parks, CapeNature), R&D organisations (e.g. the Council for Scientific and Industrial Research, CSIR) to large municipalities (eThekwini Municipality in Durban, City of Cape Town), which are users of the information generated. In addition, the CIB has a productive partnership with the South African National Biodiversity Institute (SANBI) that allows staff of the two organisations to work together on a range of activities from student co-supervision to joint implementation of invasive species legislation.

The aim to engage in research that is locally relevant and internationally competitive in order to enhance the pursuit of research excellence and capacity development has been augmented since the establishment of the CIB with the aim of maximising research impact on industry, business and society. The CIB's key performance areas (KPAs) include research, education and training, networking, knowledge brokerage and service provision. These five KPAs represent different ways of achieving impact in the research sphere—that is, they are elements of the research impact 'process'. This chapter will show how the CIB's key activities go beyond community involvement to community participation in research itself. Much of the CIB's research is conducted through post-graduate students and post-doctoral associates. Along with research excellence and education and training, DST-NRF Centres of Excellence are required to undertake networking, information brokerage and service provision, thereby maximising the impact of the CIB's work on different sectors of society and economy.

While the CIB's internationally competitive research and capacity building activities address impact at the academic level, the other KPAs are intended to ensure impact on other areas of society. Networking, information brokerage and service provision involve working with decision-makers to increase knowledge about biological invasions and seek to make invasion-related information resources

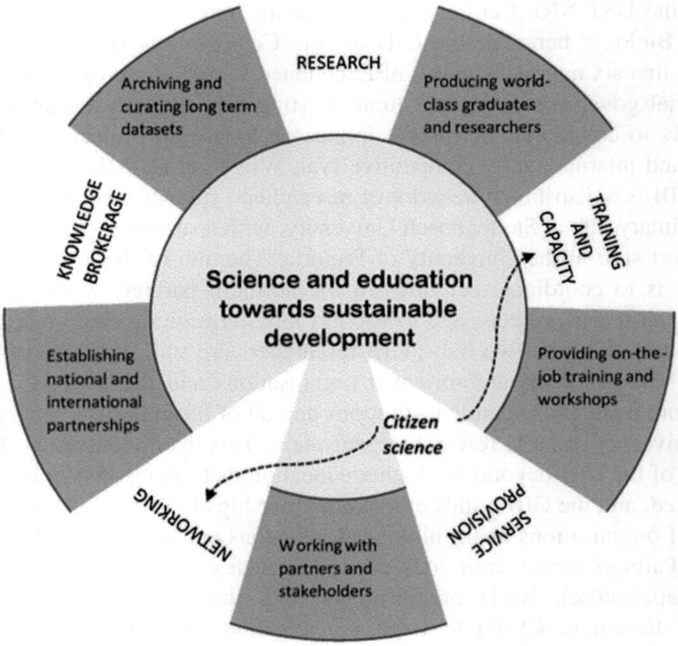


**Fig. 1** The Centre for Invasion Biology's key performance areas and major activities contributing to education

readily available to broader society. Thus, the CIB sees all its key activities as hinging on education in different sectors (Fig. 1). For example, under its service provision activities, the CIB provides policy advice and consultancy services at reasonable cost to decision-makers in government, non-government organisations (NGOs) and the private sector. The long-term working relationships built between researchers and these institutions through service provision allow mutual learning to take place, thereby advancing the field.

Education in schools: The Iimbovane Outreach Project

Biodiversity and the conservation thereof are important for the maintenance of ecosystem function and delivery of environmental goods such as food and potable water. However, mounting evidence shows that the rate at which we are losing biodiversity and degrading ecosystems is unprecedented. One of the strategies that can help reduce the loss of biodiversity is increased public education and understanding of the consequences of biological invasions. While the scientific community has a reasonable understanding of threats to biodiversity, most South Africans are not familiar with biodiversity, its loss and the consequences of biological invasions. If we want the public to support conservation and management efforts, we must be purposeful in increasing environmental literacy and putting initiatives in place that educate citizens about the environment. The landscapes where poorer urban communities are located are often heavily invaded and transformed by invasive alien plants, with virtually no natural indigenous plant diversity (City of Cape Town 2013). As a result, residents may have no direct contact with indigenous biodiversity (City of Cape Town 2013; Elmqvist et al. 2013; Miller 2005; Turner et al. 2004).

Only in the last decade has the basic education sector in South Africa become actively engaged in biodiversity education. Before 2006, biodiversity was not included in the formal school curriculum. This changed with the National Curriculum Statement for Grade 10–12 Life Sciences which now includes two chapters dealing with biodiversity, environmental change and the impacts of human activities on biodiversity. Many educators have had no formal training in the relatively young fields of biodiversity science and invasion science. Due to a lack of training in or direct experience with biodiversity, most educators struggle to develop practical projects in this subject area. Consequently, learners do not get the necessary content knowledge, or accumulate the critical thinking and practical skills needed to master life science as a subject, and therefore seldom consider taking up a career in science. With growing class sizes, a lack of access to transport and financial constraints, educators tend to move away from asking learners to collect specimens in the field. Instead, learners are told how specimens are collected, rather than being allowed to explore the process for themselves through direct experience. In most South African schools, these problems are exacerbated by the lack of access to scientific equipment such as microscopes and information and communication technology (ICT). These growing pressures on the delivery of high quality education are happening at a time when there is a growing demand for learners with critical thinking skills who are aware of human impacts on their environment and the need for sustainable solutions. The Department of Science and Technology's Ten-Year Innovation Plan articulates the need for knowledge workers who are equipped to participate in South Africa's developing knowledge economy (Department of Science and Technology 2008).

## 1.1 Serving the Education Need

The CIB runs an innovative, long-term project, Iimbovane, that combines science outreach and biodiversity monitoring, as well as ongoing relationships with partner organisations that contribute to its mandate. The Iimbovane Outreach Project grew out of the dual challenges to educate society about biodiversity and deepen public understanding of the consequences of biological invasions. The project provides support to learners and educators encountering biodiversity science at the secondary school level. Iimbovane, meaning 'ants' in isiXhosa, a widely-spoken South African language, focuses on ants as a model group for teaching biodiversity and invasion science. The project uses an experiential learning approach whereby participants accumulate knowledge and skills through direct involvement with the model group (ants). Participating learners and educators assist with the collection of ant samples and relevant environmental data at monitoring sites in their school grounds (typically highly disturbed environments) and in matched reference sites in nearby protected areas. The project uses a simple sampling protocol: arrays of pitfall traps to collect ant species with matched vegetation samples to collect species richness and abundance information. An important advantage of this collection method is the diversity of groups that the learners discover when they remove the traps after a few days. Seeing the variety of insects encourages learners to ask questions about biodiversity.

Ants were selected as a focal group because of the group's high level of diversity in South Africa and the ease of collecting specimens using pitfall traps and low-cost equipment, making this protocol repeatable for educators who wish to repeat the project for teaching purposes. In addition, the project maintains an ant identification key based on the ant species identified in the project. The identification guide, together with classroom technology, such as laptop computers, microscopes and data projectors are handed to each school that participates in the project, making it possible for learners and educators to see and work with both biological samples and data using ICT. Project activities with learners take place during school contact hours, while educators receive training and provide feedback to the project team at separate workshops. During these workshops the project staff, together with educators, develop lesson plans and assessment activities that can be used in the classroom. The activities and products are therefore not an extra-curricular burden for the educators, but compliment their teaching.

## 1.2 Educational Advantages of Iimbovane

The value of education projects such as the Iimbovane Outreach Project lies mainly in their contribution to science and biodiversity education at school level. The project improves educator capacity in the field of biodiversity; educators benefit from project workshops by gaining in-depth knowledge on biodiversity, environmental change and human impacts, curriculum areas that are challenging for them. Iimbovane Outreach Project support enables them to teach biodiversity in a more confident and thoughtful manner. The project addresses the formal requirements of the South African National Curriculum for the Life Sciences, which requires that learners develop an understanding of science and how it is undertaken and applied in society. The project consists of a 'doing' phase when students carry out the fieldwork in school grounds and a 'reflective' phase during which the data collected are assimilated and applied (Fig. 2). By providing educational resources to the educators and working with learners in the field and laboratory, Iimbovane supports these curriculum aims and engages learners directly with the scientific process. The learners are thereby exposed directly to science and the scientific process in a real setting, rather than from textbooks in a classroom setting. Learners also see what the career of a researcher involves. Most learners participating in the Iimbovane

**Fig. 2** The Centre for Invasion Biology's Iimbovane Outreach Project consists of a 'doing' phase during which participating learners plant pitfall traps to collect ants while interacting with scientists, followed by a 'reflective' phase during which learners analyse and interpret the data they have collected

Outreach Project do not receive this exposure any other way, at home or in their communities.

Direct contact with scientists working on the Iimbovane project gives learners a wider experience and understanding of what it means to practice science (Braschler 2009; Braschler et al. 2010). Scientific fieldwork also provides an opportunity for learners to work as a team and to work purposefully outdoors, learning to appreciate the natural world and link theory with observation. Perhaps one of the most significant advantages in terms of education and awareness is the learners' realisation that areas around their schools are in a poor environmental condition. Learners then become aware of pollution, habitat destruction and the extent of invasive alien species in their local environments (Ballouard et al. 2011).

*Experiences of participating learners*

> I really found the field work informative because I saw what hard work it is to collect many different specimens and to do research. One of my favourite things during the week was the lab work and the microscope work. Our school does not have many microscopes and we do not get to work with them often, which are why it was so interesting to work with them last week. I also want to do forensic science and I love the possibility of just being in a lab and doing experiments all day. The week was very, very informative and I have learned new skills because I am not an outdoorsy person but I learned that I can do it and I can survive the elements and work under different types of conditions. I've also learned so many things in the lab like learning how to identify different insects and use the microscope to do so. (Learner from Malibu High School)

I've learnt that ants are very important to us even if they're small and that we must value biodiversity. (Learner from Vusisizwe Secondary School)

The workshop was extremely helpful - working with actual microscopes was great. The fieldwork was awesome, as it made me realize how precious those animals are and also how human activities can affect animals' habitats. (Learner from Sarepta Secondary School)

### Education at tertiary level

Education at tertiary institutions is traditionally delivered in modular units that encompass broad to specialist themes as students progress through their courses. A close rapport between course material and cutting-edge research is one that is strived for in many tertiary education programmes globally. In this respect, a semester course on invasion science at Stellenbosch University was initiated in 2014. The course was initiated and is delivered annually by members of the CIB's research team, typically by four lecturers and several guest lecturers who introduce a diversity of topics and their specialist knowledge. The course provides an introduction to the exciting and important field of invasion science for senior under-graduate (final year Bachelor degree) students.

### The employment sphere

Biological invasions are widely recognised as one of the major threats to both the conservation of biodiversity and the maintenance of ecosystem services worldwide. In many parts of the world, the most challenging and time-consuming tasks of land managers and conservation biologists are those relating to controlling alien species and preventing impacts, and, increasingly, repairing systems already damaged by aliens. The interaction between invasions and other drivers of global change creates fascinating areas for research at many organisational levels, e.g. from genes to ecosystems. Invasions also represent a major challenge to the goal of sustainable development, as they affect the operations of a broad swathe of society from rural communities to major shipping companies. As such, invasion science demands insights from a wide range of disciplines. Furthermore, as an applied biodiversity field, invasion biology prepares graduates and young researchers effectively for employment. Graduates of the CIB work in a range of organisations from universities to government bodies, and private consultancies to parastatal biodiversity-focused organisations. Between 2004 and 2014, the CIB produced 211 graduates and supported 37 post-doctoral associates. Of these, most have found positions within their chosen field or continued to another advanced degree or post-doctoral position in South Africa or abroad. Alumni work in provincial environmental departments, national parks and nature reserves, South African and international universities, environmental observation networks and other research and policy-making organisations (such as the South African National Biodiversity Institute) (Table 1).

On-the-job training, or training towards advanced degrees for people already working full time, is an emerging focus for the education sector in South Africa. South Africa's government, through its Ten-Year Innovation Plan, is committed to producing 6000 PhD graduates per annum by 2025 (Department of Science and

**Table 1** Examples of organisations employing recent graduates of the Centre for Invasion Biology

| |
|---|
| National government departments (e.g. Department of Environmental Affairs, Department of Agriculture, Forestry and Fisheries) |
| Provincial nature conservation and land management agencies (e.g. CapeNature) |
| Local authorities, including metropolitan municipalities such as the City of Cape Town |
| BirdLife South Africa (an affiliate of BirdLife International) |
| Blue Science (environmental consultancy focussing on water issues) |
| Coastal Environmental Services (consultancy focusing on environmental impact assessment) |
| Council for Scientific and Industrial Research (CSIR; a government-sponsored science council conducting directed and multidisciplinary research, innovation and development) |
| Fruit Fly Africa (industry-owned service organisation using sterile insect technique for area-wide fruit fly control) |
| National Research Foundation (NRF; independent government agency mandated to promote and support research) |
| South African Broadcasting Corporation (SABC; the national broadcaster of South Africa) |
| South African Environmental Observation Network (SAEON; an environmental observation network that delivers data for scientific research and informs decision-making) |
| South African Institute for Aquatic Biodiversity (SAIAB; a national research facility dedicated to the study of aquatic biodiversity) |
| South African National Biodiversity Institute (SANBI; government agency coordinating research, monitoring and reporting on the state of biodiversity) |
| South African National Seed Organization (SANSO; sectoral industry body that represents the South African seed industry) |

Technology 2008). In 2009 South Africa produced 1380 doctoral graduates (Mouton 2011) indicating that a three- to four-fold increase in graduation rate will be required to meet the 2025 target. At the same time, many universities are experiencing a dearth of qualified applicants, and already more than 80 % of South African PhD students study part time (Mouton 2011). These students are typically more mature, experienced, and may be active in broader fields that their study itself addresses; however, they may need refresher courses to enable them to perform in their degree programme. Although there are significant obstacles to increasing part-time PhD enrolments in South Africa, including the lack of provision for part-time registrations in some universities and supervision capacity, efforts are being made to address these shortcomings (e.g. African Doctoral Academy, http://www0.sun.ac.za/ada/).

Service provision and networking

The CIB's managed network affords opportunities to extend our education work into our partner institutions. There are three main ways in which the CIB brings education into the working day of our partners: (a) through issue-based workshops (Novoa et al. 2015), (b) through field-work and research collaborations (Measey et al. 2014), and (c) through formal registration for post-graduate degrees.

Stakeholder workshops are a powerful tool for understanding the perceptions of the broad array of role players who are involved with invasive species problems. These meetings regularly include individuals from our partner network, as well as private land owners and lobby groups, and can contribute to developing research programmes and management strategies to address ‘conflict species’—invasive organisms that are both beneficial and detrimental. Many alien species are conflict-generating, as they were introduced for a particular use and then established self-sustaining and expanding populations beyond the area of introduction, where they may have both positive and negative impacts. For example, South Africa has a major problem with invasive alien trees, many of which were introduced and widely disseminated as forestry plants, ornamentals or amenity trees (van Wilgen and Richardson 2012). The CIB has held several workshops to bring stakeholders together to seek the best solutions for dealing with conflict invasive tree species and other groups of organisms, such as cacti used in ornamental horticulture. These workshops have proved to be opportunities for stakeholders to share their values and experiences, and measurable changes of attitude have been documented (e.g. Novoa et al. 2015).

Workshops are also rich environments for ascertaining the level of understanding that participants have of invasive species and invaded systems, and therefore to set the agenda for ideas that need to be taught, identify common misconceptions and highlight communication gaps. For example, invasive mesquite (*Prosopis*) species and their hybrids were was thought to be widely used by farmers and rural dwellers in South Africa for fuel and as building material, until workshop discussions showed that mesquite is in fact considered to be inferior for these purposes (see Shackleton et al. 2015). In this respect, workshops are a learning experience for all involved, although at the outset it is not easy to predict who will be the educated or the educators. However, it is usually the case that all participants leave a workshop with an appreciation of invasive species from another standpoint.

The detection of a new species of invasive frog in a national park in 2010 (Fogell et al. 2013; Measey and Davies 2011) led to the development of a control programme and its inclusion in SANParks’ Annual Plan of Operations (de Villiers et al. 2015). Control activities are now carried out by national park staff with the assistance of interns from local universities and under the guidance of CIB researchers. These activities take place annually over three days. The briefing for the field operation is carried out jointly by CIB and SANParks staff and includes information on the cause of the invasion, as well as an explanation behind the management decision and control programme. At least one intern usually takes the subject on as a project that will count toward a formal part of his or her degree. The outcome of these interactions with partner institutions has led to a better understanding of common invasive species problems, as well as increasing the support for invasion biology research in these organisations, and registration of members of staff for part-time degrees.

Currently, around 10 % of CIB post-graduate students also work for partner organisations. Although the student-supervisor role is formalised within the university context, links between higher education institutions and their partner organisations provide many opportunities for co-supervision of students by scientist and managers working outside academia, as well as exchange of information between the CIB and the partner institute. In some cases this has resulted in three-way partnerships between the CIB and two partner organisations, with positive results for all parties. Because post-graduate degrees typically take two to three years, the graduates often stay in their former positions, and there is often a long-term positive feedback toward supervising more students and strengthening ties between partners and the CIB.

The role that citizen science can play in science education

Citizen science, or the involvement of citizens in research, has grown exponentially in the last fifteen years. While public involvement in science has well-established roots, the ease of connection between scientists and the public (i.e. via the internet) and the widespread digitisation of media have opened up significant opportunities for scientists and participating citizens alike (e.g. Silvertown 2009). Principally, citizen science has enabled data to be collected on a scale that was previously unfeasible at the regional, continental or even global scale (Bonney et al. 2009). South Africa has produced a number of regional atlas projects covering the region (South Africa, Lesotho and Swaziland) which have citizen science as a key part of data collection (e.g. Bates et al. 2014; Mecenero et al. 2013; Minter et al. 2004). The CIB has been a key participant in several of these projects, which have established South Africa as a global hotspot of citizen science. These projects have, in turn, called for more information on the distribution of invasive species in South Africa to be collected, and have generated important data about the distribution and dispersal of alien species. Citizen science has also been of great importance in generating data for invasion biologists in the USA (Crall et al. 2010).

Finding ways to engage efficiently with large numbers of contributors is particularly important in citizen science projects because engaging on an individual level would be expensive and onerous. One solution relies on crowd-sourcing to provide feedback as well as enhance the roles of citizens and scientists alike (Silvertown et al. 2015). A common way of soliciting information from citizen scientists is to call for the submission of electronic photographs of organisms or landscapes such as on iSpot (http://www.ispotnature.org/communities/southern-africa), but the identification of these photographs is difficult and tedious. Crowd-sourcing appears to be an excellent solution, and 80–90 % of identifications can be made through this form of social network. In addition, the original contributor receives feedback on his/her original submission, together with comments from experts in the network. The citizen scientist can then use the identification information to learn more about the distribution and biology of the organism of interest. He or she may in turn become an expert in identification of this or other taxa. Silvertown et al. (2013) offer some important considerations for those planning to employ citizen science towards a project, including the clarity of learning objectives and target audience, the project duration and legacy, as well as the scientific rationale behind the project.

Many citizen science projects have specific education targets built into their design (Silvertown 2009). Citizens are encouraged to engage not only in data collection, but to build their knowledge with tests and quizzes. For example, participants in the North American amphibian monitoring program had to learn the calls of native frog species in their area of North America. They then had to pass an online test in order to take part as a data-collecting member of the project. Data from this project have since been used to answer hypothesis-driven questions concerning the conservation of amphibians in the eastern USA (Cosentino et al. 2014).

**Acknowledgements** This work was funded by the DST-NRF Centre of Excellence for Invasion Biology, and by donations from Rand Merchant Bank Fund and AfriSam. The Western Cape Education Department hosts and supports the work of the Iimbovane Outreach Project in schools. The authors are grateful to Mrs C.M. Momberg, CIB for collating data on the CIB's performance.

## References

Ballouard, J.-M., Brischoux, F., & Bonnet, X. (2011). Children prioritize virtual exotic biodiversity over local biodiversity. *PLoS ONE, 6*(8), e23152. doi:10.1371/journal.pone.0023152.

Bates, M. F., Branch, W. R., Bauer, A. M., Burger, M., Marais, J., Alesander, G. J., et al. (eds.) (2014). Atlas and red list of the reptiles of South Africa, Lesotho and Swaziland. Pretoria, South Africa: South African National Biodiversity Institute.

Bonney, R., Cooper, C. B., Dickinson, J., Kelling, S., Phillips, T., Rosenberg, K. V., et al. (2009). Citizen science: A developing tool for expanding science knowledge and scientific literacy. *BioScience*, 59, 977–984.

Braschler, B. (2009). Successfully implementing a citizen-scientist approach to insect monitoring in a resource-poor country. *BioScience, 59*, 103–104.

Braschler, B., Mahood, K., Karenyi, N., Gaston, K. J., & Chown, S. L. (2010). Realizing a synergy between research and education: How participation in ant monitoring helps raise biodiversity awareness in a resource-poor country. *Journal of Insect Conservation, 14*, 19–30.

City of Cape Town. (2013). State of the Environment Report 2012. http://www.capetown.gov.za/en/EnvironmentalResourceManagement/publications/Documents/State_of_Environment_Report_2012.pdf

Cosentino, B. J., Marsh, D. M., Jones, K. S., Apodaca, J. J., Bates, C., Beach, J., et al. (2014). Citizen science reveals widespread negative effects of roads on amphibian distributions. *Biological Conservation*, 180, 31–38.

Crall, A. W., Gregory, G. J., Jarnevich, C. S., Stohlgren, T. J., Waller, D. M., & Graham, J. (2010). Improving and integrating data on invasive species collected by citizen scientists. *Biological Invasions, 12*, 3419–3428.

de Villiers, F. A., de Kock, M., & Measey, G. J. (2015). Controlling the African clawed frog *Xenopus laevis* to conserve the Cape platanna *Xenopus gilli* in South Africa. *Conservation Evidence*, 12.

Department of Science and Technology. (2008). Innovation towards a knowledge-based economy: Ten-year plan for South Africa (2008–2018).

Elmqvist, T., Fragkias, M., Goodness, J. Güneralp, B., Marcotullio P. J., McDonald, R. I., et al. (2013). *Urbanization, biodiversity and ecosystem services: Challenges and opportunities, a global assessment* (p. 755). Berlin: Springer.

Fogell, D. J., Tolley, K. A., & Measey, G. J. (2013). Mind the gaps: Investigating the cause of the current range disjunction in the Cape Platanna, *Xenopus gilli* (Anura: Pipidae). *PeerJ, 1*, e166.

Measey, G. J., Annecke, W., Davies, S. J., Dorse, C., Stafford, L., Tolley, K. A., et al. (2014). Cape collaborations for amphibian solutions. *FrogLog, 109*, 46–47.
Measey, G. J., & Davies, S. J. (2011). Struggling against domestic exotics at the southern end of Africa. *FrogLog, 97*, 28–30.
Mecenero, S., Ball, J. B., Edge, D. A., Hamer, M. L., Henning, G. A., Krüger, M., et al. (Eds.). (2013). *Conservation assessment of butterflies of South Africa, Lesotho and Swaziland: Red list and atlas*. Cape Town: Saftronics, Johannesburg and Animal Demography Unit.
Miller, J. R. (2005). Biodiversity conservation and the extinction of experience. *Trends in Ecology & Evolution, 20*, 430–434.
Minter, L. R., Burger, M., Harrison, J. A., Braack, H. H., Bishop, P. J., & Kloepfer, D. (Eds.). (2004). *Atlas and red data book of the frogs of South Africa, Lesotho and Swaziland*. Washington, DC: Smithsonian Institution.
Mouton, J. (2011). Doctoral production in South Africa: Statistics, challenges and responses. *Perspectives in Education: The changing face of doctoral education in South Africa (Special Issue 3), 29*, 13–29.
Novoa, A., Kaplan, H., Kumschick, S., Wilson, J. R. U., & Richardson, D. M. (2015). Soft touch or heavy hand? Legislative approaches for preventing invasions: Insights from cacti in South Africa. *Invasive Plant Science and Management, 8*, 307–316.
Pyšek, P., & Richardson, D. M. (2010). Invasive species, environmental change and management, and ecosystem health. *Annual Review of Environment and Resources, 35*, 25–55.
Richardson, D. M. (2011). Invasion science: The roads travelled and the roads ahead. In D. M. Richardson (Ed.), *Fifty years of invasion ecology: The legacy of Charles Elton* (pp. 397–407). Oxford: Wiley-Blackwell.
Richardson, D. M., Pyšek, P., & Carlton, J. T. (2011). A compendium of essential concepts and terminology in invasion ecology. In D. M Richardson (Ed.), *Fifty years of invasion ecology. The legacy of Charles Elton* (pp. 409–420). Oxford: Wiley-Blackwell.
Shackleton, R. T., Le Maitre, D. C., van Wilgen, B. W., & Richardson, D. M. (2015). Use of non-timber forest products from invasive alien *Prosopis* species (mesquite) and native trees in South Africa: Implications for management. *Forest Ecosystems, 2*, 16. doi:10.1186/s40663-015-0040-9.
Silvertown, J. (2009). A new dawn for citizen science. *Trends in Ecology & Evolution, 24*, 467–471.
Silvertown, J., Buesching, C. D., Jacobson, S. K., & Rebelo, T. (2013). Citizen science and nature conservation. *Key Topics in Conservation Biology, 2*, 124–142.
Silvertown, J., Harvey, M., Greenwood, R., Dodd, M., Rosewell, J., Rebelo, T., et al. (2015). Crowdsourcing the identification of organisms: A case-study of iSpot. *ZooKeys*, 480, 125–146.
Turner, W. R., Nakamura, T., & Dinetti, M. (2004). Global urbanization and the separation of humans from nature. *BioScience, 54*, 585–590.
Van Wilgen, B. W., Davies, S. J., & Richardson, D. M. (2014). Invasion science for society: A decade of contributions from the Centre for Invasion Biology. *South African Journal of Science, 110*(7/8), 12p (Art. #a0074). http://dx.doi.org/10.1590/sajs.2014/a0074
Van Wilgen, B. W., & Richardson, D. M. (2012). Three centuries of managing introduced conifers in South Africa: Benefits, impacts, changing perceptions and conflict resolution. *Journal of Environmental Management, 106*, 56–68.

# Engaging Society to Fight Invasive Alien Plants in Portugal—One of the Main Threats to Biodiversity

Elizabete Marchante and Hélia Marchante

**Abstract**

Invasive alien species (IAS) are a major threat to biodiversity all over the world and Portugal is no exception. The problem is so serious that such species are recognized by the Portuguese legislation since 1999 which includes 32 species of invasive animal and plants. Nevertheless, a large proportion of the population is still unaware of the problem. Considering that citizens represent a vector of introduction and spread of IAS and, on the other hand, can play a major role in helping to prevent and control IAS, this lack of awareness can be largely detrimental. In an attempt to reduce this gap, and aiming to contribute for a more sustainable environment and society, researchers from CFE/UC and from ESAC/IPC have been raising public awareness and engaging the public with the IAS problem, namely with invasive plants. Since 2003, several strategies have been used, including the website invasoras.pt which aims to aggregate several elements of these strategies. The most recent version of the website is online since 2013 and includes as core element a citizen science platform that aims to engage the public countrywide, voluntarily, to report sightings of invasive plants. Publications about invasive plants in Portugal, including a field guide and other printed materials, workshops and social media have been used to engage the public with the WebMapping platform. Additionally, field-work projects for university students and training courses for professionals dealing with alien plants and for school teachers have been organized. In this chapter, the different strategies implemented and the results of an effort to evaluate the effectiveness

E. Marchante · H. Marchante (✉)
Department of Life Sciences, Centre for Functional Ecology,
University of Coimbra, Coimbra, Portugal
e-mail: hmarchante@gmail.com

H. Marchante
Department of Environment, Escola Superior Agrária,
Instituto Politécnico de Coimbra, Coimbra, Portugal

 
P. Castro et al. (eds.), *Biodiversity and Education for Sustainable Development*,
World Sustainability Series, DOI 10.1007/978-3-319-32318-3_8

and outreach of these various approaches are presented. Overall, awareness about IAS is increasing amongst the Portuguese population, with citizens more educated about the problem and contributing more to the prevention and control of IAS, but much more work is needed. The challenge to reach publics other than the peers or professionals related to the topic is still daunting.

## 1 Invasive Plants as a Threat to the Biodiversity Worldwide

Biological invasions mediated by human activities change ecosystems worldwide, with escalating impacts, at ecological (e.g., elimination of native species and disruption of plant communities), economic and human-health (e.g. major threats to agriculture, forestry production or fishing) levels (Gaertner et al. 2009; Jäger et al. 2009; Marchante et al. 2008a, 2015a; Mazza et al. 2014; Pejchar and Mooney 2010; Pyšek and Richardson 2010; Vilà et al. 2010).The origins of biological invasions are lost in time with some authors pointing to Late Devonian (Stigall 2010). Nevertheless, it has increased consistently under human-mediation, reaching a peak in the second half of the 20th century (Pyšek and Richardson 2010) when it gained "formal" recognition with the pioneering work of Elton (1958), often considered as the "father of Invasion Biology" (Richardson and Pysek 2008). Numerous scientists, politicians and Global Organizations (European Parliament and Council [EPC] 2014; Invasive Species Specialist Group/International Union for Conservation of Nature [ISSG/IUCN] 2008; Ministério do Ambiente 1999; Pyšek and Richardson 2010) recognize that invasive alien species (IAS; it is important to stress that not all the alien species, but only the ones that become invasive, see Table 1 for definitions) damage ecosystem services, disrupt human well-being and are amongst one of the main causes of biodiversity decline. Ecosystem services are significantly affected by invasive species, include supporting (e.g., alteration of succession of plant communities and soil and nutrient cycling), provisioning (e.g., threats to native species, alteration of genetic resources), regulating (e.g., changing pollination services and fire regimes) and cultural services (e.g., effects on ecotourism, changes in perception of landscape) (Millennium Ecosystem Assessment [MEA] 2005; Vilà et al. 2010). The alterations that are promoted at one trophic level have frequent repercussions into several other trophic levels amplifying even more the impacts of IAS (Carvalheiro et al. 2010). Humans have a key role in most of the processes of invasion, either as facilitators or opponents contributing to manage them, and as such being aware of the problem is essential.

IAS are found amongst all living organisms, including plants, animals and microorganisms (Elton 1958; Perrings et al. 2010). The present chapter focuses on different approaches implemented to raise awareness about invasive plants amongst the Portuguese population, but some brief reference to other invasive organisms in Portugal are included.

**Table 1** Main concepts and definitions used in plant invasion biology

| Concept | Definition |
|---|---|
| Exotic, alien, aloctonous, non-native, non-indigenous | Plant taxa whose presence in a given area is due to introduction, intentional or accidental, as a result of human activity |
| Casual, occasional escape, transient | Subset of alien plants that may flourish, and even reproduces occasionally in an area, but which do not form self-replacing populations, and which rely on repeated introductions to persist |
| Naturalized, sub-spontaneous | Subset of alien plant that reproduce consistently and sustain populations over many life cycles without direct intervention by humans (or in spite of human intervention); often recruit offspring freely, usually close to adult plants, and do not necessarily spread into natural, semi-natural or human-made ecosystems |
| Invasive, environmental weed | Subset of naturalized plants that produce reproductive offspring's, often in very large numbers, at considerable distances from parent plants (approximate scales: >100 m; <50 years for taxa spreading by seeds and other propagules; >6 m/3 years for taxa spreading by roots, rhizomes, stolons, or creeping stems), and thus have the potential to spread over a considerable area; frequently adversely affecting native biodiversity and/or ecosystem functioning |

*Source* Richardson et al. (2000)

## 2 Brief Overview of Invasive Alien Species in Portugal

There are over 940 alien species in Portugal (excluding species merely cultivated/domesticated), comprising terrestrial animals (>245) and plants (>670), aquatic animals (2) and plants (7) and fungi (16) (values updated after DAISIEteam 2008). Amongst these, a part is considered invasive, perceived as widespread and as causing major negative impacts (Anastácio et al. 2005; Hellmann et al. 2011; Marchante et al. 2014; Sousa et al. 2008) with 32 species of plants and animals being legally listed as invasive (Decreto-Lei n.º 565/99, 21 December) and as such forbidden to use. However, the list of species needs to be updated as several more recent IAS have not yet been included.

The more widespread and injurious invasive animals in Portugal include species such as the Louisiana crayfish (*Procambarus clarkii*), the pinewood nematode (*Bursaphelenchus xylophilus*), the Asian clam (*Corbicula fluminea*), the red Palm Weevil (*Rhynchophorus ferrugineus*) or the pond slider (*Trachemys scripta*) (Anastácio et al. 2005; Naves et al. 2006; Sousa et al. 2008).

Invasive plants are far more numerous than animals (Marchante et al. 2014) and when compared with central/northern European countries include more tree species (DAISIEteam 2008), which often results in more extreme changes at the landscape level. Over 670 alien plant species (including only casuals, naturalized and invasive) are considered to be introduced in Portugal mainland, being rather well characterized regarding introduction time and pathways, native range and taxonomy (Almeida and Freitas 2012; Almeida 1999; Marchante et al. 2014). Their native ranges include regions all around the world, with over 220 species originating from the Americas', over 270 from Eurasia and Mediterranean region, and 70 from Africa; Australasia despite being the native range of fewer species (35; Almeida 1999; Almeida and Freitas 2006; Almeida and Freitas 2012) is home to a substantial proportion of the most problematic ones, such as *Acacia* and *Hakea* species (Marchante et al. 2014).

Almost 50 % of the alien plant species were introduced as ornamentals, with agri/horticulture also accounting for a substantial number of introductions. These pathways are still "active" with new species being introduced every year. About 1/6 of the species were accidentally introduced, which corresponds to numerous cases of agricultural weeds whose seeds were acquired unintentionally with crop seeds (Almeida and Freitas 2000).

Amongst the vast taxonomic diversity (110 families) of alien plant species present in the country, several families are absent from the Portuguese native flora (e.g. *Oxalidaceae, Proteaceae, Pittosporaceae*) and many species (ca. 25 %) are grouped into a few large families—*Asteraceae*, *Fabaceae* and *Poaceae* (Almeida and Freitas 2006). Alien plant species occur in a wide range of habitats with some of the more anthropogenic and disturbed habitats having more species (>300) (Almeida 1999), reflecting the pattern of distribution predicted for Europe (Chytrý et al. 2009). Accordingly, the areas of the territory that have registered the introduction of most alien plant species are the heavily populated coastal areas, namely Estremadura (>330 species), Beira Litoral (>250 species) and Douro Litoral (>200 species), confirming the strong human factor associated with IAS problems. Nevertheless, many species are also present in other habitats, including in coastal sand dunes (48 species) and margins of inland water bodies (70 species) (Aguiar et al. 2007; Marchante et al. 2014).

The attribution of an invasive status to alien species is neither consensual, nor completely objective or static in time. Most alien species occur as either naturalized or casuals but presently with no invasive behavior (though some with risk of becoming invasive, Marchante et al. unpublished) and about 15 % of these alien species show invasive behavior at least in some areas. Based on field observations and consultation to experts, Marchante et al. (2014) considered 103 invasive (47) or potentially invasive species (24 casuals and 32 naturalized, see Table 1), based on present behavior in Portuguese territory. Of these, 29 are considered invasive by law, including most of the more problematic species, e.g., *Acacia dealbata*, *Acacia longifolia*, *Ailanthus altissima*, *Carpobrotus edulis*, *Eichhornia crassipes* or *Hakea sericea* (Fig. 1); other species not yet listed as invasive in the legislation (e.g., *Cortaderia selloana*), already reveal a large distribution along the country (Marchante et al. 2014).

**Fig. 1** Some of the worst invasive plant species in Portugal

Considering the number and the significant negative impacts of IAS, citizens need to be aware of the problem and conscientious that their action, either as a professional of areas dealing with alien and invasive plants or as an anonymous citizen, may be contributing to disperse the species but also to mitigate the impacts.

## 3 The Importance of Public Awareness and Engagement as Key Components to Deal with Invasive Plants in a More Sustainable Way

Preventing the introduction or further spread of species with high risk of becoming invasive is one of the most cost-effective management strategies (Pyšek and Richardson 2010). Several actions are considered in prevention, such as border interception of potentially problematic species, pathway and vector management, legislative frameworks and public awareness. Public awareness and education are often considered an important part of prevention but they should also be seen as essential to other phases of management (Wittenberg and Cock 2001). Awareness activities should be planned to target technicians and stakeholders but also the public in general who is an important vectors of introduction and spread of IAS (Ruiz and Carlton 2003), and may additionally play an important pro-active role in controlling the species. A well informed public may adopt a more responsible attitude (e.g., selecting native or non-problematic alien species, being aware of introduction pathways and excluding them, adopting measures to avoid being an "accidental vector" for spread of seeds) and become active (e.g., contributing to early detection programs, participating in citizen science programs with these species, controlling species in private lands) in the management of the species, with significant repercussions for mitigation of the problem. Awareness activities need to educate about environmental and economic risks involved with IAS, laws and regulations to prevent introductions of alien species (e.g., reasons for the restrictions, regulatory actions), species recognition and, additionally, propose easy, practical options that can help with management of IAS. Appealing printed materials, video presentations, talks, workshops, interactive games, use of social networks and hands on activities are all promising approaches (Marchante et al. 2010; Wittenberg and Cock 2001).

## 4 Strategies Used to Raise Public Awareness About IAS in Portugal

Short after starting to work with invasive species, a team of researchers from CFE/UC and ESAC/IPC realized the great lack of knowledge about IAS in Portugal. In order to fill this gap, diverse activities were undertaken since 2001 including workshops, training and field work events, public talks, development of printed materials, websites, etc. The different strategies implemented are described below and their outreach and effectiveness discussed.

### 4.1 Invasoras.Pt—Website on Invasive Plants in Portugal

Aims: to make available information about invasive plant species in Portugal, where they are and how can they be contained; to engage the public with the theme and publicize awareness activities and news about IAS; to make available information that can serve as reference for social networks and different approaches.

Description: in 2003 a website was developed and made available; this was the first website in Portugal with information about invasive alien plants at the country level. Later, in 2013, a new version of the website was developed (invasoras.pt) with a new approach—besides the static information about species (factsheets to facilitate the recognition of the species) it includes more dynamic and editorial style contents, comprising also connections to social media networks (https://www.facebook.com/InvasorasPt and https://twitter.com/Invasoraspt) and a citizen science platform for WebMapping of IAS in Portugal (Marchante et al. 2015b). The website in bilingual (Portuguese and English) and contacts are available for users to consult experts about invasive plants. The Facebook and Twitter pages work mostly as distribution channel of the website and WebMapping platform. The contents shared/published include, among others, weekly posts with alerts for the species which are easier to spot at that period and always links to the WebMapping platform.

Results and evaluation: the new website is available from March 2013 and in two and half later more than 106,000 visitors have accessed it (almost 500,000 pageviews); these values can be added to the more than 135,000 visitors of the previous website (Table 2). Although a large percentage of visitors are from Portugal and Portuguese speaking countries (almost 90 %), users from over 70

**Table 2** Approaches used to raise public awareness about invasive plants in Portugal by a team from CFE/UC and ESAC/IPC

| Type of activity/methodology | Target public | Public reached[a] | Time frame | Costs (€)[c] |
|---|---|---|---|---|
| *Web-page* invasoras.pt | General public | >240,000 | Available since 2003: new edition from 2013 | 17,000 |
| *Field-work projects* | University students; professionals mainly of environmental, forestry and biological sciences | >300 | 15 annual editions since 2003 (1 week each) | 30,000 |
| *Training courses* | | | | |
| Identification and control of IAP[b] | Technical publics dealing with IAP[b] | 340 | 3 editions: 2005, 2006 and 2007 (25 h each) + 12 in 2015 (7 h each) | 4700 |
| Biological invasions and environmental education | Schoolteachers | 50 | 2 editions: 2009 (25 h) and 2014 (4 h) | 1200 |
| *Printed documents* | | | | |
| Plant species technical profiles (also available online) | Technical publics dealing with IAP[b] | >3500 | Available since 2005 (out of print) | 6400 |
| Invasive plants field guide (also available online) | General public | >6700 | Available since 2009: new edition 2014 | 25,000 |
| Postcards to color | 8–12 years old | >2000 | Available since 2009 (out of print) | 2500 |
| Bookmarks collection | General public | >25,000 | Available since 2009 | 3500 |
| *Other initiatives* | | | | |
| Thematic workshops | Mainly students, but also the general public | >850 | >20, since 2008 | 5000 |
| Science and nature forums and fairs | General public and students | >2000 | >10, since 2008 | – |
| Talks | General public, students, horticultural trade, conservation experts, foresters, etc. | >5000 | >100, since 2007 | – |
| Social media: Facebook | General public | >5200 | Since march 2013 | – |
| | Sub-total (not considering the web page) | >50,000 (~0.5 % Portuguese population) | | |
| | total | >290,000 | | 91400 |

[a]Approximate numbers: [b]*IAP* invasive alien plants: [c]some values are rough estimates based on man-days to develop the activities, though such values were not in most of the cases, specially allocated to fund these tasks

countries have visited the page (plus another 80 with <10 visitors). Numerous people and institutions collaborate at the WebMapping platform (described below) and use the page e-mail address to request technical assistance on control methodologies and species identification, as well as to ask for collaboration in public awareness activities and environmental education sessions. These collaborations and contacts validate the website as an effective awareness tool. Almost 30 % of the visitors return to the site and the bounce rate is around 40 % which are good indicators that the website is working correctly. The Facebook page gathers over 5200 "likes".

## 4.2 Webmapping Platform to Geolocate Invasive Plants in Portugal

Aims: engage citizens to submit geolocation records of invasive plants, contributing proactively in the creation of geolocation dataset covering the national territory, including mainland and Archipelagos of Madeira and Azores. It also aims to engage the citizens with invasive plants problem.

Description: The WebMapping platform is available at the website invasoras.pt, since March 2013, and includes a Web App and a smartphone App (for Android devices, freely available at Google Play), which gradually build-up an online map of invasive plants (Fig. 2); it includes 46 invasive plants. Although the target public is all citizens, frequently unaware of the problem, experts and stakeholders are also targeted. The smartphone App includes simple factsheets of the invasive plants and shows images. The Web App easily access much more information included at the website namely more detailed factsheets. After acquiring the skill to recognize the invasive plants included on the platform, contributors can spot them and add the records to the WebMapping platform. Validation of the sightings is done at the back-office, by team members, and only after that the records are made available on the map. The map uses Google Maps interface and makes available the last 500

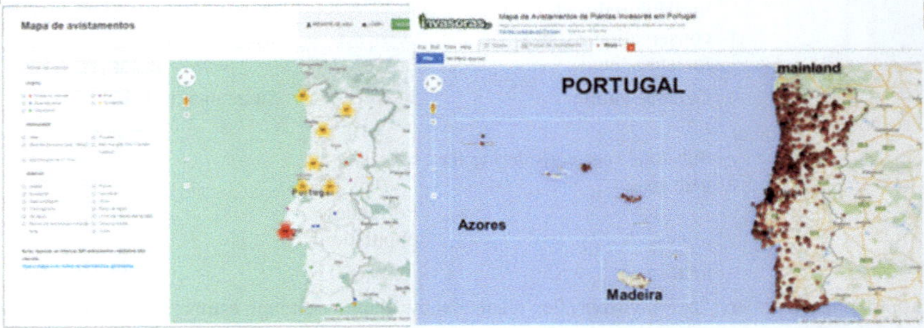


**Fig. 2** WebMapping platform to map invasive plants in Portugal (**a**) map available at invasoras.pt with the last 500 sightings; (**b**) map available at GoogleFusion with all sightings

sighting at Invasoras.pt, in order not to limit site speed; all validated sightings are visible on a Google Fusion Table, linked to the WebMapping platform, which allows free download of all data.

Results and evaluation: A total of 816 registered members (243 active) submitted 6785 geolocation records, using either the Web App (3467) or the smartphone App (3318); of these 93 % were validated, with the Web Map showing, by the end of September 2015, 6281 validated sightings. Over 75 % of the sightings validated were fully completed, i.e., included the compulsory and the optional information, suggesting that contributors are not conditioned by the quantity of information required. The number of validated sightings per month was varied from almost 900 (July 2013) to values between 50 and 100 (December 2013–March 2014). The peaks with over 500 sightings/month, registered in June–August 2013 and March 2015, correspond, respectively, to the participation of a few users particularly active, and a stronger effort on a raising awareness campaigns (through Facebook and Invasoras.pt) when some of the most expressive invasive plants in Portugal (*Acacia* spp.) were in flower.

## 4.3 Fieldwork Projects About Invasive Plants

Aims: to increase awareness amongst university students and young (or not so young) professionals, mostly from areas related to environmental, forestry and biological sciences, namely through training and collaboration on control of invasive plants in Conservation Areas.

Description: the projects include different approaches to engage the target public: (1) participation in control of invasive plant species, e.g., *Acacia longifolia, A. dealbata, A. melanoxylon, Cortaderia selloana, Tradescantia fluminensis* and *Carpobrotus edulis*, (2) brief training about IAS and Nature Conservation, and (3) small projects involving invasive plants, namely scientific experiments and public awareness activities for the general public and schools. The philosophy behind these projects is to strongly engage the target public with the theme, through learning about IAS, hands-on activities to control invasive plants and creation of a healthy and fun working/learning environment. In 2003, when the first project was organized, this type of project was quite innovative in Portugal and the public was outstandingly receptive and enthusiastic. Although activities are planned for 20 volunteers each year, the number of applicants has been always much higher, reaching more than 80 in several editions. These projects are developed in summer vacations, for one week, with volunteer groups sharing accommodation, meals, learning, working and leisure time.

Results and evaluation: Since 2003, 15 field-work projects were organized in seven places in Portugal (one per year), including Conservation Areas and other invaded areas involving over 340 volunteers, who contributed to the control of eight invasive plant species. These projects are very effective and successful in training people and raising awareness, especially among university students and young professionals. Additionally, they showed to be engaging: after participating, several volunteers became involved in invasive species projects, and some of them now work professionally with this subject. Furthermore, it has been a good way to encourage the staff of Conservation Areas and to publicize their work on the management and control of invasive plants.

In 2013 a questionnaire was sent to the previous field-work projects participants and to non-participants as well (240 questionnaires sent to each group) in order to better quantify the effectiveness of this approach. 40 % of participants and 32 % of non-participants answered the questionnaire. Briefly, the results showed that: (1) the level of knowledge about IAS is higher in people that attended the field-work projects (Fig. 3a); (2) participants are more motivated to participate/organize activities or work with IAS (Fig. 3b); (3) in general, more people heard about IAS from participants than from non-participants (Fig. 3c); and (4) participants identify correctly more IAS (Fig. 3d). Finally, participants consider themselves as having changed behaviors and attitudes regarding the problem.

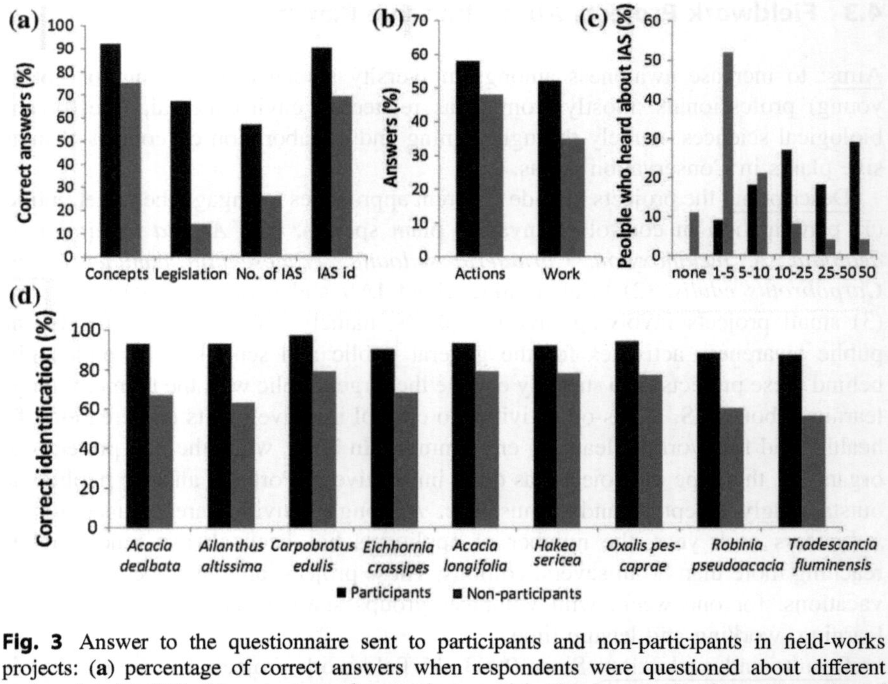


**Fig. 3** Answer to the questionnaire sent to participants and non-participants in field-works projects: (**a**) percentage of correct answers when respondents were questioned about different topics related to IAS; (**b**) percentage of respondents participating in control actions; (**c**) number of people who heard about IAS from the respondents; (**d**) percentage of correct identification of some of the worst invasive plants in Portugal

## 4.4 Training on Invasive Plants Identification and Control

Aims: to provide tools to capacitate the trainees to (1) identify and manage invasive plants present in Portugal (technical courses for professionals dealing with alien and invasive plants) and (2) develop educational projects and activities about invasive species (courses for school teachers).

Description: the courses involved theoretical and practical sessions, and field trips to areas invaded by different species. Three courses (25 h each) about identification and control of invasive plants were organized in 2005, 2006 and 2007 and 12 shorter ones (7 h) in 2014 and 2015. The target publics were technicians from municipalities and nursery industry, conservation and forestry experts, researchers, and other technical staff who deal with alien and invasive species. In 2009 and 2014 a different course was offered to school teachers, as they are in a privileged position to disseminate information among young people. The program was adapted from the technical course focusing more on the theory behind biological invasions and considering environmental education projects and activities that could be developed and used in school classes.

Results and evaluation: ca. 340 technicians and 50 teachers attended the courses. This approach has proved to be very effective in changing attitudes. Some technicians have actively integrated the knowledge gained in the courses in their regular activities, namely in programs to control invasive species or excluding invasive species from their lists of working species. Some of the teachers developed programs to be applied in their schools and as a consequence many students have heard about this theme and many have been involved in hands-on activities.

## 4.5 Printed Documents About Invasive Plants

Aims: to develop printed documents that can be used to raise awareness about invasive plants.

Description: the different activities organized and the contact with the public highlighted the need of printed documentation about invasive plants that could both support hand-on activities and also be available as sources of information for specific target-audiences. To fill this gap, different documents were produced, targeting different publics: (1) Plant species technical profiles (2005)—about identification and control of the 29 plant species considered invasive by the Portuguese law, plus three other species with invasive behavior (Marchante et al. 2005). The target-public was the technical staff dealing with invasive plants, and the plant species technical profiles were made available both online and printed; the printed version was distributed to professionals working with alien plants and private and public entities responsible for the management of areas invaded by alien plants; (2) Invasive plants field guide (2008 and 2014): in 2008 the first field guide of invasive alien plants in continental Portugal was published (Marchante et al. 2008b). In 2014 this was updated incorporating a risk assessment for every species and enlarged to include more species and invasive plants from Azores and Madeira (Marchante et al. 2014). More than 80 plants species were included (over 100 in the 2nd edition), including invasive plants and other with invasive potential (casuals and naturalized). The guide includes an introduction to biological invasions and invasive plant species; (3) Booklet with postcards to color (2008): although IAS can be a theme somewhat complex to young children, it is important to raise awareness from an early age. A small booklet with 13 postcards (some of the worst invasive plant species in Portugal) was developed targeting school children. It included a fixed part (to keep, with simple information) and postcards to detach which are drawings of invasive plants; the reverse is an ordinary postcard—the idea is that each child can learn a bit about invasive plants, personalize the card, coloring it, and write a message to friends and family about IAS, working as vectors of the information. Postcards were initially made for children from 8 to 12 years old, but worked also fine with younger as well as older students; (4) Bookmarks collection (2008): 13 bookmarks were made on a selection of some of the worst invasive plants in Portugal. Each bookmark includes simple information about invasive plants in general, information about a specific invasive plant and the link of the website where more information and contacts can be looked after. These are targeted to the general public, and used for different publics and activities. The aim was to have available a simple, appealing (and cheap) publication that can be given to everyone.

Results and evaluation: the technical profiles about invasive plants are available in a platform where it is the fifth most downloaded document amongst several thousand, with almost 3000 downloads since July 2007; the printed version (500 copies) is out-of-print. Frequent requests for the printed version and consultation concerning control of different invasive plant species are received. Two thousand free copies of the 1st field guide were printed and are now out-of-print; the reviews/criticisms to this first edition were very good and the new edition has incorporated them. A total of 4750 copies were printed of the 2nd edition of the field guide and copies may be requested at Invasoras.pt. Both editions (on paper) were distributed, mainly under direct request, to several official entities and people interested in the theme, reaching very distinct publics; it was also distributed to public and school libraries, being available to people all over the country. The first edition is available for download at https://pombalina.uc.pt/ and the second edition is available at this site and also at issuu.com (http://issuu.com/plantasinvasoras). The bookmarks were (still are) mostly distributed to entities dedicated to science communication and environmental education but also to conservation areas, schools and the general public in nature and science festivals and other events. Postcards were mainly used with school children and activities organized for this specific public. As much as possible, the printed documents were used together with different initiatives organized in order for them to be understood in context.

### 4.6 Other Activities

Aims: to raise awareness about invasive plants and to communicate results of research projects to different publics.

Description: thematic workshops were organized, mainly targeting school students, including different activities, such as short talks, hands-on activities for the control of invasive plants, interactive games and invasive plant identification games (Schreck et al. 2013). Further dissemination of information about invasive plants was attained through participation in several environmental conferences, forums, conference and talks, etc., targeting very diverse publics: the general public, school children and students, university students, foresters, horticultural trade, conservation experts, ENGOs, etc.

Results and evaluation: since 2007, more than 100 talks were given, 20 workshops and hands-on activities were carried out, and several science and nature forums and fairs for different publics were joined. The contexts and publics of these initiatives were very diverse. As a result, over the past few years and all over the country many citizens became aware about invasive plants. Effectiveness of the workshops organized for schools was accessed through questionnaires sent to schools, one year later, targeting students who attended the workshop as well as a control group who did not attend it. Results showed that, after one year, the participants in the workshop knew significantly more about invasive species and recognized more invasive plant species than non-participant students (Schreck et al. 2013).

## 5 Final Considerations

After more than 12 years communicating about invasive alien plants in Portugal, our perception is that awareness about biological invasions has increased, although lack of awareness is still a substantial reality. There is still a lot to be done and reaching the unaware citizens, outside the scientific or technical world, is particularly difficult. Nevertheless, information on IAS is nowadays more frequent in the media and many people and institutions have contributed, and are committed to continue, to raise public awareness. The diversified methodologies and strategies used by the team of CFE/UC and ESAC/IPC are slowly contributing to change mentalities and attitudes, making the public better educated on the topics of invasive plants and biological invasions. This public can then have an important role in the prevention, early-detection and control of invasive species.

Our perception is that approaches including hands-on (e.g., field-work projects) or interactive activities (e.g., WebMapping platform or Facebook) and that involve the participants for a longer time are more engaging and efficient in raising awareness about invasive plants (Schreck et al. 2013). The estimated number of people reached by the different activities/approaches is higher than 290,000 (or >50,000 if the webpage is exclude, Table 2). However, the main contribution to this number is the website, which effectively contributes to raise awareness and provides information, but which is probably less effective in making people changing their attitudes about alien and invasive plants than other activities. Evaluation of effectiveness is not always easy and as such a stronger effort and investment needs to be made in order to better evaluate the activities/approaches used to communicate on IAS. Nevertheless, funding for communication is often scarce and so it is important that it can be used in the most efficient way, targeting approaches that are more effective in changing attitudes and engaging the public with this subject. The collaboration of experts on communication is also of utmost importance if a well-coordinated and effective campaign is to be promoted.

We are committed to this challenge of engaging the public with IAS and will continue to do so along with our research activities. For that, we are planning to diversify activities in the field, establishing protocols with local and regional administrative agencies, implementing new tools on the website and initiating a pilot early-detection program.

## References

Aguiar, F. C., Ferreira, M. T., Albuquerque, A., & Moreira, I. (2007). Alien and endemic flora on reference and non-reference sites from Mediterranean type-streams of Portugal. *Aquatic Conservation: Marine and Freshwater Ecosystems, 17*, 335–347.

Almeida, J. (1999). *Flora exótica subespontânea de Portugal continental (plantas vasculares). Catálogo das plantas vasculares exóticas que ocorrem subespontâneas em Portugal continental e compilação de informações sobre estas plantas.* Master thesis, University of Coimbra, Coimbra.

Almeida, J., & Freitas, H. (2000). A flora exótica e invasora de Portugal. *Portugaliae Acta Biologica, 9*, 159–176.

Almeida, J., & Freitas, H. (2006). Exotic naturalized flora of continental Portugal—A reassessment. *Botanica Complutensis, 30*, 117–130.

Almeida, J., & Freitas, H. (2012). Exotic flora of continental Portugal—A new assessment. *Bocconea, 24*, 231–237.

Anastácio, P. M., Parente, V. S., & Correia, A. M. (2005). Crayfish effects on seeds and seedlings: Identification and quantification of damage. *Freshwater Biology, 50*, 697–704.

Carvalheiro, L. G., Buckley, Y., & Memmott, J. (2010). Diet breadth influences how the impact of invasive plants is propagated through food webs. *Ecology, 91*(4), 1063–1074.

Chytrý, M., Pyšek, P., Wild, J., Pino, J., Maskell, L. C., & Vilá, M. (2009). European map of alien plant invasions based on the quantitative assessment across habitats. *Diversity and Distributions, 15*, 98–107.

DAISIEteam. (2008). European invasive alien species gateway. Accessed on September 10, 2015 http://www.europe-aliens.org

Elton, C. S. (1958). *The ecology of invasions by animals and plants*. London: T. Methuen and Co.

European Parliament and Council. (2014). Regulation (EU) No 1143/2014 of 22 October 2014 on the prevention and management of the introduction and spread of invasive alien species. *Official Journal of the European Union, 317*, 35–55

Gaertner, M., Den Breeyen, A., Cang, H., & Richardson, D. M. (2009). Impacts of alien plant invasions on species richness in Mediterranean-type ecosystems: A meta-analysis. *Progress in Physical Geography, 33*(3), 319–338.

Hellmann, C., Sutter, R., Rascher, K. G., Máguas, C., Correia, O., & Werner, C. (2011). Impact of an exotic $N_2$-fixing *Acacia* on composition and N status of a native Mediterranean community. *Acta Oecologica, 37*(1), 43–50.

ISSG/IUCN. (2008). Invasive species specialist group, species survival commission (SSC) of the IUCN. Available at http://www.issg.org/

Jäger, H., Kowarik, I., & Tye, A. (2009). Destruction without extinction: long-term impacts of an invasive tree species on Galápagos highland vegetation. *Journal of Ecology, 97*(6), 1252–1263.

Marchante, E., Freitas, H., & Marchante, H. (2008a). *Guia prático para a identificação de Plantas Invasoras de Portugal Continental.* Coimbra: Imprensa da Universidade de Coimbra.

Marchante, E., Kjøller, A., Struwe, S., & Freitas, H. (2008b). Short and long-term impacts of *Acacia longifolia* invasion on the belowground processes of a Mediterranean coastal dune ecosystem. *Applied Soil Ecology, 40*, 210–217.

Marchante, H., Marchante, E., & Freitas, H. (2005). *Plantas Invasoras em Portugal—fichas para identificação e controlo.* Coimbra: Ed. dos autores.

Marchante, H., Marchante, E., Freitas, H., & Hoffmann, J. H. (2015a). *Temporal changes in the impacts on plant communities of an invasive alien tree.* Acacia Longifolia: Plant Ecology. doi:10.1007/s11258-015-0530-4:.

Marchante, E., Marchante, H., Morais, M., & Freitas, H. (2010). Combining methodologies to increase public awareness about invasive plants in Portugal. In *Paper Presented at the 2nd International Workshop on Invasive Plants in Mediterranean Type Regions of the World*, Trabzon, Turkey. 2–6 August.

Marchante, H., Morais, M., Freitas, H., & Marchante, E. (2014). *Guia Prático para a Identificação de Plantas Invasoras em Portugal.* Coimbra: Imprensa da Universidade de Coimbra.

Marchante, H., Morais, M. C., Gamela, A., & Marchante, E. (2015b). Using a WebMapping platform to engage volunteers to collect data on invasive plants distribution. *Transactions in GIS*,. doi:10.1111/tgis.12198.

Mazza, G., Tricarico, E., Genovesi, P., & Gherardi, F. (2014). Biological invaders are threats to human health: An overview. *Ethology Ecology and Evolution, 26*(2–3), 112–129.

Ministério do Ambiente. (1999). Decreto-Lei n° 565/99, de 21 dezembro. In: Diário da República —I Série—A. 295: 9100–9114

Millennium Ecosystem Assessment. (2005). *Ecosystems and human well-being: biodiversity synthesis.* Washington, DC.: World Resources Institute.

Naves, P. M., Camacho, S., & De Sousa, E. M. (2006). Entrance and distribution of the pinewood nematode Bursaphelenchus xylophilus on the body of its vector Monochamus galloprovincialis (Coleoptera: Cerambycidae). *Entomologia Generalis, 29*(1), 71–80.

Pejchar, L., & Mooney, H. (2010). The impact of invasive alien species on ecosystem services and human well-being. In C. Perrings, H. Mooney, & M. Williamson (Eds.), *Bioinvasions and globalization: Ecology, economics, management, and police.* New York: Oxford University Press.

Perrings, C., Mooney, H., & Williamson, M. (2010). *Bioinvasions and globalization. ecology, economics, management, and policy.* New York: Oxford University Press.

Pyšek, P., & Richardson, D. M. (2010). Invasive species, environmental change and management, and health. *Annual Review of Environment and Resources, 35*, 25–55.

Richardson, D. M., & Pysek, P. (2008). Fifty years of invasion ecology—The legacy of Charles Elton. *Diversity and Distributions, 14*(2), 161–168.

Richardson, D. M., Pyšek, P., Rejmánek, M., Barbour, M. G., Panetta, F. D., & West, C. J. (2000). Naturalization and invasion of alien plants: concepts and definitions. *Diversity and Distributions, 6*, 93–107.

Ruiz, G. M., & Carlton, J. (2003). *Invasive species: Vectors and management strategies.* Washington DC: Island Press.

Schreck Reis, C., Marchante, H., Freitas, H., & Marchante, E. (2013). Public perception of invasive plant species: Assessing the impact of workshop activities to promote young students' awareness. *International Journal of Science Education, 35*(4), 690–712.

Sousa, R., Nogueira, A. J., Gaspar, M., Antunes, C., & Guilhermino, L. (2008). Growth and extremely high production of the non-indigenous invasive species *Corbicula fluminea* (Müller, 1774): Possible implications for ecosystem functioning. *Estuarine, Coastal and Shelf Science, 80*, 289–295.

Stigall, A. L. (2010). Invasive species and biodiversity crises: Testing the link in the late devonian. *PLoS ONE, 5*(12), e15584.

Vilà, M., Basnou, C., Pyšek, P., Josefsson, M., Genovesi, P., Gollasch, S., et al. (2010). How well do we understand the impacts of alien species on ecosystem services? A pan-European, cross-taxa assessment. *Frontiers in Ecology and Environment, 8*(3), 135–144.

Wittenberg, R., & Cock, M. J. W. (2001). *Invasive alien species: A toolkit of best prevention and management practices*. Wallingford, Oxon, UK: CAB International.

# Part III
# Engaging Society in Biodiversity Conservation and Sustainability

# Bugs and Society II: Testing Two Communication Strategies for Public Engagement in the Azores

Isabel R. Amorim, Ana Moura Arroz, Rita São Marcos, Paulo A. V. Borges and Rosalina Gabriel

**Abstract**

Two initiatives to raise biodiversity awareness towards Azorean endemic species, an outdoor exhibition *Açorianos há milhões de anos* (Azoreans for millions of years) and a web contest to name insects *Chama-lhe Nomes!* (Pick a Name!), are presented in this chapter. Both communicational strategies targeted non-traditional audiences, relied on the Portuguese and Azorean cultural identity and on anthropic verisimilitude of situations involving insects. The context, principles, assumptions and multidisciplinary approach involved in the development of the public awareness activities were presented in detail in chapter "Bugs and Society I: Raising Awareness About Endemic Biodiversity". Apart from having symbolically occupied the urban public domain, the outdoor exhibition triggered positive reactions in more than three-quarters of the observers, prompting them to seek more information about the insects, to want to see them alive, to photograph them, etc. On the other hand, the web contest attributed common names to 12 endemic species of insects and motivated over one hundred people to carefully consider their photos and descriptions, engaging in sheer naturalistic pleasure. Less favourable were the post-observation

Isabel R. Amorim and Ana Moura Arroz—contributed equally to this work.

I.R. Amorim (✉) · P.A.V. Borges · R. Gabriel
Departamento de Ciências Agrárias, cE3c—Centre for Ecology, Evolution and Environmental Changes/Azorean Biodiversity Group and Universidade dos Açores, 9700-042 Angra do Heroísmo, Açores, Portugal
e-mail: isabel.ma.rosario@uac.pt

A.M. Arroz (✉) · R. São Marcos
Departamento de Ciências da Educação, cE3c—Centre for Ecology, Evolution and Environmental Changes/Azorean Biodiversity Group and Universidade dos Açores, 9700-042 Angra do Heroísmo, Açores, Portugal
e-mail: ana.mm.arroz@uac.pt



P. Castro et al. (eds.), *Biodiversity and Education for Sustainable Development*, World Sustainability Series, DOI 10.1007/978-3-319-32318-3_9

recognition indicators of urban exposure, since only less than one-third of the interviewees correctly identified the insect' group represented on the banner or could place the origin of the animal. Nevertheless, the means assigned to the evaluation of the intervention were not enough to undertake a full assessment of the impacts produced in the knowledge, attitudes and behaviours of viewers and users. Still, the resonance in the media was considerable, in terms of the number of news stories and the diversity of vehicles and audiences reached at regional (outdoor exhibition) and national levels (web contest). Communication devices like the ones presented are expected to raise biodiversity awareness and empower people regarding its preservation in the Azores. Moreover, the critical analysis of these initiatives is expected to provide guidelines that maximise the transferability of communication strategies to other social settings.

## 1 Introductory Note

Biodiversity loss is one of the most neglected environmental risks, with a low level of priority among public concern (Cardinale et al. 2012). Thus, in order to raise awareness to the problem of endemic biodiversity disappearance, this issue was framed in the light of risk governance (Arroz et al. 2016, this book), using insects from the Azores, the taxonomic group with the highest number of endemic species in the archipelago.

This chapter presents two of the activities that were part of the Azorean intervention *Bugs and Society*, and discusses their results. The activities differed considerably, namely on the: (i) degree of involvement of the target population—to inform versus to involve; (ii) audiences—city dwellers versus Facebook users; (iii) proposed tasks—to appreciate/contemplate versus naming insects; and (iv) grammars used—outdoor banners vs. an online contest. However, the same principles and purposes underlie both activities, which also rely on immaterial values that characterise the Portuguese/Azorean identity as a strategic option to bridge the gap between insects and humans.

The specific goals of this work were to: (i) design communicational devices to promote endemic biodiversity, framed by indigenous ethnoenthomology and by the Azorean and Portuguese core identities; (ii) present the process and evaluate the results of two communication activities in mainstreaming biodiversity; and (iii) provide guidelines about the message design, the regulation and the evaluation of communication strategies in order to maximise its transferability to other social contexts.

## 2 Raising Public Awareness: The Exhibition *Açorianos há Milhões de Anos*—Azoreans for Millions of Years

### 2.1 Communication Goals and Strategies

To raise public awareness regarding conservation biodiversity, an urban outdoor exhibition was conceived to catch city dwellers' attention while going about their daily activities. Because insects represent a big portion of the species that can only be found in the Azores, but are among islanders least favourite organisms (Gabriel et al. 2007, 2009, 2012a,b), endemic insects were the protagonists of this initiative. Besides providing accurate scientific information on the particular species, the exhibition sought to defy the dominant representations about insects in order to:

- draw attention to the existence of species that can only be found in the Azores;
- raise awareness among local citizens for the need to protect endemic species; and
- bring these species to the city, giving the general population a chance to see them, as these insects are difficult to observe in their natural habitat.

This initiative aimed to spark curiosity, promote interest and incite casual conversations, bringing insects into the public arena, thus facilitating an invisible (latent) participation of the people (sensu Pasquino's system of political participation, 2002).

The communication strategy draws on previous research on the perspectives of Azoreans on insects (Gabriel et al. 2007, 2009, 2012b). Those studies revealed that people were unaware of the fact that several of the species that exist in the archipelago can only be found there, felt aversion (*dislike*) towards insects because they are *disgusting*, a *nuisance*, *scary*, *useless*, *ugly* and *irrelevant*, and were not very concerned about insect conservation.

To deconstruct this negative public opinion on insects, several team brainstorming meetings were held in order to prepare a briefing for communication designers to produce the *copy* for this specific initiative (Box 1), bearing in mind that the purpose, the function and the legitimation criteria of communicating science are unlike the ones of making science (Dahlstrom 2014).

**Box 1: Excerpt of the designers briefing describing positive attributes of the selected insect species** Key ideas to convey concerning the selected Azorean endemic insect species:

1. They are unique and only exist in the Azores;
2. They are special, all having important ecosystem functions (*insects are useful*):
    a. Pollination—guarantee the survival of most plant species, assuring seed and fruit production;
    b. Regulation of natural populations—by feeding on other organisms (predation and/or herbivory) insects control those populations, preventing them from attaining pest status (e.g., other insects, other groups of arthropods, plants);
    c. Forest structure—insects feed on living and dead plants, shaping the landscape;
    d. Decomposition and nutrient cycling—insects are responsible for fast cycling of nutrients, particularly in forest ecosystems; without them, a huge amount of nutrients would remain unavailable for other organisms;
    e. Food source—many other animals feed on insects.

3. They are true *Azoreans*, they were on the islands long before Portuguese settlers arrived in the 15th century: they have been around for millions of years;
4. They are "under siege" on the islands and their existence depends on the conservation of their natural habitats (native Azorean forest), which currently is reduced to about 2.5 % of its original area. Extinction is forever; once they disappear from the Azores, unique biological diversity has been definitely erased from Earth;
5. They may and should be celebrated for their eccentric and idiosyncratic beauty;
6. They are not a threat! These insects are not dangerous, not poisonous, not disease vectors and would not invade our homes; conversely, their existence is threatened by human-mediated habitat destruction;
7. They are also *animals*. When compared to other animals (e.g., birds, mammals) insects attract very little attention and are less valued; insects are the aesthetically challenged Mother Nature's "ugly ducklings"!

The operationalised strategy (Table 1) involved, in terms of iconic and symbolic forms of representation, the creation of hyper-realistic depictions of endemic insects, exploiting the capability of extreme macrophotography, in order to make unavoidably visible what is naturally tiny and socially ignored. The extreme magnification revealed traits in a larger than life format, otherwise impossible to see with the naked eye. The change of scale resonated, along the lines of modern expressionism, with ugliness *coming from the depth of what is alien to human experience, disgusting, unsettling…* (Fortuna 2009, p. 193), somewhat near to the notions of the ugly/sublime described by Kant (Kuplen 2015).

The "portrait" was the visual grammar chosen for the banners, to draw the observer's attention to "a face", which was intended to foster human empathy with the living beings represented (Merleau-Ponty 1945). Additionally, the unusualness of the images inspired an alternation of the perception of the *figure* and *ground* (sensu Gestalt) and between *studium* and *punctum*—the structural elements of a picture (Barthes 1980). The former element refers to the generalised interest, pleasure or concern one might take in a photo (Iversen 1994), sparking interest and curiosity. The latter corresponds to a unique detail, "jumping out of the photo", coming across the *studium*, grounded by the individual experience of the observer: it may be the eccentric feathery look of the scales, a peculiarly shaped tongue or the prognathism of the mandible, or some other personally meaningful detail.

The weirdness and, in some cases, the latent threat of the images mirror the perceptions of the population towards insects. The logic behind the communication strategy was to take advantage of those dominant public perspectives and, using the *copy* developed by the designers, catapult the exact opposite message in order to deconstruct and demystify current views on insects. This deconstruction of mental

**Table 1** Message design used for the deconstruction of the negative public opinion on insects and for the promotion of Azorean endemic species in the outdoor exhibition *Açorianos há milhões de anos*—Azoreans for millions of years

| ARGUMENT | TACTIC | OPERATIONALIZATION | | | | | |
|---|---|---|---|---|---|---|---|
| | |  | | | | | |
| ACCURACY | SCIENTIFIC NAME | *Hipparchia azorina azorina* | *Drouetius borgesi borgesi* | *Hemerobius azoricus* | *Ascotis fortunata azorica* | *Trigonophthalmus borgesi* | *Trechus terceiranus* |
| BEAUTY<br>ugly/sublime | HYPERREALISM<br>extreme macro photos | 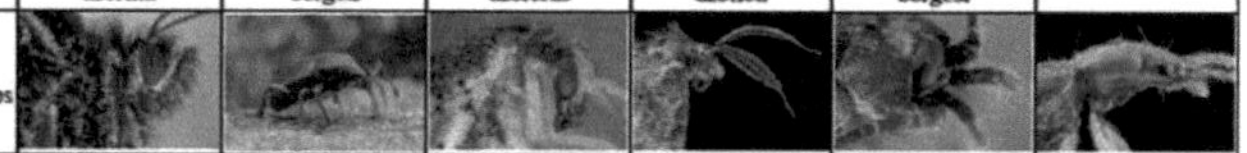 | | | | | |
| IDENTITY<br>recognition by others is required | HUMANIZATION<br>usage of cognomen and portraits | *A Desejada*<br>The Desired | *O Artista*<br>The Artist | *O Justiceiro*<br>The Righteous | *A Formosa*<br>The Beautiful | *O Restaurador*<br>The Renovator | *O Bravo*<br>The Brave |
| | | most photographs only captured the insect's head | | | | | |
| | CLAIMING OWNERSHIP<br>originally from the Az | *… for millions of years* | | | | | |
| SPECIAL ROLE<br>utility and/or resilience | PUBLIC OPINION DECONSTRUCTION<br>positive attributes | *A precious butterfly spreading life flying from flower to flower…* | *A weevil that shapes the landscape…* | *An insect that protects the forest from pests…* | *A flamboyant forest moth displaying beauty wherever it goes…* | *A jumping bristletail that recycles forest rubbish…* | *A beetle that withstands the hardship of living in caves…* |
| EXCLUSIVITY<br>endemic | CERTIFICATION<br>"certificate of origin and exclusivity" stamp | INSETOS ÚNICOS DOS AÇORES | | | | | |
| THREAT<br>risk of extinction | | | | | | | |

Macro photographs: Javier Torrent; other photographs: Enésima Mendonça, except *H. azorina azorina* (Martin Gascoigne-Pees), *A. fortunata azorica* (Axel Hausmann) and *T. borgesi* (Paulo A. V. Borges)

images and false beliefs was further reinforced by discursive analogies that decoded complex scientific concepts and by creative highlights of the importance of insects in the ecosystem (specific functions), therefore raising awareness for the need to protect them.

The purpose of using Portuguese royal cognomen for the insects was to foster a link between them and historical figures inhabiting the Portuguese cultural imaginary as well as to attribute an easily recognised social role to each one. Using historical references also tends to take people into the past, and although dealing with very different time scales, this facilitates the acknowledgement that the depicted endemic species have been in the Azores for a very long time: millions of years.

Finally, a "stamp" was used on the insect photographs to "certify" the scientific veracity of the conveyed information: geographic exclusivity—"only found in the Azores"; risk of extinction—"in the last 500 years, 97 % of these species natural habitats were destroyed", and natural heritage relevance—"a treasure in peril". This "certification stamp" summarises some of the conservation arguments most valued by biology experts.

The strategic options that guided the conceptual and argumentative logic used for the (de)construction of negative social perspectives on insects resulted in the urban outdoor exhibition *Azoreans for millions of years* (Arroz et al. 2016).

## 2.2 Insect Banners on the City Streets of Angra do Heroísmo: Claiming Territory in the Public Arena

The outdoor exhibition *Azoreans for millions of years* comprised two sets of large banners (4 m × 2.75 m) representing six insect species endemic to the Azores (Table 1; Fig. 1; Arroz et al. 2013a; Gabriel et al. 2016), which are poorly known and lack common names, but have important roles for proper ecosystem functioning.

Streets were deliberately occupied with giant photographs (focus stacking of extreme macro photos) of these species. Instead of hosting a biodiversity exhibition in a traditional science venue, the exhibition was moved outdoors into the streets, claiming a public territory for these endemic species that may face the risk of extinction in the near future. This was no ordinary territory, it was the historical centre of Angra do Heroísmo city, a UNESCO World Heritage Site, i.e., neglected natural heritage was displayed side by side with widely acknowledged built heritage. Furthermore, bringing these *sui generis* insects into the city allowed people to become acquainted with some elements of the unique Azorean natural heritage, both the particular species on display and the threatened habitats where they could be found (native forest and caves).

The 12 banners were strategically displayed on white building walls along the main streets of Angra do Heroísmo (Arroz et al. 2013b), as to create an endemic insect discovery path across the city. The exhibition was on display from April until

**Fig. 1** Communication device—banner—for raising public awareness of the Azorean endemic moth *Ascotis fortunata azorica*, Pinker, 1971

June 2013. Other urban areas, within and outside of the Azores, have since showed interest in hosting *Azoreans for millions of years.*

## 2.3 Outdoor Exhibition Evaluation

To assess public response to the outdoor exhibition Azoreans for millions of years in the city of Angra do Heroísmo, an audience study was undertaken during the three weeks that followed the opening of the exhibition (3rd April 2013), as planned in the logic model (Tables 2, 3 and 4; Arroz et al. 2012a).

According to the observations made by six trained field assistants, 5752 people walked by the banners during 219 observation periods of 10 minutes (14 days, 9–17 h), an average of 26 (±6) persons per period. When compared with the number of visitors of the Science Centre of Angra do Heroísmo[1] for the same month (n = 302; unpublished data), this is an impressive figure and validates the option of using the streets as a vehicle for science communication.

On average, 3.0 % (174) of the observed people explicitly showed some kind of reaction to the insect banners, 41 of these city dwellers were approached, and 37 agreed to be interviewed. All interviewees were residents of Terceira Island and the sample was balanced in terms of sex (20 women), age group (17 under 25,

[1]Science Centre of Angra do Heroísmo—CCAH—http://ccah-oaa.blogspot.pt

**Table 2** Evaluating the Knowledge dimension of the outdoor exhibition *Açorianos há milhões de anos* (Azoreans for millions of years): Analytic model and main results

| Sub-dimension indicator | Question or item | Main results |
|---|---|---|
| Previous information | | |
| Knowledge of endemic fauna | 16. List animals that only occur in the Azores. | People naming endemic animals: 7 (18.9 %)—the birds: *Buteo buteo rothschildi* (Swann, 1919) and *Pyrrhula murina* (Godman, 1866) |
| | 17. What about insects? Do you know any? Which one(s)? | People naming endemic insects: 1 (2.7 %)—"certain butterflies" |
| Relevance of insects in the ecosystem | 18. In your opinion, is it possible to live without insects? | Yes: 0 (0.0 %); No: 33 (89.2 %); DNK/NA: 4 (10.8 %) |
| Understanding the information | | |
| Representation of the purpose of the exhibition | 4. What do you think these banners are all about? | Correct: 13 (35.1 %) (Conservation of species: 7; Information about endemic biodiversity: 5; nature conservation: 1); Incomplete: 8 (21.6 %) (Information about biodiversity: 7; Information about nature: 1); Incorrect: 6 (16.2 %) (Alert to pests: 3; others: 3); DNK/NA: 8 (21.6 %) |
| Highlights: identify | 5. What caught your attention? | Image: 28 (75.7 %); Scale: 4 (10.8 %); Text: 3 (8.1 %); DNK/NA: 2 (5.4 %) |
| Highlights: reason | 6. Why? | Hyperrealism and beauty/ugliness: 6 each (16.2 %); colour and scale and unusual: 4 each (10.8 %); novelty: 2 (5.4 %); location and sensation: 1 each (2.7 %); DNK/NA: 9 (24.3 %) |
| Recognition of information | | |
| Acknowledgement of the outdoor banners | 1. Have you noticed the large banner on the wall that you just walked by? | Yes: 37 (90.2 %); No: 4 (9.8 %) |
| Selection and identification of a previously seen banner | 7a. The BRAVE is… a butterfly; a mosquito; a moth; a spider | Correct: 9 (24.3 %); incorrect: 23 (62.2 %); DNK/NA: 5 (13.5 %) |
| Ability to recall the distribution range of the insect | 7b. The BRAVE exists only in: Europe; Portugal; the Azores; the Portuguese islands | Correct: 11 (29.7 %); Incorrect: 14 (37.8 %); DNK/NA: 12 (32.4 %) |

(continued)

**Table 2** (continued)

| Sub-dimension indicator | Question or item | Main results |
|---|---|---|
| Ability to recount—freely reproduce the intervention (banner or exhibition) | 3. What have you just seen? | Insect: 32 (86.5 %); DNK/NA: 5 (13.5 %) |
| Self-evaluation of the impact of the initiative | | |
| Perceived knowledge acquisition | 29. Do you think this initiative was able to teach you something? | No affect: 8 (21.6 %); Minor affect: 5 (13.5 %); Neutral: 16 (43.2 %); Moderate affect: 5 (13.5 %); Major affect: 0 (0.0 %); DNK/NA: 3 (8.1 %) |

*DKN* do not know; *NA* no answer

**Table 3** Evaluating the Engagement dimension of the outdoor exhibition *Açorianos há milhões de anos* (Azoreans for millions of years): Analytic model and main results

| Sub-dimension indicator | Question or item or parameter | Main results |
|---|---|---|
| Attractiveness | | |
| Number of passers-by | Record of the number of passers-by that walked by the banners | Number of people passing by the banners every hour—average: 158; max: 402; min: 6<br>Estimated by 219 observation periods of 10 min each, in 14 days [April 4–30, 2013], from 9 to 17 h |
| Number of passers-by that observed the banners | Record of the number of passers-by that observed (stopped and/or looked) the banners | People reacting: 174/5752 (3.0 %) (Glances: 148 [85.1 %]; Admires: 14 [8.0 %]; Does something: 12 [6.9 %]) (Observation periods as above) |
| Assessment of the degree of attractiveness | 8. Did you appreciate what you just saw? | Not at all: 0 (0.0 %); Yes, to some extent: 2 (5.4 %); Neutral: 19 (51.4 %); Yes, to a large extent: 13 (35.1 %); Yes, to a very great extent: 2 (5.4 %); DNK/NA: 1 (2.7 %) |
| First impressions | 2. What were the first words that came into your mind when you looked at it? 1st; 2nd; …; nth word | Persons answering: 35 (94.6 %); DNK/NA: 2 (5.4 %)<br>Total of words: 56; Number of different words: 35; Words per interviewee—average: 1.5; max: 4; Most common word: Insect (10; 27.0 %) |

(continued)

**Table 3** (continued)

| Sub-dimension indicator | Question or item or parameter | Main results |
|---|---|---|
| Representation of the effect of the exhibition (emotional responses) | 9a. For you, this banner is: ugly 1-2-3-4-5 beautiful | Level 1: 0 (0.0 %); Level 2: 2 (5.4 %); Level 3: 9 (24.3 %); Level 4: 3 (8.1 %); Level 5: 4 (10.1 %); DNK/NA: 19 (51.4 %) |
| | 9b. For you, this banner is: useless 1-2-3-4-5 useful | Level 1: 1 (2.7 %); Level 2: 1 (2.7 %); Level 3: 6 (16.2 %); Level 4: 1 (2.7 %); Level 5: 7 (18.9 %); DNK/NA: 21 (56.8 %) |
| | 9c. For you, this banner is: repugnant 1-2-3-4-5 attractive | Level 1: 1 (2.7 %); Level 2: 4 (10.8 %); Level 3: 11 (29.7 %); Level 4: 5 (13.5 %); Level 5: 5 (13.5 %); DNK/NA: 11 (29.7 %) |
| | 9d. For you, this banner is: difficult 1-2-3-4-5 simple | Level 1: 2 (5.4 %); Level 2: 2 (5.4 %); Level 3: 4 (10.8 %); Level 4: 5 (13.5 %); Level 5: 4 (10.8 %); DNK/NA: 20 (54.1 %) |
| | 9e. For you, this banner is: alarming 1-2-3-4-5 soothing | Level 1: 2 (5.4 %); Level 2: 2 (5.4 %); Level 3: 2 (5.4 %); Level 4: 9 (24.3 %); Level 5: 3 (8.1 %); DNK/NA: 19 (51.4 %) |
| Interest | | |
| Number of banners seen | 11. Is this the first banner of the exhibition that you see? | Yes: 15 (40.5 %); No: 21 (56.8 %); DNK/NA: 1 (2.7 %) |
| | 12. Which ones have you seen so far? | All banners were mentioned; max number of banners mentioned: 4/6; min: 0 |
| Repercussions | 13. Have you heard about this? | Yes: 10 (27.0 %); No: 25 (67.6 %); DNK/NA: 2 (5.4 %) |
| Source of information | 14. If you have heard about the exhibition, how did you learn about it? | Television: 0 (0.0 %); Radio: 0 (0.0 %); Newspapers: 3 (8.1 %); Family and/or Friends: 3 (8.1 %); Magazines: 0 (0.0 %); Internet: 3 (8.1 %); Others: 1 (2.7 %) |
| Reason for observing | 15. You looked at this banner because… | You walked by and noticed it: 32 (86.5 %); You came here on purpose to look at it: 2 (5.4 %); You searched the city for banners: 1 (2.7 %); DNK/NA: 2 (5.4 %) |
| Intention to extend the experience beyond the viewing time | 10. Now that you have looked at this banner, you feel like you want to… | Know more about the insect: 11 (29.7 %); Talk about the banner: 9 (24.3 %); Look at a live specimen of this insect: 3 (8.1 %); Take a picture and Look for other banners: 2 each (5.4 %); Do nothing in particular: 7 (18.9 %); DNK/NA: 1 (2.7 %) |

(continued)

**Table 3** (continued)

| Sub-dimension indicator | Question or item or parameter | Main results |
|---|---|---|
| References on general and specialized media | Number of news | Broadcasted: 3 tv + 1 radio; Press: 7 regional + 1 national; Online: 16 sites; Science divulgation: 2 articles; Others: 1 (ART) |

*DKN* do not know; *NA* no answer; *max* maximum; *min* minimum

**Table 4** Evaluating the Attitudes dimension of the outdoor exhibition *Açorianos há milhões de anos* (Azoreans for millions of years): Analytic model and main results

| Sub-dimension indicator | Question or item | Main results |
|---|---|---|
| Attitudes towards insects | | |
| Biophilia | 20. Do you like any particular insects? | Yes: 15 (40.5 %); No: 19 (51.4 %); DNK/NA: 3 (8.1 %) |
| | 21. Which one(s)? | Butterflies: 10 (27.0 %); Bees, Flies, Mosquitos, Crickets and Mantises: 1 each (2.7 %); All species: 2 (5.4 %) |
| | 22. Why/Why not? | Like insects 16 (43.2 %)—Beautiful: 4; Cute and Interesting: 3 each; Useful: 2; Liking all animals: 2; Other reasons: 2; DNK/NA: 0<br>Dislike insects: 18 (48.6 %)—Ugly: 4; Disgusting: 2; Dirty: 1; Other: 2; DNK/NA: 9 |
| Biophobia | 23. Are there any insects that you do not like? | Yes: 28 (75.7 %); No: 4 (10.8 %); DNK/NA: 5 (13.5 %) |
| | 24. Which one(s)? | Cockroaches: 8 (21.6 %); Mosquitos: 5 (5.4 %); Flies: 4 (10.8 %); Bees and Butterflies and Termites: 1 each (2.7 %). Other (non-insects): 6 (16.2 %) |
| | 25. Why/Why not? | Disgusting: 6 (16.2 %); Scary: 5 (13.5 %); Dangerous: 4 (10.8 %); Annoying and Bad 2 each (5.4 %); Destructive and Dirty and Uninteresting: 1 each (2.7 %); Other: 2; DNK/NA: 12 (32.4 %) |
| Importance of protecting endemic species | | |
| Perceived seriousness of biodiversity loss | 19. How do you think that species extinction will affect life on the planet? | No affect: 0 (0.0 %); Minor affect: 1 (2.7 %); Neutral: 3 (8.1 %); Moderate affect: 23 (62.2 %); Major affect: 7 (18.9 %); DNK/NA: 3 (8.1 %) |

(continued)

**Table 4** (continued)

| Sub-dimension indicator | Question or item | Main results |
|---|---|---|
| Degree of support | 26. Do you find important to protect animals and plants that only exist in the Azores? | Not important: 0 (0.0 %); Slightly important: 0 (0.0 %); Neither important nor unimportant: 3 (8.1 %); Important: 29 (78.4 %); Very important: 3 (8.1 %) |
| Priority level assigned to nature conservation | 27. How much of the Azorean government budget for the environment do you think should be invested in protecting species that only exist in the Azores? | No budget (0 %): 1 (2.7 %); Low budget (20 %): 8 (21.6 %); Medium-low budget (40 %): 10 (27.0 %); Medium-high budget 60 %: 3 (8.1 %); High budget (80 %): 3 (8.1 %); Total budget (100 %): 1 (2.7 %); DNK/NA: 11 (29.7 %) |
| Self-evaluation of impacts | | |
| Degree of attitudinal change due to the exhibition | 28. Has this initiative changed your opinion on the need to protect insects that only exist in the Azores? | No affect: 7 (18.9 %); Minor affect: 3 (8.1 %); Neutral: 17 (45.9 %); Moderate affect: 7 (18.9 %); Major affect: 1 (2.7 %); DNK/NA: 2 (5.4 %) |

*DKN* do not know; *NA* no answer

11 between 25 and 45, and 13 over 45 years old), educational attainment (18 did not finished high school; 22 did, and 1 had a university degree) and area of residence (18 urban); additionally, six persons had experience with environmental organisations. As regards cultural habits (adapted from Virtanen 2007), more than half of the interviewees had not been to a museum (54.1 %), exhibition (64.9 %), theatre (67.6 %) or archaeological site (83.8 %) during the previous year and going to music events, the cinema and natural parks were the only cultural activities that more than a quarter of the people had performed regularly (three times or more) during that period. Concerning biological interests, most interviewees recognised that extinction of species is a serious problem (81.0 %), and that animals and plants that only exist in the Azores should be protected (86.5 %), but less than half (40.5 %) liked at least one insect (Table 4). About a quarter of the interviewees (27.0 %) already knew that the exhibition was up and running before being interviewed, mostly through their families, newspapers and the internet (Table 4).

About a third of the interviewees (35.1 %) correctly identified the main purpose of *Azoreans for million years* (e.g. inform about endemic biodiversity and conservation of species) while a fifth (21.6 %) answered in a vague, imprecise way (e.g. information about biodiversity and nature) and the remaining either failed to identify the purpose (16.2 %) or did not answer this question (21.6 %) (Table 2).

The large majority of people (75.7 %) were drawn to the image. For most, the *punctum* was the portrayed insect, but the unusualness, the colours and the lettering also attracted the attention of many. A minority mentioned the text—particularly the kings' cognomen—and the size of the banners (Table 2). The reasons invoked to explain the attraction to specific banner elements mainly included hyperrealism and beauty/ugliness of the insects, colour, scale of the banner/insect and unusualness (e.g., shape of the insect's antenna) (Table 2).

Knowledge on Azorean endemic animals was rather limited. A few persons mentioned the common buzzard of the Azores (10.8 %) and the Azorean bullfinch (8.1 %)—both birds. When asked about endemic insects, a single person (2.7 %) acknowledged "some butterflies". Nevertheless, insects were recognized as very relevant in the ecosystems (89.2 %) (Table 2). The majority of people (86.5 %) recognized that they had just seen an insect and 29.7 % rightly stated that the species were originally from the Azores while 24.3 % correctly identified the kind of insect portrayed (beetle, butterfly, moth, etc.) (Table 2). These results are actually better than the self-efficacy report of the interviewees, since only 13.5 % acknowledged moderate/major affect from the exhibition.

Less than half of the interviewees (40.5 %) stated that they appreciated the exhibition to a large/very large extent, while about half of them (51.4 %) adopted a neutral position, which reveals a large zone of indifference or latitude of non-commitment (Table 3). The self-recognised impact of the exhibition in their attitudes towards conservation of endemic insects was suboptimal with about one-fifth (21.6 %) recognising that it had contributed to change their attitude (Table 4). The silver lining came when people were asked about what they would do after having seen the banners. A large majority (78.4 %) expressed interest either in the insects (e.g. learning more about them, seeing them alive), or the exhibition (e.g. in talking about it, photographing the banners) (Table 3).

The verbalisation of "what comes to mind" after looking at the banners resulted in 56 words (35 different; average of 1.5) (Table 3). The word "insect", besides being the most frequent, it was always the first mentioned. Another word always among the first to be mentioned was "alien", although with a much lower frequency (8.1 %). The expectation that first impressions would relate to the beauty and function of the insects was not supported, as people used a more descriptive approach in this question.

To investigate the feelings evoked by the exhibition, a semantic differential of five pairs of qualities was used: ugly/beautiful; useless/useful; repugnant/attractive; difficult/simple; and alarming/soothing. The overall impression towards the banners was diversified, with the qualities directly associated with the portrayed insect, "beauty" and "attractiveness", producing the highest neutral responses (Table 3). Different banners elicited different reactions from the public. The moth, *Ascotis fortunata azorica* (Fig. 1), was considered attractive by all viewers and this was also the favourite insect of the contest put forward by the *Campus alive* blog[2] users, although moths were never referred to as the most-liked insects (neither in this study nor in the preliminary study with 223 persons from Terceira Island; unpublished). Conversely, the butterfly *Hipparchia azorina azorina* (Table 1) was the

[2]Campus alive—*Há Vida no Campus*—http://havidanocampus.blogspot.pt

least associated with favourable qualities, which is interesting since butterflies generally rate high on public preferences and, in this particular study, were mentioned by 10 persons (27.0 %) as their favourite insects (Table 4). It is suspected that the fact that only the hairy head of *Hipparchia* was portrayed in the banner may have worked against this species. Having no wings in the picture, one of butterflies' most distinctive features may have clashed with the idea of "beauty" and "cuteness" that are invoked by most people to justify their butterfly preference among insects (Table 4). Although the insect *Hemerobius azoricus* was not deemed beautiful or attractive, the banner was considered the most soothing. This was possibly due to the greenish hue of the picture background, while, in the other banners, it was black, grey or brown.

The impact evaluation of the exhibition on public knowledge and attitudes towards endemism, insects and nature was planned, and a preliminary study engaging 223 persons living in Terceira Island was made prior to the launching of the exhibition. Nevertheless, only 37 people were interviewed during the exhibition and, thus, no substantive conclusions could be drawn on what impact the exhibition may have had on Azorean people concerning endemic insect species.

At the time the exhibition was launched, a press release was sent out to the media to promote and explain the exhibition to people living in or visiting Angra do Heroísmo. As a result, *Azoreans for millions of years* was featured on three television news programmes (both regional and national RTP), the radio (local), seven newspapers (local) and 16 online sites, including the Science Centre of Angra do Heroísmo (CCAH[3]), the Portuguese Society of Entomology (SPEN[4]), and the Portuguese Ministry for Education and Science (programme *Mundo na Escola*[5]) websites. A newspaper reference (Lima 2013), used the *Azoreans for millions of years* exhibition to draw attention to the meagre public funds allocated to termite control in the Azores; he entitled his piece *A Desgraçada* (The Unfortunate), using the same cognomen rationale used in the exhibition. Other unforeseen results were the many instances where people, mainly tourists, inquired at the local tourism office and city hall about the outdoor insect exhibition. They sought general information on the exhibition, but also had specific questions regarding particular banners/species. To address this need, bilingual (Portuguese and English) informative A2 size posters were produced to be displayed at the science centre, tourism office and city hall in Angra do Heroísmo, which included the following sections: (i) What is this exhibition all about?; (ii) What to find on each banner?; (iii) Why insects?; (iv) Why in the city?; (v) When and where to visit the exhibition?; and (vi) Authorship and acknowledgements. Equally positive was the response to an online reference to the exhibition on the international forum "skyscraper", which received 838 messages and 573 likes. Finally, the exhibition *Azoreans for millions of years* also caught the attention of biologists, and two articles were published in different journals devoted to science divulgation: Gaspar et al. (2014) in Ecologi@

[3]CCAH—http://ccah-oaa.blogspot.pt

[4]SPEN—https://www.facebook.com/sociedadeportuguesadeentomologia/

[5]Mundo na Escola—https://www.facebook.com/MundoNaEscola/

—the journal of Portuguese Ecological Society, and Rego et al. (2015) in Ide@—the journal of the *Sociedad Entomologica Aragonesa*. In both articles, the exhibition was presented as a creative way to promote and celebrate natural heritage and endemic insects, in particular.

## 3 Promoting Public Participation: The Creative Naming of Endemic Insects on Facebook—*Chama-lhe Nomes!* (Pick a Name!)

### 3.1 Specific Aims and Strategy

To take advantage of the fact that digital media provide a far-reaching means of mass communication and are particularly effective among young people (Press and Livingstone 2006), a communicational device was developed as part of the Azorean *Bugs and Society* intervention using the free online social networking service Facebook. In particular, this activity aimed to: (i) raise awareness of the biodiversity found in the Azores, namely insects; (ii) make people aware of the fact that the particular species portrayed only exist in the Azores; and (iii) engage people in biodiversity conservation.

Since "people care about what they know" (Balmford et al. 2002, p. 2367b), it is important, although not enough, to share knowledge on conservation initiatives.

Besides people being unaware of the existence of most insect species, and disliking most of the ones they know, another factor contributing to this insect blindness is the fact that the majority of them lack common names. For example, for the 50 fly species that are restricted to the Azores, there are only four common names (*mosca*, *mosca-da-fruta*, *mosca-do-vinagre*, *mosquito*) and for all of the 429 fly species that occur in the Azores (Borges et al. 2010), only eight have common names.

However, why is it so important to have a name? The act of naming provides an identity to the entity being named. Often, a person's name is the first thing others learn about an individual, and people lacking a name are not socially recognised (Armstrong and Fontaine 1989). Concomitantly, the opportunity to name an entity represents a form of empowerment for the namer. For example, in the Book of Genesis (2: 19b), when Adam was given the chance to name animals: "brought them to the man to see what he would call them; and whatever the man called a living creature, that was its name". Furthermore, the act of naming also shapes the perception of the namer (Armstrong and Fontaine 1989) and, for this activity, people were challenged to abstract salient traits of each insect species from the information provided, to name them.

This communication activity was set on an identity framework, which aimed to bring people and insects closer together in an effort to foster public engagement in insect conservation. Table 5 summarises the message design process, including underlying arguments and tactics adopted.

**Table 5** Message design (imagery and text) to raise awareness of 12 Azorean endemic insect species in the Facebook page *Chama-lhe Nomes!*—Pick a Name! contest.

| ARGUMENT | TACTIC | OPERATIONALIZATION | | | | | |
|---|---|---|---|---|---|---|---|
| IDENTITY name = existence | HUMANIZATION "formal" identity card (name + photo) |  Example of a species Identity Card | | | | | |
| | EXCLUSIVITY Azorean endemic | *Só dos Açores!* - Only found in the Azores! | | | | | |
| | BELONGING reinforcing in-group | Use the colours of the Azorean regional flag (blue and white) in the identity card | | | | | |
| | INFORMING Photograph (sp. 1 to 6) |  | | | | | |
| | Scientific name (sp. 1 to 6) | *Pinalitus oromii* Ribes, 1992 | *Athous pomboi* Platia & Borges, 2002 | *Pseudechinosoma nodosum* Hustache, 1936 | *Calacalles droueti* Crotch, 1867 | *Cedrorum azoricus* Borges & Serrano, 1993 | *Tipula macaronesica* Savchenko, 1961 |
| | Species' (1-6) local range, habitat and ecosystem functions* | Where to find me: On tree canopies in the laurisilva forest. Did you know that: I feed by sucking sap from plants | Where to find me: At night, on the laurisilva forest ground. Did you know that: I only exist in Santa Maria islands and I can jump suddenly | Where to find me: At night, on the laurisilva forest litter. Did you know that: I feed on plants and take part on nutrient recycling | Where to find me: On *Euphorbia stygiana* spurge plant. Did you know that: I am an herbivore and only feed on one plant species | Where to find me: At night, on the ground and under pebbles in the laurisilva. Did you know that: I am a predator and the Genus I belong to only exists in the Azores | Where to find me: In laurisilva and exotic forests. Did you know that: As a larvae I feed of roots and as an adult I only live for a day |
| | Photograph (sp. 7 to 12) | 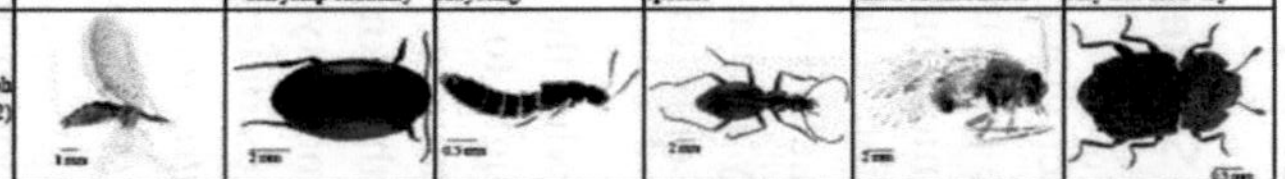 | | | | | |
| | Scientific name (sp. 7 to 12) | *Hemerobius azoricus* Tjeder, 1948 | *Hydroporus guernei* Régimbart, 1891 | *Atheta dryochares* Israelson, 1985 | *Pseudanchomenus aptinoides* Tarnier, 1860 | *Elipsocus azoricus* Meinander, 1975 | *Tarphius azoricus* Gillefors, 1986 |
| | Species' (7-12) local range, habitat and ecosystem functions* | Where to find me: At night, on tree canopies in the laurisilva forest. Did you know that: My mouthparts are adapted for piercing and sucking and I feed on other small animal | Where to find me: At night, in altitude streams and ponds. Did you know that: I am an aquatic predator | Where to find me: At night, on tree trunks in the laurisilva forest. Did you know that: I feed on other small animals | Where to find me: At night, on the ground and under pebbles in laurisilva. Did you know that: I am a predator and I only exist on Pico and São Miguel islands | Where to find me: In the laurisilva forest, on tree canopies, leaf litter, under pebbles. Did you know that: I feed on algae, fungi and plants and take part on nutrient recycling | Where to find me: In the laurisilva and exotic forests, on tree canopies and leaf litter. Did you know that: I have a strong dorsal shield and I feed on fungi |
| ENGAGEMENT foster participation | ATTENTION-GRABBING double entendre name | *Chama-lhe nomes!* (Pick a name!) | | | | | |
| | PUBLIC RECOGNITION become part of History | The Azorean Biodiversity Group selects a winning common name for each species and publishes the authorship of those names. | | | | | |

Photographs: Enésima Mendonça, except *Tipula macaronesica* (Paulo A. V. Borges)

*More detailed information concerning each species was provided on the Facebook page, outside the ID card

Animal anthropomorphisation facilitates how people relate to animals and develop empathy towards them, which is crucial to promoting concern for species and, therefore, has the potential to become a powerful tool for biological conservation (Root-Bernstein et al. 2013). In order to take advantage of this fact, identity cards were assigned to 12 Azorean endemic insect species (Table 5—Argument Identity). Following the rationale of serious games (e.g. Ritterfeld et al. 2009), the ID cards were created to be used in a fun contest intended to promote the public acquaintance with local endemic species, i.e. while providing entertainment, the primary goal of the Facebook contest *Pick a Name!* was to raise awareness of insect species that only occur in the Azores. The ID cards contained the species photograph, scientific name, and biological and ecological information, and were designed to resemble official identity cards. The "official look" of the ID cards, the colour scheme used (Azorean flag colours) and the emphasis on geographic exclusivity, were all part of the strategy used to narrow the gap between people and insects and to foster a sense of responsibility towards the "safe keeping" of these species.

In the *Pick a Name!* contest, participants were challenged to propose common names for 12 insect species and the advertised prize for the people suggesting the "winning" common names was to have their own names associated with the Natural History of the Azores. Public recognition, for both insects and participants, was, therefore, at the centre of the strategy used. As mentioned earlier, the opportunity to name an entity is a privilege and, in biological sciences, it is up to the taxonomist describing a new species to decide on the specific epithet that follows the genus name. Although taxonomists must follow nomenclature rules, they have considerable freedom in the choice of scientific names, some being more descriptive (e.g., *Homo sapiens* Linnaeus, 1758), and others more imaginative (e.g., *Ytu brutus* (Spangler, 1980)—beetle). It is also common to have species named after the region where they occur (e.g., *Nyctalus azoreum* (Thomas, 1901)—bat endemic to the Azores) and, more recently, to have species named in recognition of individual or corporate benefactors in exchange for conservation and/or research funding (e.g., GoldenPalace.com monkey, *Callicebus aureipalatii* Wallace, Gómez, Felton, & Felton, 2006) (Tree Foundation[6]). Numerous species have also been named after personalities (e.g., *Scaptia beyonceae* Lessard, 2011—horse-fly named after the singer Beyoncé; *Prethopalpus attenboroughi* Baehr & Harvey, 2012—spider named after Sir David Attenborough) and fictional characters (e.g., many species named after the work of J. Tolkien[7]), which has proven to be very successful in drawing attention to the particular species.

*Pick a Name!* gave the general public the power to name endemic insects that lacked a common name. The goal of this contest was to empower both the namer and the named: the former could aspire to have his/her name linked to the natural history of the Azores and the latter would no longer remain anonymous. In order to be able to suggest a common name, participants had to devote time to "study" the ID cards and extra information provided for each species, and, based on the species

[6]http://treefoundation.org/2013/10/13/name-a-species-save-a-forest-new-tardigrade-species/

[7]http://tolkiengateway.net/wiki/Species_named_after_Tolkien's_works

traits, to think of an appropriate name. The expectation was that the time participants spent on each species would increase biodiversity awareness and scientific literacy, and would ultimately promote individual responsibility towards biodiversity conservation.

To draw attention to this initiative, the online contest was given a double entendre name—*Chama-lhe Nomes!* (Pick a Name!): the name describes, simultaneously, the activity of naming, but also refers to name-calling, a widely held, although socially disapproved behaviour.

## 3.2 Naming an Insect: An Engaging Initiative

The Azorean Biodiversity Group (University of the Azores) launched the contest Pick a Name! on the Facebook page *Chama-lhe Nomes!* (www.facebook.com/Chama.lhe.Nomes; Amorim et al. 2013), where 12 endemic insects—part of the unique biological heritage of the Azores—lacking common names were showcased in an appealing and interactive format. As mentioned above, each species was assigned an ID card including a photograph, scientific name and concise information on species morphology, habitats and behaviour, using language appropriate to the general public (Table 5). From July to December 2012, creative Facebook users with a knack for names were challenged to come up with appropriate common names for those 12 insect species that only occur in the Azores, based on the species information provided (image and text).

A scientific committee (the authors of this chapter) selected the most appropriate common names for each species, which were posted on the Facebook page *Chama-lhe Nomes!* (Pick a Name!), along with the names of the "winning" participants that suggested them. Similarly to scientific names that are followed by the name of the taxonomist that described the species, the common name adopted for each of the 12 insect species is followed by the name of the participant. This information will soon also be featured in the updated version of the online Azorean Biodiversity Portal (http://azoresbioportal.uac.pt/; Borges et al. 2010b). The criteria to choose the "winning" common names included creativity and non-error inducing (e.g., not to assign the common name ant to a wasp, just because it may look like one), regardless of the frequency of the suggested names.

Prior to the *Pick a Name!* online contest with endemic insects, a "small trial" was organised with bryophytes, where members of the Azorean Biodiversity Group were challenged to suggest common names for species that occur in the Azores. One of the most successful names "fragrant snakeskin liverwort" (Azorean Biodiversity Portal—azoresbioportal.uac.pt) was adopted for the species *Conocephalum conicum* (L.) Dumort., since it describes both the scent and the look of the plant. This trial showed the great engagement potential of the initiative, since several people were eager to participate, as well as its learning potential, because it was necessary to apprehend the most striking features of the species in order to be able to suggest common names that are easy to recognise the species by and to remember.

To promote the *Pick a Name!* online contest, the Azorean Biodiversity Group advertised this initiative through the Portuguese Guild of Biologists and locally through environmental organisations and science centres, namely *Associação Os Montanheiros*, *Museu Carlos Machado*, *Amigos dos Açores*, *Centro de Ciência de Angra do Heroísmo—Observatório do Ambiente* and *ExpoLab—Centro de Ciência Viva*. Additionally, a press release on this initiative was also distributed to the media.

## 3.3 Facebook Contest Evaluation

Bearing in mind the logic model established for the Azorean intervention *Bugs and Society* (Arroz et al. 2016) results from this initiative were assessed using Facebook metrics data and other indicators.

First, the main short-term goal of this initiative—assigning common names to 12 Azorean endemic insects—was a success. The 132 participants suggested 776 common names for the insect species portrayed on the Facebook page *Chama-lhe Nomes!* (Pick a name!). Participants suggested names for all insect species, although some species elicited more names than others did (min = 52; max = 89), with a slight decrease in the number of names suggested from the first to the last species presented in the contest.

Another indicator of this initiative's success was site traffic statistics showing that, from July to December 2012, a daily average of 438 new visitors interacted with the Facebook page *Chama-lhe nomes!* (Pick a Name!), with the maximum number of people that saw any content associated with the page per day reaching almost 22,000 people. Moreover, 35 science and environmental online sites visited and/or promoted the Facebook page. This online contest caught the attention of several media, resulting in several interviews, magazine and newspaper articles (N = 20), and mentions on national and regional TV and radio stations (N = 5), which, in turn, reinforced the interest in this initiative (Fig. 2). For instance, after the LUSA news agency released a piece on the *Pick a Name!* contest on October 29th, the number of names submitted peaked.

The visibility attained by this initiative was crucial to draw attention to insects in general, but most importantly, to species that only occur in the Azores, and ultimately, to the conservation of Azorean biodiversity.

In addition to the data analysis concerning Facebook page visitors and audiences reached through the media, it is also important to consider the actual names proposed by the participants. According to Gurung (2003), in order to name insects, taxonomists mostly focus on morphological criteria, while non-specialists (e.g., the Tharu) privilege other norms, such as locomotion and human impact. The 12 insect species selected for this initiative did not have autochthonous known names, were small and inhabited recondite native forests—so, the probability that people had seen them before was, at best, very low. Consequently, participants' efforts to come up with an adequate name for each creature were, as expected, predominantly based

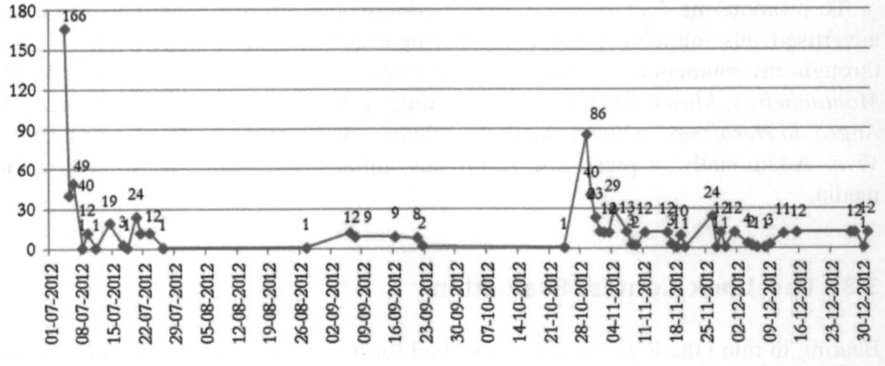

Fig. 2 Number of names suggested, from July to December 2012, to the online contest launched on the Facebook page *Chama-lhe Nomes!* (Pick a Name!)

on the information provided on Facebook (95.1 %): the proposed names could be traced either to the photograph alone (12.0 %), to text alone (51.2 %) or to both together (31.2 %), although these proportions differed among species.

Not only did participants use the information provided on the ID cards (88.1 %), but they also attributed great importance to the information provided in additional text (50.4 %). Criteria used to name the species included, colours (4.5 %), range distribution (27.6 %), resemblance to other animals (27.3 %), relation/resemblance to plants (18.5 %), geology (3.3 %), and others that were difficult to categorise. Also very interesting is the fact that, for some species, namely, *Callacales droueti*, *Pinalitus oromii* or *Atheta dryochares*, the same common name, or slight variations of it, were proposed by several participants.

The *Pick a Name!* contest involved few financial costs, as message design was performed by the Azorean Biodiversity Group multidisciplinary research team using photographs and scientific information already available in the research group, although it mobilised a great number of collaborators. A straightforward conclusion that can be drawn is that a low budget activity, even without tangible rewards for the participants, can be appealing to many people. Initiatives like this have, therefore, a great potential to increase people's awareness of biodiversity and may be used to engage citizens in the conservation of a main asset of a region's natural heritage—its endemic species.

# 4 Discussion

## 4.1 Using Scola's *Brutti, Sporchi e Cattivi* of Biodiversity to Promote Nature Conservation

Two initiatives of the Azorean intervention *Bugs and Society* were presented in this chapter based on the principles and strategies proposed by Arroz et al. (2016). The main goal was to raise public awareness to nature conservation, using the taxonomic group with the highest number of unique species in the Azores, but which, like elsewhere, has been traditionally ignored—insects.

The use of the community cultural *ethos* (values, icons) as a unifying factor was expected to be of great heuristic value for raising awareness of biodiversity loss, particularly regarding the loss of biodiversity unique to the Azores. However, to identify the values, knowledge and practise that define being Portuguese and Azorean was an elusive task, hindering the communication approach. Currently, which of the following is the most successful at embodying Portuguese national identity: D. Sebastião?[8] Egas Moniz?[9] Cristiano Ronaldo?[10] In the outdoor exhibition *Azoreans for millions of years,* kings' cognomens and the functions insects play in the ecosystems were used to narrow the gap between people and these animals. However, few people (5.4 %) mentioned either of those references, and the most relevant elements of the messages conveyed on the banners turned out to be the insect photographs (imagery power of macro-photography) and the reference to the taxonomic group of each insect. It seems, therefore, that message strategies, languages and vehicles must be diversified in order to deconstruct different people's resistances using arguments contingent to different target groups.

However, which specific resistances need to be deconstructed? Nature, biodiversity or endemicity conservation were among the concerns verbalised by people interviewed regarding the outdoor exhibition, and were reflected in the common names suggested for the *Pick a Name!* online contest. Additionally, when specifically questioned on the importance of protecting species that only exist in the Azores and on the resources that should be allocated for conservation, nine out of 10 interviewees answered that it was very or extremely important to protect endemic species, and roughly half answered that 40 % or more of the Azorean budget for the environment should be invested in the conservation of those species.

Nevertheless, because insects are among the least likeable creatures, both at local (Gabriel et al. 2012a) and global levels (e.g., Gurung 2003), all "good intentions" seem to vanish when these are the species in need of protection. In fact, social representations of insects are dynamic and vary according to the individuals' situation, education level and worldview (Amorozo et al. 2002). Besides, the bigger fauna, and especially animals that bear resemblances to humans, receive more attention compared to insects (Batt 2009). Moreover, there is a trend for people to favour exotic over local fauna (Ballouard et al. 2011). A good example of this phenomenon relates to the flagship species that lead conservation efforts (e.g., pandas, tigers), where most of the funding comes from people that live on a different continent and have never seen, and most likely will never see, them in the wild.

For the insect species portrayed in the Azorean intervention, *Bugs and Society*, neither the fact that they only exist in the Azores nor the geographic proximity with

[8]*The Desired*—Portuguese people have mythically longed for this King, who disappeared in the battle of Alcácer Quibir (1578), to end national crises.

[9]Laureate of the Nobel Prize in Physiology or Medicine in 1949.

[10]In 2015, he was awarded his fourth Golden Boot as Europe's leading football goal scorer.

the target public succeeded in promoting biophilia: the local population does not seem to value nor care more for these species just because they are endemic and exist in the same (island) territory.

Certainly contributing to the bad reputation of insects are people's own experiences with some of the most common insects (e.g., Japanese beetle, termite, flee, fly, cockroach): they destroy crops and buildings, sting, bite and spread diseases. Over time, humans have evolutionarily and culturally learned to avoid insects (Lockwood 2013; Butler 1991; Curtis 2013), despite the fact that the great majority of insects lives in native habitats, with very little chance of direct interactions with humans, and that insects are crucial for ecosystem functioning and human survival, i.e., "If insects were to vanish, the environment would collapse into chaos" (Wilson 2006, p. 33).

Another example of the lack of popularity that insects enjoy among the general public is reflected by the existence of fewer insect enthusiasts' websites, with fewer followers, compared to other taxonomic groups' websites. For example, the Facebook page of the largest insect conservation society in Europe—Butterfly Conservation—has 75,678 likes[11] compared to the 606,514 likes on the bird oriented—The National Audubon Society—page.

To change the overall aloofness and reluctance to acknowledge the problem of biodiversity loss, in particular, the loss of endemic species, and even more specifically, the loss of insect species, is, therefore, a challenging task. This challenge may be even more difficult in Portugal where scientific literacy is low compared to other European countries (EC 2010), and where people's perceived distance to nature (local "recreational or green areas") is higher compared to other European countries (PT-19.7 % vs. EU27-12.5 %) and even to Nordic Countries (3.5 %), where the climate is much more severe (Eurofound 2013).

Two communication activities dealing with the complexity of biodiversity loss have been described in this chapter. These activities differed in terms of engagement—to inform (outdoor exhibition—*Azoreans for millions of years*) and to involve (Facebook contest—*Pick a Name!*)—and used distinct languages and tactics, but shared common principles, and were designed to target groups not necessarily engaged with science and/or nature, and used Azorean endemic insects as the protagonist.

It is important, however, to reflect upon the decision of having used insects as the taxonomic group to raise awareness of biodiversity. To halt biodiversity loss and to foster empathy towards invertebrates are important goals, but the decision of combining both might have hindered the ultimate goal of nature conservation. For example, some nature advocates may have such a strong dislike for insects that they are not receptive to an intervention, which, although intended to promote endemic biodiversity, is based on those organisms.

There are a couple of local LIFE programmes (LIFE Priolo[12]; Gil et al. 2016), where the actions towards protecting specific target species, or groups of species

[11]December 10th 2015.

[12]http://life-priolo.spea.pt/en/

(birds in both cases), have positively impacted the conservation of other species within the same ecological community and are examples of successful nature conservation endeavours in the Azores. Should then other, more charismatic organisms, and not insects—the *brutti sporchi e cattivi* of biodiversity—be used to communicate about the insidious threats of biodiversity loss?

The use of non-insect species was not an option for this particular intervention, as it was developed within a research project addressing insect speciation on islands[13]. Further, because insects are, in fact, the group with the highest number of species that can only be found in the Azores. Additionally, the research team was eager to embrace the challenge of promoting unsightly species.

## 4.2 The Importance of Evaluating Outputs and Outcomes

Unfortunately, instances where science communication interventions were systematically evaluated are more the exception than the rule. There are a few examples in Portugal (Campos et al. 2009; Delicado 2010; Delicado et al. 2013; Ceríaco and Marques 2013; Schreck Reis et al. 2013), and the work edited by Filippoupoliti (2010) is an international reference.

However, even in studies where the evaluation of results and impacts are clearly stated as goals, data is sometimes missing, superficial, and subjective (Part 3 "Activating evaluation tactics" in Filippoupoliti 2010, pp. 298–571) or does not support the inferences made by the communicator:

> Nevertheless, the curators believe (though no summative evaluation or audience study was carried out) the majority of visitors did enjoy the exhibition and that the main problem was the inability to attract the public, rather than visitors' dissatisfaction with the visit (Delicado 2010, p. 479).

There are also situations where the evaluation of communication activities was assigned to independent parties (e.g., Connolly 2010; Sneider 2010; White 2010); however, in these cases, the formative evaluation of the initiatives tends to give less relevance to communication results compared to the overall achievements of the intervention within which the communication initiative was developed. Examples of best practise (e.g., Connolly 2010; Magill 2010; Schreck Reis et al. 2013) show that evaluation, either when part of all stages of the intervention, or at least present at one of the stages (e.g., Fox and Phillips 2010; White 2010), is extremely important in valuing the work developed, in the monitoring and management of future interventions and in supporting decisions concerning mobilised resources.

When a communication strategy is conceived to include evaluation, from conception to implementation, it does not imply that outputs should become the only indicators by which to assess the success of the intervention. Even though outputs, as well as outcomes, are very useful to analyse, to discuss and to promote the

[13]What can the Macaronesian islands teach us about speciation? Acase study of *Tarphius* beetles and *Hipparchia* butterflies. FCT-PTDC/BIA-BEC-104571/2008.

quality of the work being developed, allowing for changes in the way people reflect on what is being done and how it is being done, the relevance of an intervention extends well beyond those (Korn 2008).

The Azorean *Bugs and Society* intervention was planned and evaluated according to a logic model (Arroz et al. 2016), which has proven quite productive, both during the conception stage of the intervention—by allowing for the clear definition of goals, target groups and intervention processes—as well as during the gathering and analysis of outputs and outcomes, that occurred at different periods in time. The Logic Model requires high creativity in proposing expected indicators and strategies to evaluate, which implies a diversification of outcomes concerning knowledge, attitudes and behaviours, and it demands the consideration of outputs such as public participation and satisfaction. Nevertheless, this evaluation model is quite time and resource consuming and requires trained personnel for data collection.

Given the few studies concerning nature conservation interventions that evaluate their initiatives, there is no solid baseline upon which to compare the results of *Bugs and Society*, and it is very difficult to infer the impact of its results. There are, however, two exceptions in the Azores. The *SOS Térmitas*[14] (Termites SOS) intervention, dealt with a risk that directly affected citizens' built heritage, and had less than 10 % participation among the people that lived in areas affected by termites (Arroz et al. 2012a, b). The well-known *SOS Cagarro*[15] (Corys's shearwater SOS) campaign promoted by the Azorean government since 1995 and publicly evaluated since 2006, shares some indicators concerning the number of people involved in bird rescues (DRA 2014), and although rescues doubled in five years (2009–2013), this represents an increase of just 120 participants in all of the nine Azorean islands. In conclusion, the participation level in the Azores is remarkably low, regardless of institutional support or the fact that the risk may concern personal property. Based on this data, the initial expectations for the *Bugs and Society* intervention regarding participation were clearly overly optimistic. For example, it was expected that more than half of the people that walked by the insect banners of the *Azorean for millions of years* outdoor exhibition would have appreciated the initiative and apprehended that the species on display were Azorean endemics. However, only 35.1 % of the interviewees liked the exhibition and only 29.7 % realised that the showcased insect species only existed in the Azores (previous sections).

The evaluation of the intervention *Bugs and Society* revealed both strengths and weaknesses. The research team, as mentioned above, had unrealistic expectations concerning the results, but there were also unexpected boons, such as the initiative *Pick a Name!* making the front page of a major Portuguese newspaper or being incorporated into a comic sketch in a popular TV talk show.

[14]https://www.facebook.com/sos.termitas; http://sostermitas.angra.uac.pt

[15]http://www.azores.gov.pt/gra/dram-soscagarro

## 4.3 Ideas Worth Sharing

Some of the learning experiences and knowledge gathered during the *Bugs and Society* intervention are transferable to other situations, particularly in what concerns suitability, multidisciplinarity, accountability and extensiveness (Arroz et al. 2016).

Suitability

- Context matters: it is crucial for risk communication to take into account local idiosyncrasies, stakeholders, resistances, and assets, and to be able to establish a trust and credibility bond between communicators and target groups;
- The mandatory initial steps in any risk communication endeavour are to understand community attitudes and practises, which requires conducting previous research on the specific target group perspectives and social representations, and to conduct a front-end evaluation that will guide message design; and
- The compromise between science communicators and target groups, regarding their cultural capital and references, is complex and dynamic, and it requires continuous mutual assimilation and adjustment. Message design cannot be undermined by neither academic erudition nor "populism"—the final product must inform and educate without being perceived as condescending.

Multidisciplinarity

- Because both *form* and *content* are key elements in communication design, the diversity of epistemological profiles within the Azorean Biodiversity Group research team was critical to achieving the expected outputs.

Accountability

- Regulatory processes and efficiency/effectiveness evaluations must be incorporated into all science and risk communication endeavours, despite independent external evaluation;
- Evaluation results should be shared to enhance the ability to improve future interventions;
- Evaluation must be continuous and extend in time in order to assess short-, medium- and long-term impacts of a communication initiative; and
- Evaluation procedures must be extremely sensitive in order to detect subtle changes in peoples' reactions to the message being conveyed and in behaviours towards risk.

Extensiveness

- The impact of a communication intervention can be amplified by developing declinations of the main activities, as long as the different tactics and languages used are congruent, i.e. developed within a common strategy to achieve the proposed goals;

- Mass media channels are relevant for the success of a communication intervention. Investing in a media marketing strategy (e.g., newspapers, magazines, TV, radio, internet) that covers key moments of the intervention is important to spark interest among the targeted group, as a positive correlation has been established between presence on social media and public participation;
- Communicators should foster strong bonds with all intervention partners in order to engage partners in promoting the intervention;
- Communicators should invest in a repository that will aggregate the collective memory of the processes, products and impacts of the intervention over time (e.g., book, website);
- Exploring non-traditional vehicles to communicate about science has proven successful to reach a wider and more diverse target groups; and
- The public engaging potential of an initiative is not necessarily dictated by its budget (e.g., Facebook contest *Pick a Name*!).

An obvious conclusion from all that was presented in this chapter (and also, Arroz et al. 2016) is that a tremendous amount of work is needed to successfully tackle wicked problems, namely biodiversity loss. Finally, although the intervention *Bugs and Society* might not have accomplished all of its initial goals, it was certainly a step towards engaging people to think about these issues, particularly regarding a group of unsightly species that only occur in the remote archipelago of the Azores.

**Acknowledgments** We would like to thank several persons and institutions without which this work would not have been possible. FUNDING: Project PTDC/BIA-BEC/104571/2008 (FCT, Portugal), Project ATLANTISMAR—M2.1.2/I/027/2011 (DRCT, Azores), IRA was funded by Portuguese National Funds MCTES/FCT Portugal fellowship SFRH/BPD/102804/2014; PARTNERS: University of the Azores (UAç), Os Montanheiros, Amigos dos Açores (AA), Sociedade Portuguesa de Entomologia (SPEN), Ordem dos Biólogos (OB), Museu Carlos Machado (MCM), Expo-Lab Lagoa - Açores, Câmara Municipal de Angra do Heroísmo (CCAH), Associação Regional de Turismo (ART), Secretaria Regional do Ambiente e do Mar (SRAM), Ecoteca da Terceira and the owners of private buildings and managers of public institutions that made the outdoor exhibition possible; PHOTOGRAPHERS: J. Torrent (JT), E. Mendonça (EM), PAV Borges (PAVB), Axel Hausmann (AH) and M Gascoigne-Pees (MGP); PARTICIPANTS and COLLEAGUES: all the participants in pilot-studies, surveys and interviews, people involved in logistics and colleagues from the Azorean Biodiversity Group, especially C. Gaspar, for fruitful discussions and general help.

## References

Amorim, I. R., Gabriel, R., Arroz, A., São Marcos, R., & Borges, P. A. V. (2013). *Chama-lhe Nomes! As redes sociais na divulgação científica de informação taxonómica*. Encontro sobre Biodiversidade e Conservação de Invertebrados em Portugal (EBCI 2013). Faculty of Sciences, University of Lisbon, Lisbon, Portugal, 3–5 July. http://hdl.handle.net/10400.3/2377

Amorozo, M. C. M., Ming, L. C. & Silva, S. P. (Eds.). (2002). Métodos de coleta e análise de dados em etnobiologia, etnoecologia e disciplinas correlatas. Rio Claro: Coordenadoria de Área de Ciências Biológicas - Gabinete do Reitor - UNESP/CNPq.

Armstrong, C., & Fontaine, S. I. (1989). The power of naming: Names that create and define the discipline. *Writing Program Administration, 13*, 5–14.

Arroz, A. M., Amorim, I. R., Gabriel, R., São Marcos, R., Gaspar, C., & Borges, P. A. V. (2012a). *De marginal a bem patrimonial: processos de divulgação científica na exposição 'Tesouros de seis patas'*. XV Congresso Ibérico de Entomologia. University of the Azores, Angra do Heroísmo, Portugal, 1–6 September. http://hdl.handle.net/10400.3/2395

Arroz, A., Amorim, I., São Marcos, R., Gabriel, R., & Borges, P. A. V. (2013a). *Contornar 'um erro de Deus': complementaridade artístico-científica na desmistificação e incorporação social de um património natural único*. Colóquio Internacional Artes e Ciências em Diálogo. University of the Algarve, Faro, Portugal, 17–18 January. http://hdl.handle.net/10400.3/2393

Arroz, A. M., Gabriel, R., Amorim, I., São Marcos, R., & Borges, P. A. V. (2016). Bugs and society I: Raising awareness about endemic biodiversity. In P. Castro, U. M. Azeiteiro, P. Bacelar Nicolau, W. Leal Filho, & A. M. Azul (Eds.), *Biodiversity and education for sustainable development*. Dordrecht: Springer. (this volume).

Arroz, A., Gabriel, R. Amorim, I. R., São Marcos, R., Torrent, J., & Borges, P. A. V. (2013b). *Insetos pela cidade: uma intervenção urbana em etnoentomologia*. Encontro sobre Biodiversidade e Conservação de Invertebrados em Portugal (EBCI 2013). Faculty of Sciences, University of Lisbon, Lisbon, Portugal, 3–5 July. http://hdl.handle.net/10400.3/2383

Arroz, A. M., São Marcos, R., Neves, I. C., Guerreiro, O., Gabriel, R., & Borges, P. A. V. (2012b). *Relatório final da campanha: SOS TERMITAS -Unidos na Prevenção*. Angra do Heroísmo, University of the Azores. At http://cita.angra.uac.pt/ficheiros/publicacoes/1331251532.pdf

Balmford, A., Clegg, L., Coulson, T., & Taylor, J. (2002). Why conservationists should heed Pokémon. *Science, 295*, 2367.

Barthes, R. (1980). *A câmara clara*. Lisboa: Edições. 70.

Ballouard, J.-M., Brischoux, F. & Bonnet, X. (2011). Children prioritize virtual exotic biodiversity over local biodiversity. PLoS ONE, 6(8): e23152.

Batt, S. (2009). Human attitudes towards animals in relation to species similarity to humans: a multivariate approach. *Bioscience Horizons, 2*(2), 180–190.

Borges, P. A. V., Costa, A., Cunha, R., Gabriel, R., Gonçalves, V., Martins, A. F., et al. (Eds.). (2010a). *A list of the terrestrial and marine biota from the Azores*. Princípia: Cascais.

Borges, P. A. V., Gabriel, R., Arroz, A., Costa, A., Cunha, R., Silva, L., et al. (2010b). The Azorean biodiversity portal: An internet database for regional biodiversity outreach. *Systematics and Biodiversity, 8*, 423–434.

Butler, T. C. (Ed.) (1991). Entry for 'Insects'. Holman Bible Dictionary. At http://www.studylight.org/dictionaries/hbd/view.cgi?n=3036

Campos, P., Dória, A., & Sousa, M. (2009). Interactivity for museums: Designing and comparing sensor-based installations. In *Proceedings of the 12th IFIP TC 13 International Conference on Human-Computer Interaction* (pp. 612–615). Berlin, Springer.

Cardinale, B. J., Duffy, J. E., Gonzalez, A., Hooper, D. U., Perrings, C., Venail, P., et al. (2012). Biodiversity loss and its impact on humanity. *Nature, 486*, 59–67.

Ceríaco, L. M. P., & Marques, M. P. (2013). Deconstructing a Southern Portuguese Monster: The effects of a children's story on children's perceptions of geckos. *Herpetological Review, 44*(4), 590–594.

Connolly, K. (2010). From setting to subject: Plants are up to something. In A. Filippoupoliti (Ed.), *Science exhibitions: Communication and evaluation* (pp. 330–365). Edinburgh: MuseumsEtc.

Curtis, V. (2013). *Don't look, don't touch, don't eat: The science behind revulsion*. Chicago: The University of Chicago Press.

Dahlstrom, M. F. (2014). Using narratives and storytelling to communicate science with nonexpert audiences. *PNAS, 111*(4), 13614–13620.

Delicado, A. (2010). Between environment and science: An exhibition about butterflies. In A. Filippoupoliti (Ed.), *Science exhibitions: Communication and evaluation* (pp. 464–485). Edinburgh: MuseumsEtc.

Delicado, A., Gago, M. M. & Cortez, A. (2013). A visita a uma exposição científica vista pelos/as professores/as: elementos para uma análise. Educação, Sociedade & Culturas, 40, 187–207.

DRA—Direcção Regional dos Assuntos do Mar. (2014). *Relatório final S.O.S. Cagarro. Campanha 2013*. Secretaria Regional dos Recursos Naturais. Horta, Governo dos Açores. At http://www.azores.gov.pt/gra/dram-soscagarro

EC—European Commission. (2010). *Europeans, science and technology, January–February 2010*. Special Eurobarometer 340/Wave 73.1—TNS Opinion and Social.

Eurofound. (2013). *Third European quality of life survey—Quality of society and public services*. Luxembourg: Publications Office of the European Union.

Filippoupoliti, A. (Ed.). (2010). *Science exhibitions: Communication and evaluation*. Edinburgh: MuseumsEtc.

Fortuna, M. (2009). *A obra de arte além de sua aparência* (2ª ed.). São Paulo: Annablume Editora.

Fox, K., & Phillips, K. (2010). Redeveloping the science and life gallery at Melbourne Museum. In A. Filippoupoliti (Ed.), *Science exhibitions: Communication and evaluation* (pp. 364–387). Edinburgh: MuseumsEtc.

Gabriel, R. (Ed.). (2012). *Abordagens do ambiente em contexto educativo*. Cascais: Princípia.

Gabriel, R., Amorim, I. R., Arroz, A. M., São Marcos, R., & Borges, P. A. V. (2012a). *O lugar dos insetos na nova Arca de Noé*. Comunicação apresentada no XV Congresso Ibérico de Entomologia. University of the Azores, Angra do Heroísmo, Portugal. http://hdl.handle.net/10400.3/2394

Gabriel, R., Arroz, A. M., Amorim, I. R., São Marcos, R., & Borges, P. A. V. (2012b). Perspetivas acerca do património natural dos Açores. In P. A. V. Borges, I. R. Amorim, A. M. Arroz, B. C. Emerson, C. Faria & R. Gabriel, F. Pereira (Eds.), *Final Report of the project " What can the Macaronesian islands teach us about speciation?"* (pp. 89–197). FCT—PTDC/BIA-BEC-104571/2008. Angra do Heroísmo, University of the Azores.

Gabriel, R., Arroz, A. M., São Marcos, R., Borges, P. A. V., & Amorim, I. R. (2016). Promover o património natural: A exposição Açorianos há milhões de anos. *Pingo de Lava*, 40.

Gabriel, R., Borges, P. A. V., & Silva, E. (2007). A biodiversidade. In E. Silva & R. Gabriel (Eds.) *As atitudes face ao ambiente em regiões periféricas* (pp. 157–189). Angra do Heroísmo, Fundação para a Ciência e Tecnologia & Universidade dos Açores. http://hdl.handle.net/10400.3/1890

Gabriel, R., Silva, A. C., Borges, P. A. V., & Arroz, A. M. (2009). *Conservação da biodiversidade: atitudes das crianças face aos invertebrados no âmbito de um projeto do 1º Ciclo do Ensino Básico*. Comunicação apresentada no 10º Congresso em Psicologia Ambiental. Lisboa, Portugal.

Gaspar, C., Borges, P. A. V., & Gaston, K. J. (2014). Towards systematic conservation planning in the Azores. *Ecologi@, 7*, 40–49.

Gil, A., Ceia, R., Coelho, R., Teodósio, J., Sampaio, H., Veríssimo, C., et al. (2016). The Priolo Atlas: A citizen science-based census initiative for supporting *Pyrrhula murina* habitat conservation and restoration policies in São Miguel Island (Azores, Portugal). *Ecological Engineering, 86*, 45–52.

Gurung, A. B. (2003). Insects—A mistake in God's creation? Tharu farmers' perception and knowledge of insects: A case study of Gobardiha Village Development Comitee, Dang-Deukhuri, Nepal. *Agriculture and Human Values, 20*, 337–370.

Iversen, M. (1994). What is a photograph? *Art History, 17*, 450–464.

Korn, R. (2008). Transforming museums—To what end? In *Proceedings for Transforming Museums Conference*, Seattle, WA.

Kuplen, M. (2015). The sublime, ugliness and contemporary art: A Kantian perspective. *International Journal of Philosophy, 1*, 113–141.

Lima, A. (2013). A desgraçada. *Diário Insular, 20786*, 24.

Lockwood, J. (2013). *The infested mind: Why humans fear, loathe, and love insects*. Oxford: Oxford University Press.

Magill, J. (2010). Chips for everyone: Exploring engineering engagement through practical interactive simulation. In A. Filippoupoliti (Ed.), *Science exhibitions: Communication and evaluation* (pp. 388–411). Edinburgh: MuseumsEtc.

Merleau-Ponty, M. (1945). *La phénoménologie de la perception*. Paris: Gallimard.

Pasquino, G. (2002). *Curso de Ciência Política*. Estoril: Princípia.

Press, A., & Livingstone, S. (2006). Taking audience research into the age of new media: Old problems and new challenges. In M. White & J. Schwoch (Eds.), *The question of method in cultural studies* (pp. 175–200). Oxford: Blackwell.

Rego, C., Boieiro, M., Vieira, V., & Borges, P. A. V. (2015). La biodiversidad de artrópodos terrestres en Azores/The biodiversity of terrestrial arthropods in Azores. *Proyecto S.E.A. Ibero Diversidad Entomológica. Revista IDE@—SEA*, 5a/5b, 1–24.

Ritterfeld, U., Cody, M., & Vorderer, P. (Eds.). (2009). Serious games: Mechanisms and effects. New York & Oxon: Routledge.

Root-Bernstein, M., Douglas, L., Smith, A., & Veríssimo, D. (2013). Anthropomorphized species as tools for conservation: Utility beyond prosocial, intelligent and suffering species. *Biodiversity and Conservation, 22*, 1577–1589.

Schreck Reis, C., Marchante, H., Freitas, H., & Marchante, E. (2013). Public perception of invasive plant species: Assessing the impact of workshop activities to promote young students awareness. *International Journal of Science Education, 35*, 690–712.

Silva, E., & Gabriel, R. (2007). *As atitudes face ao ambiente em regiões periféricas*. Angra do Heroísmo: Fundação para a Ciência e Tecnologia & Universidade dos Açores.

Sneider, C. (2010). An essential element in exhibit design. In A. Filippoupoliti (Ed.), *Science exhibitions: Communication and evaluation* (pp. 485–519). Edinburgh: MuseumsEtc.

White, K. (2010). Beautiful science: Ideas that changed the world. In A. Filippoupoliti (Ed.), *Science exhibitions: Communication and evaluation* (pp. 298–329). Edinburgh: MuseumsEtc.

Wilson, E. O. (2006). *The creation: An appeal to save life on Earth*. New York: Norton & Company.

# Communicating Biodiversity Conservation Research Through Dialogue and Mutual Learning in Rural and Indigenous Communities

Isabel Ruiz-Mallén

**Abstract**

Dialogue and mutual learning between civil society and researchers involved in natural resource management have been increasingly advocated as a means of improving public understanding of science, biodiversity conservation, and local well-being. In rural areas in developing countries, however, science communication and environmental education strategies for disseminating biodiversity conservation research have traditionally used methods based on top-down, one-way approaches that have limited local engagement in research and undermined feedback generation between local people and researchers. This chapter examines a participatory process of developing a communication strategy for an environmental conservation research project in Southeastern Mexico. By analyzing data from interviews and focus groups with stakeholders from six rural and indigenous communities, opportunities and challenges on how to collaboratively plan a communication strategy aiming to both disseminate the research and foster mutual learning are identified and discussed. Such participatory approach increases the social relevance of the research and improves both research results and dissemination products.

I. Ruiz-Mallén (✉)
Internet Interdisciplinary Institute (IN3), Universitat Oberta de Catalunya, Barcelona, Spain
e-mail: iruiz_mallen@uoc.edu

I. Ruiz-Mallén
Institute of Environmental Science and Technology (ICTA), Universitat Autònoma de Barcelona, Barcelona, Spain

 
P. Castro et al. (eds.), *Biodiversity and Education for Sustainable Development*, World Sustainability Series, DOI 10.1007/978-3-319-32318-3_10

## 1 Introduction

Mainstream trends in education and communication for biodiversity conservation have been traditionally dominated by the information -or knowledge—deficit model. Such model assumes that lay people are receivers of limited or oversimplified information on environmental issues leading to misconceptions and unsustainable attitudes that can only be overcome by providing correct knowledge (Bickford et al. 2012). This model is thus based on the assumption that providing meaningful information will rise environmental awareness and lead to pro-environmental behaviors. The interconnectedness between knowledge, attitudes, and behavior at both individual and collective level in practice, however, is built in a more complex way than through a linear relationship. It is shaped by a myriad of psychological and socio-economic factors, such as motivations, values, emotions, institutional rules and norms, political and economic interests, and cultural traits (Kollmuss and Agyeman 2002; Sturgis and Allum 2004). Adopting a simplistic approach that neglects the influence of such internal and external factors in the public understanding of science is the reason why many conventional environmental communication strategies and education programs, which are only based on delivering information and knowledge, become ineffective in building transformative capacity and producing long-term behavioral changes (Schultz 2002; Jiménez et al. 2014).

New approaches for environmental education and communication advocate for more inclusionary tools and reflective processes to deal with the above mentioned limitations. This is the case of Education for Sustainable Development (ESD). ESD requires participatory methodologies for motivating and empowering people to take action more consciously (UNESCO 2015). Such approach not only promotes knowledge acquisition on an specific topic (e.g., biodiversity conservation), but transversal skills such as critical thinking, creativity, and collaboration in decision-making through the promotion of communication and education strategies encouraging deliberation among different actors involved in sustainability processes. The current European Union research strategy (2014–2020) framed in the Responsible Research and Innovation approach (RRI) also advocates for promoting dialogue and mutual learning between scientists and civil society as the strategy to improve public understanding of science and foster people capacity to make informed decisions that contribute to a more democratic society (Sutcliffe 2011). RRI has a commitment with science education, sustainability, inclusiveness and diversity, and reflexivity.

In the fields of conservation research, and specifically when working with indigenous and rural communities in developing countries, a call for dialogue and mutual learning between local people and scientists has been encouraged through the use of participatory research methods to both enrich the scientific practice and contribute to biodiversity conservation and social well-being (Berkes et al. 2006). Even though communities' participation and voice is becoming more relevant and an increasing number of conservation researchers and practitioners advocate for involving them in the research process and decision-making (Castillo et al. 2006; Rist et al. 2011), scientists conducting research in this field rather include inclusive tools in their communication strategy. Communication strategies are generally based on the production of scientific articles in peer-reviewed journals and communications in conferences. These actions are recognized as scientific activities by the academic evaluation system and, consequently, are useful in promoting and consolidating researchers' academic careers.

Differently, education and communication actions addressed to non-academic audiences do not contribute to researchers' career and consequently are not a priority for them. Researchers habitually experience constrained research agendas, narrow training in communication and education skills, and lack of recognition and funding for conducting outreach activities (Torres-Albero et al. 2011). They cannot invest enough time to involve local actors in planning communication tools. Resulting actions and materials are thus decontextualized from local actors' reality, needs, interests, and concerns, negatively influencing their motivation for learning and taking consciously action (Ruiz-Mallén et al. 2014). How to develop and embed a process of dialogue and mutual understanding in the design of a communication strategy still remains a challenge.

This chapter examines a participatory process of collaboratively developing a communication strategy for a biodiversity conservation research project in Southeastern rural Mexico between researchers and local stakeholders. In doing that, the chapter discusses the opportunities and challenges of planning and implementing participatory communication and education actions in the field of natural resource management and conservation research for both research and conservation goals. Creating dialogue and enhancing mutual learning are highlighted as two main principles to enhance the societal impact of scientific communication.

## 2 A Mexican Experience of Participatory Communication

The case presented here is framed within a two-year research project conducted by a Mexican-European consortium of six universities and one non-governmental organization (NGO) in six rural and indigenous communities living in or around protected areas in Southeastern Mexico (CONSERVCOM, funded by the Fondo de Cooperación Internacional en Ciencia y Tecnología UE-Mexico FONCICYT 94395). The aim of the CONSERVCOM project was to understand how forest management and conservation is influenced by three types of conservation initiatives with different scenarios of local participation in decision-making. First, government-managed protected areas, such as biosphere reserves and national parks where communities have no power in decision-making. Second, indigenous peoples' and community conserved territories and areas (ICCA), such as ecotourism projects and community conservation areas, which are areas managed by and for communities. Third, Payments for Ecosystem Services programs (PES) implemented in areas held by communities who follow rules designed by the national government.

Participant communities are located in the states of Veracruz, Oaxaca, Campeche, and Quintana Roo (Table 1). Five of them are placed within or around a government-managed protected area, three of them are managing ICCAs and other three are involved in PES. Land tenure is communal in five of them; the other one is an indigenous community of small-landowners whose members maintain customary institutions based on collective decision-making over natural resource management (i.e., assemblies).

The CONSERVCOM project used an interdisciplinary methodological approach to identify and analyze social-ecological factors and processes influencing the effectiveness of conservation initiatives, such as land use and land cover change, livelihoods, local engagement in decision-making and management, and local knowledge, perceptions and attitudes towards conservation (Reyes-García et al. 2013; Méndez-López et al. 2014; Ruiz-Mallén et al. 2014).

**Table 1** Main socio-economic characteristics of the participant communities and conservation initiatives

| Community (state) | Land tenure | Ethnicity | Population | Main activity | Conservation initiatives |
|---|---|---|---|---|---|
| Tonalaco (Veracruz) | Communal | Mestizo | 1202 | Forestry | Cofre de Perote national park<br>Payment for ecosystem services |
| Santa Cruz Tepetotutla (Oaxaca) | Communal | Chinantec | 452 | Subsistence agriculture | Community-based conservation area *La tierra del faisan* (ecotourism)<br>Payment for ecosystem services |
| Xmaben (Campeche) | Communal | Yucatec-Mayan | 1300 | Subsistence agriculture | Calakmul biosphere reserve<br>Payment for ecosystem services |
| La Mancolona (Campeche) | Private | Tzeltal | 432 | Subsistence agriculture | Calakmul biosphere reserve<br>Community-based conservation area *La Raiz del Futuro* |
| Felipe Carrillo Puerto (Quintana Roo) | Communal | Yucatec-Mayan | 25,744 | Services | Sian Ka'an biosphere reserve<br>Community-based conservation areas *Much Kanan Ka'ax* (ecotourism) and *La Sabana*<br>Payment for ecosystem services |
| Chunyaxché (Quintana Roo) | Communal | Yucatec-Mayan | 920 | Subsistence agriculture | Sian Ka'an biosphere reserve |

The CONSERVCOM methodology also included a communication strategy to be developed from the beginning of the project. Such strategy was divided according to the target audiences into academic and non-academic communication plans. The academic communication plan was addressed to researchers, practitioners, and other experts working in the fields of biodiversity conservation and rural development. The CONSERVCOM research team elaborated a list of communication tools to reach this audience, e.g., website, peer-reviewed publications, thesis, and conferences. Some of these tools were also included in the non-academic communication plan, which aimed to reach involved stakeholders in management decision-making of the studied conservation initiatives (i.e., local communities, NGOs, protected area managers, policy makers). More specifically, and differently from the vast majority of communication strategies related to research projects, the CONSERVCOM communication plan addressed to a non-academic audience also included a myriad of interactive tools that were agreed and developed between local communities and researchers, such as exhibitions and environmental education programs. In what follows, I examine how such non-academic communication strategy was decided and developed through a participatory research approach.

## 3 Three Stages in the Participatory Process for Developing a Communication Strategy

To participatory build the non-academic communication strategy within the CONSERVCOM project, the methodology was structured in three stages according to the following objectives: (1) Understanding of the communication context, (2) Fostering interaction and exchange, and (3) Validating the tools. Each stage and objective was guided by a research question as shown in Fig. 1.

Free, prior, and informed consent from the local authorities of the six communities was obtained through verbal or written collaboration agreements between them and the project team in order to conduct the research and disseminate the results.

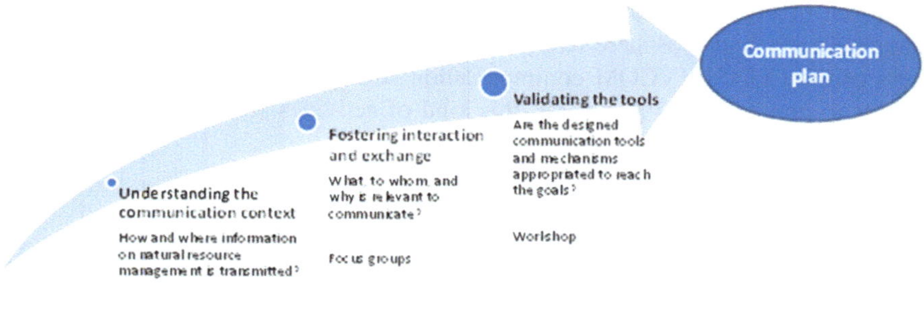


**Fig. 1** Methodological approach used to develop the participatory communication strategy

## 3.1 Stage 1: Understanding of the Communication Context

Prior to develop any communication strategy aiming to disseminate the results of a research on biodiversity conservation to non-academic audiences, including local communities, it is necessary to identify and understand the ways of knowledge transmission usually employed by local people to get informed on natural resource management and conservation issues in and outside their communities. On this basis, an ethnographic study based on participant observation and interviews was conducted in each selected community to explore how and where information and knowledge on natural resource management and conservation issues were transmitted at local level. A total of 130 adults previously identified as key informants were asked on the places where such information was disseminated in their communities (e.g., community meetings, schools, church) and who were involved in the process of knowledge transmission (including actors, social groups, and institutions). Such information was collected by four trained researchers, who were assisted by local translators when participants did not speak Spanish but the indigenous language (i.e., Tzeltal, Maya-Yucatec, Chinantec). As a result, a description of stakeholders and spaces of communication was obtained in each community, which allowed for further identifying appropriated tools to communicate the project results (Table 2).

In general, local informants identified the local authority, community groups, associations of local producers, and NGOs as key actors in disseminating information on local natural resource management and conservation issues within the local context. Informants from those communities with ecotourism cooperatives or managing a community-based conservation area for ecotourism also mentioned ecotourism guides as key actors in such dissemination process. Interestingly, biosphere reserves and natural parks were not mentioned as spaces of communication or learning in any of the cases. Although five participant communities were located within or around protected areas, local people did not use to visit such areas due to lack of transportation facilities, but also because of their lack of involvement in managing these areas that are exclusively administrated by the government (Ruiz-Mallén et al. 2014). Local communities' lack of recognition of nearby protected areas as spaces of communication questioned the mainstream dissemination approach of organizing environmental education and communication activities in national parks and biosphere reserves. Consequently, and to ensure the effectiveness of the CONSERVCOM communication strategy to disseminate the research results to non-academic audiences, this kind of activities were not included in the communication plan.

## 3.2 Stage 2: Fostering Interaction and Exchange

Once the stakeholders and spaces of communication for the transmission of information and knowledge on biodiversity conservation were characterized in each

**Table 2** Stakeholders and spaces of communication in selected communities

| Communities | Stakeholders | Spaces |
|---|---|---|
| Felipe Carrillo Puerto | Local authority<br>Regional and federal government<br>Local cattle ranching and beekeepers associations<br>Local and regional NGOs<br>International agencies (PNUD)<br>Academics | Community meetings<br>School<br>House of culture<br>Churches |
| Chunyaxché | Local authority<br>Federal government<br>Ecotourism cooperatives<br>Regional NGO<br>International agencies (PNUD) | Community meetings<br>School<br>Church |
| La Mancolona | Local authority<br>Regional and federal government<br>Ecotourism guides<br>Local cattle ranching, beekeepers and agroforestry associations<br>Regional NGO | Community meetings<br>Ecotourism area<br>Tree nursery<br>School<br>Municipality<br>Churches |
| Xmaben | Local authority<br>Regional and federal government<br>Local cattle ranching and beekeepers associations<br>Local NGO | Community meetings<br>School<br>Municipality<br>Churches |
| Santa Cruz Tepetotutla | Local authority<br>Federal government<br>Ecotourism guides<br>Local beekeepers and agroforestry associations<br>International agencies (MIE)<br>Academics<br>Local and regional NGOs | Community meetings<br>Ecotourism area<br>School<br>Church |
| Tonalaco | Local authority and community groups (water, cleaning, etc.)<br>Federal government<br>Local fishing and farming associations<br>Academics | Community meetings<br>School<br>Church |

community, CONSERVCOM researchers fostered a process of dialogue with communities in order to identify and discuss preferred mechanisms and tools to disseminate the research results. A series of focus groups were conducted during organized exchange meetings between the six selected communities in two of them: (1) Xmaben, Campeche, with the participation of three communities also located in the Yucatan Peninsula (Felipe Carrillo Puerto, Chunyaxché, and Mancolona), and (2) Santa Cruz Tepetotutla in Oaxaca with participants from Tonalaco, Veracruz. A total of 49 people participated in both meetings (21 in Xmaben and 28 in Tepetotutla). Each community agreed those community members attending them.

During the six focus groups, participants of each community were invited to discuss with researchers on the contents of the CONSERVCOM project that could be of potential interest and the reasons for and ways of disseminating such contents (e.g., radio, internet, educational materials, community meetings). In each focus group, the dialogue was moderated by a facilitator and based on prior knowledge about the communication context in each community.

In general, focus groups participants identified three types of audiences: (1) local adults, (2) local children, and (3) people abroad (Table 3). They shared a common interest to disseminate the CONSERVCOM project findings on both social and ecological community aspects enforcing biodiversity conservation at local level, such as the history of the community, land use and cover change, and traditional knowledge on the diversity and uses of plants and animals. They also highlighted the need of increasing local people's awareness on cultural practices related to conservation and strengthening communities' environmental values. Suggested dissemination tools targeted to local people mostly consisted of low-cost materials such as school drawing exhibitions, posters, and ecological itineraries. They also included some tools that are not usually suggested by researchers, such as songs and radio spots.

A more expensive but relevant tool identified by focus group participants was an itinerant exhibition about the project findings to be implemented, in an interactive way, in participant communities. In each village, people attending the exhibition could write or record messages and leave handicrafts to be shared with members of other communities.

Furthermore, participants highlighted the need of disseminating the results on their livelihoods, local ecological knowledge, and sustainable management activities beyond their communities to advertise their productive activities and handicrafts and to enhance their image as "rural people who take care of nature". Including such information in the project website and elaborating leaflets and videos were agreed as actions to potentially communicate such information to broad audiences.

From the final list of agreed communication tools, the research team selected nine of them according to the availability of financial resources and personnel to implement the actions in each community (additional funding was granted by the Programa de Cooperación Inter-Universitaria e Investigación Científica of the Ministerio de Asuntos Exteriores y Cooperación A/023406/09 and A/030044/10, and the Fundació Autònoma Solidària-UAB). These tools were: (1) land-use change maps of each community, (2) itinerant exhibition among communities, (3) drawings exhibition at schools, (4) website with local productive activities, (5) environmental education program on the ecotourism project of Felipe Carrillo Puerto, (6) calendars of the Sian Ka'an biosphere reserve, (7) illustrated leaflets, (8) national forum on community conservation and (9) video on the exchange meetings. A research team member was assigned to be responsible for elaborating the material or conducting the activity corresponding to each tool.

**Table 3** Communication contents and tools agreed by communities and researchers

| Audience | Content | Tools | Communities |
|---|---|---|---|
| Local adults | Land-use change map of the community | Poster | All |
| | History, biodiversity and conservation projects | Book | All |
| | History, biodiversity and conservation projects from other communities involved in conservation | Itinerant exhibition | All |
| | Why conservation is important | Radio program and spots | Mancolona |
| Local children | History, biodiversity, and conservation projects | Illustrated book | All |
| | History, biodiversity and conservation projects from other communities involved in conservation | Itinerant exhibition with pictures and drawings | All |
| | Children's perception of their community | Drawings exhibition at the school | Felipe Carrillo |
| | | | Puerto, Chunyaxche |
| | Sian Ka'an biosphere reserve biodiversity and management | Environmental education program and calendar | Felipe Carrillo |
| | | | Puerto, Chunyaxche |
| | Importance of biodiversity | School presentation and song | Mancolona |
| | Forest management practices and activities | Illustrated school notebooks | Tonalaco |
| | Biodiversity and traditional knowledge on plants and animals uses | Itineraries in the community-based conservation area | Santa Cruz |
| | | | Tepetotutla |
| People abroad | Ecotourism project and local associations of producers activities and products | Website | All |
| | History, biodiversity conservation and management and traditional ecological knowledge of the community | Illustrated leaflet | All |
| | Community-based conservation projects and activities | Workshop | All |
| | Exchange meetings between communities | Video | All |
| | Sustainable management of community forests and fire prevention | Posters | Tonalaco |
| | Community values | Articles in newspapers | Santa Cruz Tepetotutla |

## 3.3 Stage 3: Validating the Tools

In order to check the appropriateness of the designed communication tools according to local people's expectations, a consultation process for validation was conducted with representatives of local communities who attended the national forum on community conservation at the end of the research project. The forum took place in the city of Campeche with 99 participants, 16 of them were local people, 42 academics, 33 NGOs and eight government representatives.

As mentioned above, the forum was planned as a communication tool to disseminate the CONSERVCOM project findings to the academia, NGOs, and government institutions working on biodiversity conservation and community-based natural resource management as well as to members of selected communities. In order to take advantage of this communication tool, community representatives attending the workshop were invited to validate the content of those tools that were already designed at that moment: land-use change maps, the video, and illustrated leaflets.

The validation process allowed researchers to identify misunderstandings in the description of research results that were further amended in order to ensure that the information provided was concise. For example, a community member realized that one of the historical events included in the leaflet of his community was wrongly dated, whereas a member of another community noted that a picture included in the corresponding leaflet showed a group of women who were not from the community. Moreover, local people validation of the dissemination materials also allowed for obtaining information that complemented the research. For instance, members from one of the communities realized that the land use change map of their community did not include an area of PES, which had not been identified in a previous research fieldwork, thus giving the CONSERVCOM researchers the opportunity of amending the analysis.

The forum itself was evaluated by using a participatory assessment mural. This method is based on a 4-point Likert scale but implemented collectively (Güell 2004). Around 60 % of workshop participants were involved in the validation that was voluntary. Participants were overall satisfied with the forum. More than 90 % of them acknowledged that the event achieved their expectations of sharing experiences on community-based conservation, and only four people highlighted that their ideas were not discussed (Table 4).

The validation of the environmental education program of the ecotourism project in the community of Felipe Carrillo Puerto followed a different process. This communication tool was collaboratively designed between community members working on the local ecotourism project and an undergraduate student who further assessed the implementation of this program with local secondary students (Fig. 2). The program was well received by the students. It increased students' knowledge on flora and fauna species of the reserve, their uses, and sustainable forest management practices. The evaluation of the program also showed its value as an environmental education tool that addressed local people's interests in enhancing young people's awareness on conservation issues (Gonzalez-Ventosa 2011).

**Table 4** Results of the participatory evaluation of the final workshop

| Statement | Number of answers | Totally agreed | Agreed | Disagreed | Totally disagreed |
|---|---|---|---|---|---|
| I only came to listen | 56 | 7.1 | 10.7 | 26.8 | 55.4 |
| I came to work on new experiences | 60 | 55.0 | 33.3 | 11.7 | 0 |
| I came to share experiences | 64 | 46.9 | 39.1 | 9.4 | 4.6 |
| My ideas have been discussed | 62 | 22.9 | 70.9 | 6.2 | 0 |
| Now I know more about community conservation | 66 | 27.3 | 60.6 | 12.1 | 0 |
| Now we have new proposals to improve community conservation | 68 | 38.2 | 39.7 | 22.1 | 0 |
| The workshop has facilitated the exchange of experiences | 69 | 52.2 | 40.6 | 7.2 | 0 |

**Fig. 2** Community member conducting the environmental education program to secondary students in Felipe Carrillo Puerto (Picture by I. Gonzalez-Ventosa)

Unfortunately, it was not possible to validate other communication tools (i.e., calendar, itinerant exhibition) with the corresponding community members due to lack of time and resources in the latest stage of the research project.

## 4 Concluding Remarks

The participatory experience presented here shows that involving local communities as active actors in planning and developing communication and education tools and actions not only contributes to enhance the impact of the resultant communication strategy in terms of improving public understanding of science and the societal impact of the research itself, but also can enhance the relevance of the research and validate the analysis.

The communication strategy developed within the CONSERVCOM project received feedback from local people, engaged them and other stakeholders in a dialogue about the past, present, and future of the relationship between their communities and their environment, and generated new knowledge on the effectiveness of natural resource management strategies for biological conservation through fostering mutual learning. Specifically, community meetings exchange for discussing the contents and tools to be used for the dissemination between CONSERVCOM researchers and local communities are clear examples of types of actions that can be conducted to generate dialogue and deliberation.

In addition, mutual learning between local communities and CONSERVCOM researchers during the final forum was relevant to identify research gaps in data collection. Community members' collaboration in research also empowered them as knowledgeable actors who could be able to contribute to the research process. Moreover, dialogue could also be enhanced by integrating communication and education actions as an inherent part of the research proposal, which may allow for achieving institutional and financial support. Such approach includes stakeholders' visions, values, and interests in the research process, and thus implies a clear political choice (Rist et al. 2011). It may also require extra time and financial and personnel resources. But making such efforts might allow for avoiding potential conflicts on further management decisions since local stakeholders are aware of and actively involved in the whole research process since the beginning.

Creating spaces for dialogue and deliberation and promoting mutual learning thus emerge as main principles to enhance the impact of scientific communication. Developing such participatory process, of course, involves certain challenges. Rural and indigenous communities are not homogeneous units, but constructed around power relationships that may shape access to information and the ability to act on it. Researchers should promote equality in community members' involvement in the design of the communication strategy to grant priority to the interests and concerns of the most vulnerable groups. Other challenges are related to communication barriers, lack of resources, and limited time for developing participatory communication because of the schedule of research projects. As the experience of the CONSERVCOM project has shown, however, such challenges can be overcome if anticipated. The described participatory approach can thus be used and adapted by researchers working in the field of natural resource management and conservation who are interested in emphasizing the interaction with stakeholders for the benefit of both science and society.

## References

Berkes, F., Reid, W. V., Wilbanks, T. J., & Capistrano, D. (2006). Conclusions: Bridging scales and knowledge systems. In W. R. Reid, F. Berkes, Wilbanks, T. J. & D. Capistrano (Eds.) *Bridging scales and knowledge systems: Concepts and applications in ecosystem assessment*. Washington: Island Press (pp. 315–331).

Bickford, D., Posa, M. R. C., Qie, L., Campos-Arceiz, A., & Kudavidanage, E. P. (2012). Science communication for biodiversity conservation. *Biological Conservation, 151*, 74–76.

Castillo, A., Torres, A., Velázquez, A., & Bocco, G. (2006). The use of ecological science by rural producers: A case study in Mexico. *Ecological Applications, 15*(2), 745–756.

González-Ventosa, I. (2011). *Elaboración, análisis y evaluación de una propuesta de Programa de Educación Ambiental para la reserva ejidal Much' Kanan K'áax, de Felipe Carrillo Puerto, Quintana Roo*. México. Environmental Science Degree. Barcelona: Universitat Autonoma de Barcelona.

Güell, N. (2004). Análisis de Foros Participativos. Propuesta y Ensayo de un Método e Instrumentos de Análisis, en el Caso del Primer Foro Público de la Estrategia Catalana de Educación Ambiental. Environmental Science Master. Barcelona: Universitat Autònoma de Barcelona.

Jiménez, A., Iniesta-Arandia, I., Muñoz-Santos, M., Martín-López, B., Jacobson, S., & Benayas, J. (2014). Typology of public outreach for biodiversity conservation projects in Spain. *Conservation Biology, 28*(3), 829–840.

Kollmuss, A., & Agyeman, J. (2002). Mind the gap: Why do people act environmentally and what are the barriers to pro-environmental behavior? *Environmental Education Research, 8*(3), 239–260.

Méndez-López, M. E., García-Frapolli, E., Pritchard, D. J., Sánchez González, M. C., Ruiz-Mallén, I., Porter-Bolland, L., et al. (2014). Local participation in biodiversity conservation initiatives: A comparative analysis of different models in South East Mexico. *Journal of Environmental Management, 145*, 321–329.

Reyes-Garcia, V., Ruiz-Mallen, I., Porter-Bolland, L., Garcia-Frapolli, E., Ellis, E., Mendez, M. E., et al. (2013). Local understandings of conservation in Southeastern Mexico and their implications for community-based conservation as an alternative paradigm. *Conservation Biology, 27*(4), 856–865.

Rist, S., Boillat, S., Gerritsen, P. R. W., Schneider, F., Mathez-Stiefel, S., & Tapia, N. (2011). Endogenous knowledge: Implications for sustainable development. In U. Wiesmann & H. Hurni (Eds.), *Research for sustainable development: Foundations, experiences, and perspectives*. Perspectives of the Swiss National Centre of Competence in Research (NCCR) North-South, University of Bern (Vol. 6, pp. 119–146). Bern, Switzerland: Geographica Bernensia.

Ruiz-Mallén, I., Newing, H., Porter-Bolland, L., Pritchard, M. D., García-Frapolli, E., Méndez-López, M. E., et al. (2014). Cognisance, participation and protected areas in the Yucatan Peninsula. *Environmental Conservation, 41*(3), 265–275.

Schultz, P. W. (2002). Knowledge, information, and household recycling: Examining the knowledge-deficit model of behavior change. In T. Dietz & P. Stern (Eds.), *New tools for environmental protection: Education, information and voluntary measures* (pp. 67–82). Washington: The National Academy Press.

Sturgis, P., & Allum, N. (2004). Science in society: Re-evaluating the deficit model of public attitudes. *Public Understanding of Science, 13*, 55–74.

Sutcliffe, H. (2011). *A report on responsible research and innovation*. London: MATTER Business Group.

Torres-Albero, C., Fernández-Esquinas, M., Rey-Rocha, J., & Martín-Sempere, M. J. (2011). Dissemination practices in the Spanish research system: Scientists trapped in a golden cage. *Public Understanding of Science, 20*(1), 12–25.

UNESCO. (2015). Education for Sustainable Development (ESD). At: http://www.unesco.org/new/en/education/themes/leading-the-international-agenda/education-for-sustainable-development/

# Education for Sustainability in the Context of Community Forestry

Rita Serra, Patrícia Ferreira, Iryna Skulska, Mayrén Alavez-Vargas, Anailton Salgado, João Arriscado Nunes and Raúl Garcia-Barrios

**Abstract**

Community forestry in Portugal is emerging as a promising form of multifunctional forestry that combines scientific and technical knowledge with the participation of the local residents in decision-making. These forests are governed by collective property arrangements (*baldios*) based on millenarian traditional usufruct rights of a local community of commoners (*compartes*). Participation is open to all the new residents regardless of their gender, activity or status. However, during the 20th century the connection between the commoners and the commons was severely disrupted by the Forest Services by compulsory afforesting the lands with tree species unknown to the local populations and provoking the decline of collective agro-pastoral practices, which was most severe close to urban areas. We describe our experience with a community of *compartes* that recently gained back control of its common lands and initiated a project to revitalize a degraded forest in the mountains after four decades of co-management with the State. We also detail the specificities and challenges that in the context of community forestry are crucial for the residents to authentically construct and take part in a commons land narrative, and the educational activities we have jointly developed to activate meaningful engagement in collective practices, intergenerational responsibility and active citizenship.

R. Serra (✉) · P. Ferreira · I. Skulska · M. Alavez-Vargas · A. Salgado · J.A. Nunes
Centro de Estudos Sociais, Universidade de Coimbra, Coimbra, Portugal
e-mail: miscara@gmail.com

R. Garcia-Barrios
Centro Regional de Investigaciones Multidisciplinarias, UNAM, Cuernavaca, Mexico

A. Salgado
Universidade da Floresta, Cruzeiro do Sul, Acre, Brazil



P. Castro et al. (eds.), *Biodiversity and Education for Sustainable Development*,
World Sustainability Series, DOI 10.1007/978-3-319-32318-3_11

# 1 Introduction

Community forests are a global phenomenon that re-emerged during the 20th century since the 80s after two centuries of strong repression by nation-states (Bullock and Hanna 2012). Traditionally, community forests were multifunctional spaces that served the needs of local populations which had usufruct rights to collect firewood, bushland, mushrooms, berries and a multitude of non-wood forest products. However, the emergence of modern silviculture during the 18th century simplified the forest to maximize the production of wood, forests were placed under the direct rule of the state and the local communities were dispossessed from their common lands (Scott 1998). This triggered violent conflicts between the state and local populations and the impoverishment of the people whose subsistence was based on forest livelihoods (Peluso 1992). During the 20th century, nation-states were severely criticized for their intents to control the society through top-down approaches that frequently failed to improve the human condition (R. García-Barrios and L. García-Barrios 2008). During the 80s, nation-states adopted neoliberal policies that handed over the pretensions to control society to the market forces, and were met with fierce opposition by social movements all over the globe (Klein 2007). A number of diverse and heterogeneous actors from civil society struggled for an alternative reform of the institutions of the state to strengthen civil participation in decision-making as a way to reduce inequalities and increase social and environmental justice (Santos 2006). It was in this context that local struggles over the common lands intersected with national and international struggles to increase local control over natural resources. The devolution of common lands and forests to local communities earned the support of global organizations such as FAO and was backed by studies from the scientific community that affirmed the capacity of local actors to sustainably manage natural resources without state intervention (Ostrom 1990). Although initially community forests were considered a phenomenon primarily found in developing countries, recent studies are revealing its true global dimension by recognizing the experiences of North America and Europe (Bullock and Hanna 2012).

As the result of many intersecting processes and heterogeneous constructions (Taylor 2010), community forests generate high but diverse expectations of achieving environmental, economic and social benefits. The word community evokes warm feelings of social actors "doing the right thing" working for the common good. Jeffery Burley, of the Oxford Forestry Institute, described it as "more or less equivalent and reflect Abraham Lincoln's view of democracy—government of the people, by the people, for the people" (Sarre 1994). However, when it comes to define what community forestry is, the word community plays no role at all. The definition of FAO[1] makes no assumptions on the nature of the

[1]Community forestry was initially defined by FAO as, "any situation which intimately involves local people in a forestry activity. It embraces a spectrum of situations ranging from woodlots in areas which are short of wood and other forest products for local needs, through the growing of trees at the farm level to provide cash crops and the processing of forest products at the household,

institutional arrangements or social relationships behind "the people" governing them, taking a pragmatic stand that good forest governance can come in all shapes and formats. Nevertheless, empirically grounded studies revealed that this stand is socially naïve. Defining community is crucial to guide instruments and policies that can actually make community forest projects thrive and reconnect the forest ecosystem management with the people's well-being (Flint et al. 2008). For the authors just cited, community emerges through communication and interaction among people who care about each other and about the place they live in. It is a phenomenon that arises through the concerted actions of residents of a local society who come to share values, concerns, interests, and actions that are expressed in prevailing local narratives. Community is marked by its multiple and often conflicting perspectives, but when the conditions are right—where there is common space, shared way of life, and collective actions, local citizens are able to overcome differences and special interests to recognize the common good. This interactional approach draws attention to forces that block or retard the emergence of the interactional community in particular settlements that keep communities in a neotenic condition (García-Barrios and Serra, in press) that allows them to persist and even reproduce but prevents them from achieving institutional maturity and purpose authenticity. Without such shared resources, the bottom-up processes essential to involving local people in decision-making are stymied and community-based initiatives cannot occur. Communities are neither permanent nor immutable: they last as long as the people in an area continue to care about each other and the place, and express this caring in the actions they take to enhance general well-being.

In Portugal mountain places are constantly faced with the migration of young people for urban areas in search for jobs. Locally available quality jobs are crucial to provide an income and support lively communities. By increasing the dependency of the population on forest resources, more people may come to recognize community forests as a common good and accept the responsibility to manage them. This implies keeping community forests as spaces capable of supporting not only leisure activities but a livelihood, deepening the connection between the forests and local people. This is particularly relevant in times of economic crisis as the one we are living in the Eurozone. One of the possible ways to achieve this objective is through community forest enterprises (CFE). The model of CFEs is quite successful in Mexico, and aims to place commercially-oriented companies under the control of the commoners to finance common good purposes, especially social care services to the population (Tejera-Hernández and García-Barrios 2008). Experiences of CFEs also exist in Europe in regions with a strong culture of autonomy as the Trento Valley, in Italy (Jeanrenaud 2001), and may mitigate the social damages caused by the current demise of the State to provide social care services at the local level. However, CFEs require a great increase of the

(Footnote 1 continued)
artisan or small industry level to generate income, to the activities of forest dwelling communities" (http://www.fao.org/docrep/u5610e/u5610e04.htm).

commoners' responsibilities closing off the possibility for oscillatory movements of getting in and out of participation.

In this chapter, the authors' argument goes as follows: (1) the processes supporting the ongoing formation of place-oriented communities were interrupted; (2) community forestry have a potential to prompt anew the formation of communities, but only under the right conditions; (3) educational approaches oriented to the commons can sparkle social interactions and the sustainability of community forest initiatives.

## 2 Community Forests in Portugal

In Portugal, community forests are the result of common lands that were compulsory afforested by the State Forest Services during the Estado Novo dictatorship (1926–1974) and devolved to the local populations after the Carnation Revolution (1974). These common lands are named *baldios*, a word that is also used as a depreciating term meaning waste, useless or abandoned space.

*Baldios* represent close to 12 % of the Portuguese forests. Although the afforestation of the country was mainly the result of state forest policies, the state has the lowest proportion of forests in Europe (2 %). Most of the forests are privately owned (86 %) and managed by industrials or small owners associations. The Portuguese forest is increasingly market-oriented and *Eucalyptus* has become the dominant species.

*Baldios* occupy an area close to 500 thousand hectares and are localized mainly in mountain regions (Fig. 1).

At the moment, 1441 *baldios* are registered in the country, with an average area of 412 ha. Mountain forests can support multifunctional forests and a *plethora* of activities such as recreation, tourism, hunting, fishing and pastoralism, together with diversified timber and non-timber forest products (Lopes and Cristóvão 2010). *Baldios* can be very relevant for mountain economies, and their aggregated economic value is estimated at 70 million euros (CNVTC 2010). They are non-profit institutions that support qualified local jobs, social enterprises and charities.

*Baldios* are managed by the commoners. The commoners are local residents who have legally recognized rights to usufruct and to administrate the lands in common. The community of commoners can decide to manage the lands exclusively or in co-management with the State. It can also delegate administration to other entities, such as local parishes or municipalities. Currently, the majority of the *baldios* are co-managed with the state (Table 1).

The commoners face huge challenges to organize themselves to manage community forests. Due to reasons explained in the next section, the processes supporting the ongoing formation of place-oriented communities were interrupted, separating the people from the forests and blocking the possibility to overcome local conflicts to give birth to community-based initiatives.

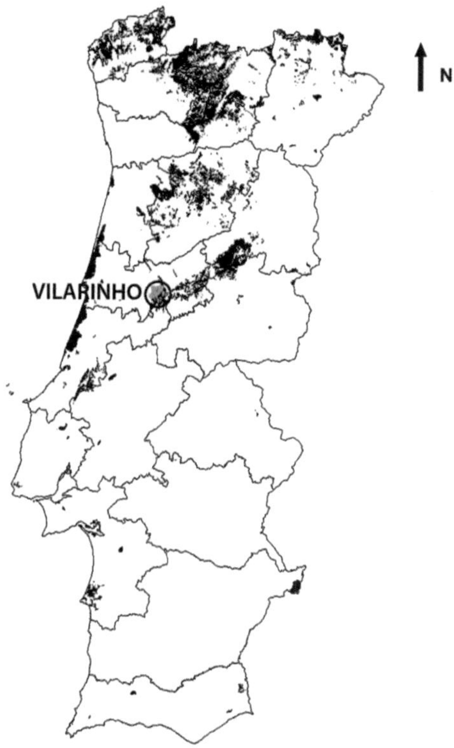


**Fig. 1** Distribution of the *baldios* in mainland Portugal (adapted from the BALADI—National Federation of *Baldios*. Source information based on the *baldios* submitted to the forest regime with a management plan). The location of the Vilarinho case-study in the Lousã Mountains is indicated

**Table 1** Management regimes of *baldios*

| Management regime | Number of *baldio* units[1] | % of total |
|---|---|---|
| Co-management with the State | 861 | 60 |
| Exclusive management by commoners | 246 | 17 |
| Delegation of competences in parishes | 334 | 23 |
| Total | 1441 | 100 |

[1] Data provided by the Institute for the Conservation of Nature and Forests (ICNF) in 2013

## 3 Disconnecting the People from Old Forests

The process of separating residents from the local community forests lasted over two centuries. The reason it took so long was that community forests sustained collective practices indispensable to the domestic economies of rural populations. Traditionally, forests were multifunctional agro-silvo-pastoral systems that provided fertilizers, fodder and fuel to subsistence agriculture activities of the peasants and to the poorest and landless members of the community. Thus, the disconnection of local people from their commons required a breakdown of the collective practices performed in such places. We highlight three fundamental stages in the historical process: (1) impediment of collective actions in common lands; (2) emergence of alternative forms of satisfying domestic needs based on exchange; (3) elimination of traditional needs based on the extinction of subsistence ways of life.

Impediment was achieved by dispossessing local communities of their rights to the commons through two distinct avenues: privatization and nationalization. Both were aimed to allocate the resource on the hands of agents with the right resources and capabilities. In one case the lands were parceled and distributed between the well-off members of the community at the expense of the poor and landless. In some places, the individualization of rights to land occurred by granting tree tenure to the planter (Nunes and Feijó 1990), as was the case with fruit trees, namely chestnuts and olive trees that are registered until today in the Ministry of Finances. While tree planting by former commoners was a way to materially support well-off individual claims to land rights, the afforestation of the common lands by the State had the same purpose to be acquired by other means (Brouwer 1995). The State forest services planted fruitless trees intended for wood production which local communities were unfamiliar with. The new forest regime intended to place the trees under the care of professionals that followed scientific prescriptions to maximize wood production. To the eyes of the State foresters, local populations had to be excluded from the forest space and traditional uses were forbidden. The afforestation of the commons by the Estado Novo dictatorship occasionally prompted violent conflicts between the State and the local populations who were defending their subsistence way of life.

In some situations, communities found alternative ways to satisfy domestic needs of fertilizers, fodder and fuel through exchange practices between farmers and grazers, such as grazing sheep in private lands to fertilize them for agriculture. In other situations, mass migrations took place, depopulating rural areas in favor of a concentration in urban areas in Portugal and abroad. The social and economic transformations gradually extinguished subsistence ways of life to give birth to commercial agriculture, industrial and service-based economies. When any of these situations become dominant, the forest becomes a distant landscape; the old collective practices that sustained a shared way of life are turned into folklore or disappear completely from the memory and local narratives.

However, when the authoritarian Estado Novo regime reached its end through the military coup which started the Carnation Revolution in 1974, the memory many mountain communities had of the compulsory and violent afforestation of the common lands was still fresh. The socialist agenda of the first democratically elected government intended to emancipate peasants through an agrarian reform. The devolution of the commons fitted perfectly the agenda and allowed to gain the support of the peasants up north, where smallholdings were the dominant reality.

## 4 Reconnecting the People with New Forests

The devolution of the commons was the first step to restore historical justice to the local communities dispossessed during the dictatorial regime. For the first time in history, the commoners (named *compartes*) gained legal rights to usufruct and administration of their lands. Rights are held in common by local residents regardless of their gender, activity or status and are inalienable, non-heritable, and cease to exist when one leaves the place. The devolution process marked a new beginning for the common lands, but the restoration of collective practices and communities had a long path to take (Klein and Stok 1986).

Once a limited group of *compartes* organized to claim the commons, the forest services had to return the administration of the lands. However, the State failed to capacitate the local communities to manage the forests. Whenever old uses were abandoned, the *compartes* chose to co-administrate the forests with the State, which kept control of the forest management decisions without involving the local residents.

As the connection between the commoners and the forests failed to be restored, the economic interests of private agents increased. The successive law reforms opened the way for the residents to delegate administration to other entities (such as the parishes and councils) and to establish concessions to external agents to explore the forests and natural resources in the commons (*Eucalyptus* monocultures, quarries, windmill farms, etc.). The rents are distributed more or less democratically to local institutions by the residents participating in the assemblies for solidarity purposes, infrastructure provision and maintenance and cultural activities.

The experience of the commoners with the co-management and the companies exploring their natural resources was, in some cases, disastrous. Due to the new public management reforms since the 80s, the Forest Services were greatly incapacitated to face the emerging problems of forests, such as invasive species, large fires and pests. It was only after the catastrophic fires of 2003 and 2005 that a national program was launched to protect the forest against fires, with the actual participation of the private and community sectors. For the first time, the *compartes* were capacitated with technical knowledge and specialized teams of forest workers co-funded by the State.

The slow recognition that forests can create local jobs and economies and that poor management can cause environmental damage and risks led some commoners to manage their forests autonomously. When *compartes* take into their hands the responsibility of managing the forests in the long-term, they must face the challenge of continuously engaging the residents in community forest projects if the management of forests is to be genuinely community-based.

## 5 Vilarinho, a Case Study

Vilarinho is an ancient settlement (the earliest mention of it in documents dates back to 1360) located in the Lousã Mountains in Central Portugal (Fig. 1). It was one of the parishes extinguished during the contested 2013 administrative reform of the territory. Its population is close to 3000 residents, most of them living in the town of Lousã. The region is trapped between rural and urban dynamics, and the landscape of the rural settings is dominated by forestry, as farming is mostly reduced to smallholdings for family needs (Tavares et al. 2014). In 2012, a group of Vilarinho commoners initiated the management of close to 1000 ha of community forests without the intervention of the State. This was the result of a series of historical intersecting processes.

Vilarinho was among the first mountain communities to reclaim the common lands afforested by the State in 1976, after the Carnation Revolution. Most of the forest was composed by plantations of pines and other conifers, such as Douglas-fir. Without the technical knowledge to manage the forests, they opted for the co-management regime with the State forest services. However, over the next three decades, the forest entered into a serious process of degradation due to recurrent fires, tree diseases, in particular the pine nematode, and the proliferation of invasive *Acacia* species. Fires are a real danger to the local populations, and during the catastrophic fires of 2005, a firefighter who was also a commoner lost his life on duty.

Dissatisfied with the management of the Forest Services, the managing council of *baldios* elected by the assembly of commoners called for several interventions to prevent the degradation process. These calls were received by the regional administration of the Services with indifference. Taking advantage of the national Program for Forest Defense against Fires, the managing council hired a specialized team of forest fire fighters and aimed to initiate forest recovery. In order to establish their own management plan, they requested the immediate end of the co-management regime, which was legally recognized by the court in 2012.

Without delay, the commoners hired a forest engineer full-time to initiate the revitalization of the forest with professional support. The goal was to produce a multifunctional forest, capable of supporting multiple goods and activities, such as timber production, hunting, apiculture and non-timber forest products, together with mountain sports and leisure activities, while keeping open to the commoners the collection of firewood, brushes and other traditional uses. Currently, the *baldios* of Vilarinho provide other sources of collective income besides the community forest, namely rents from a wind park and an airfield. However, community forestry is the single activity that supports local jobs, as it employs 9 workers full-time.

The community forest project implies making considerable investments that will only be enjoyed by the future generations. There is a permanent tension between social investments in local associations, in particular for the care of the elderly and the young and the care of the forest. From the near 3000 residents with legally recognized rights to usufruct and administrate the *baldios*, less than 2 % participate in the general assemblies. Most of them are not forest users and are unaware of the existence of the community forests of their form of management and of their rights and responsibilities as *compartes*.

We crossed paths with Vilarinho in the scope of the research project SCRAM,[2] while we were searching for emergent socio-ecological arrangements and the revitalization of Portuguese forests. A protocol was established for collaborative research between the Centre for Social Studies and the Community of *compartes* from Vilarinho, aiming to produce resources relevant to the community forestry context. One of the key issues addressed were educational activities for the *compartes* that could open up spaces for communing and taking part in the ongoing common lands narrative.

## 6 Education for the Commons

How can educational activities contribute to community forest projects? This question must be answered in the wider context of how community forestry may contribute to the emergence of place-oriented communities, especially when the processes allowing communities to develop are blocked. According to the interactional approach followed by Flint et al. (2008), communities emerge through the ongoing communication that takes place between local residents when doing things together. Through shared actions and ways of life, local residents may come to share values that allow them to recognize the common good and overcome internal conflicts. The approach followed by the above mentioned authors is somewhat in line with the pedagogical creed of the American pragmatist John Dewey, who understood education as the unconscious process though which individuals gradually come to share the intellectual and moral resources which humanity has succeeded in getting together (Dewey 1897). Dewey believed this process could only be prompted by immersing the individuals in problematic social situations that would stimulate them to act as members of a unit with regard for the welfare of the group to which they belong. As it is increasingly recognized by academics and activists (Bollier and Helfrich 2014), the commons are more than material resources, and the result from the commoning activities of the commoners in their daily life (Dardot and Laval 2014). In this case the crucial aspect is to know if community forestry qualifies as an activity residents can actually authentically participate in, that is, take part in the common lands meaningful narrative, mutually shaping and being shaped by it (Nunes et al. 2014).

[2]SCRAM—Crises, risk management and new socio-ecological arrangements for forests: a perspective from science and technology studies, hosted by the Centre for Social Studies, University of Coimbra, Portugal.

We took an educational approach considering that residents may authentically participate in community forests by also engaging in social practices which could support common meaningful narratives (MacIntyre 1984). We tested the potential of two distinct practices for reconnecting local residents with the forest ecosystem management: mushroom picking and forest governance.

## 7 Communing with Mushrooms

The initial interest of the commoners of Vilarinho in wild mushrooms was to assess the economic potential of the species flourishing in the common lands. The Portuguese law considers wild mushrooms as property of the landowner, yet open access is traditionally granted to pick mushrooms for personal consumption. Mushroom commercialization, however, motivated owners to enforce their right. In Vilarinho one pine plantation was frequently raided by non-residents for commercial purposes without the consent of the commoners, who aimed to devise a plan for managing the mycological resources. Additionally, a few commoners had mushroom picking habits and aimed to know how to prevent mushroom poisonings and diversify their knowledge on edible species.

The first mushroom-related activity in Vilarinho was a training course in mushroom biology, ecology and the identification of poisonous and edible species. The course was attended by all those interested in mushrooms, including commoners and non-commoners. Although the potential for commercial exploitation turned out not to be viable, wild mushroom species provided great opportunities for storytelling, sharing experiences and sparking the curiosity of the participants about the diversity of the natural world. As these mushrooms are not domesticated species, their emergence is out of our control (Tsing 2012). Mushroom picking is a seasonal practice that takes people to walk in the woods and wander in search of hidden treasures among the falling leaves of autumn. Human-mushroom encounters require the development of an intimate ecological knowledge of the species and the territory, wandering to find the best places. One must cultivate the virtue of patience, respect the interspecies relationships and learn the tacit rules to safely meet again every year.

We took the specificities of the humans-wild mushrooms relationship as an entry point to develop educational activities related to the governance of *baldios*. The aim was to make visible the connections between the emergence of wild mushroom species and the care of the forest. This implied turning responsible wanderers into responsible commoners.

The first activity was a mycological walk in the common lands, entitled "*baldios* and their friends—a walk to discoveries" that is being repeated yearly. The walk is open to everybody, but we introduced the difference of *compartes* and non-*compartes* in signing-up for it. Among the participants were local residents unaware of the existence of *baldios*, activists, academics and students supporting the commons, as well as people searching an escape from urban routines. The participants form a heterogeneous group and meet regularly every year. The walk has both structured and un-structured activities. It has a previously defined trail, an interspecies narrative through the landscape and a meal prepared in the woods. During the walk and lunch time, informal conversations can take place between the participants, including the managers and workers of the community forest. The walk was complemented with a one-week workshop based on an experimental curriculum that uses the ecology of mushrooms, especially mushroom-tree relationships, to connect mushrooms with forest management and forest governance.

## 8 The Mini-Commoners *Baldios* Governance

The degrading image of *baldios* as wastelands, spaces abandoned or prone to conflicts, drives away the people from participating in the governance of *baldios,* regardless of its open institutions. The fact that community forestry is an activity that only makes sense in an intergenerational perspective and that the youngest generations are unaware of its existence endangers the sustainability of the initiatives. This prompted us to devise educational activities for children, related to the governance of community forests.

A "Summer School for mini-commoners" was created with children in the age-range 6–12, who were at a daycare center of Vilarinho during the summer holiday. We established a program in collaboration with the managing council of the *baldios*. It included two weeks of activities related to the governance of community forests. We implemented the school in collaboration with the parish of Vilarinho and the association holding the daycare center (Association for the Protection of the Elderly and the Young—ADIC). The experimental pedagogical approach was based on four intercepting axes: (1) territory; (2) community; (3) local institutions; (4) communing. The program included outdoor activities such as (i) a guided visit to the common lands introducing the children to the key forest resources and management problems; (ii) guided tours to the key beneficiary institutions of *baldios*; (iii) interviews with the elderly to recover memories about common lands; (iv) interviews with the managers of the community forests to learn about the formal governance structures, administrative positions established by law and the democratic procedures for collective decision-making. During the summer school, the children were engaged in joint activities such as musicing the *baldios* song and elaborating a mini-dictionary for community forest governance. The School for mini-commoners aimed to provide quality experiences to connect the children's social context with the community forests. Thus, communication with the families was a constant concern.

The central aspect of the School was the dramatization of an assembly of *compartes*. The preparation of the assembly followed strictly current legal proceedings; the children chose their roles and elaborated a discussion based on a realistic scenario: the clear cut of a plantation with pine nematode disease and the decisions on the species to plant and the social investments necessary to satisfy the needs of the local population. The assembly was attended by the children's families, who were actual *compartes* and participated in the dramatization by voting the decisions.

The Summer school showed a potential for transformative learning among all the participants, including adults. As stated by Freire (1996), whoever teaches learns in the act of teaching, and whoever learns teaches in the act of learning. The children were free to raise the questions that adults were too embarrassed to ask. The sight of the children as future managers of the community forests regardless of their socio-economic status increased the self-esteem of the families whose children face learning difficulties in the formal school, leading to public statements like: "after all, our children are bright"! Simultaneously, the children were saying: "our parents also need a School for commoners"!

## 9 Key Reflections and Future Challenges

Modern community forestry presents specific challenges for community engagement. It is an activity that requires a centralized managing structure with professional support, previously under state control and later under local control, which must plan the uses of the forest in the long-term. This creates a challenge for those intimately engaged with the forest, as the community forest project is ultimately controlled by local residents not engaged in any forest practice.

The approach to use wild mushroom picking as a forest practice is being successful in producing communing spaces and responsible wanderers in the forests. Non-commercial mushroom picking is a leisure activity and a naturalistic delight that occasionally offers the occasion for a good meal with family and friends. However, it is not simple to turn responsible wanderers into responsible commoners. The opposite argument could indeed be made, that leisure activities in the forest are an escape from the constraints and responsibilities of daily life. Still, mycological walks and workshops allow people to meet the community forest project and have an experience, even if limited, of what communal activities look like. The fact that mushroom based initiatives are open to locals and non-locals allowed the establishment of alliances with outsiders that support the community forest cause and increased the self-esteem of the locals by attracting people from all over the country to meet their community forest.

The approach to consider forest governance as a practice implies citizens willing to participate in political life. Unlike mushroom picking, however, politics is not regarded as a pleasant activity, but one that is reluctantly engaged in.[3] By devising educational activities with children we aimed to democratize politics and make it

[3]Yanis Varoufakis, a professor of economics and former Minister of Finance of the Greek government, claimed that "anyone who wants to be a Minister should be disqualified from that position". He considers that politics needs "reluctant politicians" that take administrative roles as a sacrifice and a genuine act of public service, one enjoy doing only for the altruistic pleasures of contributing to the community (conference accessible online at https://www.youtube.com/watch?v=2a3ZJE-mu3I).

simple. This brought into the light the social embarrassment of most commoners for being unaware of how the *baldios* institutions function, as family members were invited to learn about their rights through a game for children. Both children and their parents considered the school for the mini-commoners a success and an initiative that should be repeated over time and replicated in other places. Family members highlighted the importance of a place-based education for the children to learn what stays out of school curricula and for engaging in collective activities. "During the school they formed a group", a mother told us. Parents could sense their children flourishing in a different educational setting enriched by outdoor activities, problematic situations faced in daily life and meeting community role models engaged in forestry-related professions and practices. However, when we asked the parents if they intended to participate in the assemblies from now on, their answer was negative. Parents regraded themselves as a lost generation for virtuous politics but requested investment in their children to capacitate them to fulfill this much needed role.

The president of the managing council, the forest engineer and the president of the Vilarinho parish were restless in their collaboration and support to plan and implement the mycological activities and the Summer school. The support was materialized in countless hours preparing and monitoring the activities. Although the education activities to engage new commoners in community forest governance are considered very important, they constantly face two interrelated challenges: (1) to support community forest jobs and (2) to improve common land management.

As we hope to have shown throughout the text, community-forming processes and the collective appropriation of forests as common goods are highly challenging. Nevertheless, when the goal is to link good forest management with the community's well-being, one has no other alternative than "to die trying".

Community forest projects can provide benefits not only for the commoners but for all of society, through local, sustainable management of resources. Adequate policies are required, though, to facilitate the emergence of the complex community-forming processes. In Portugal, the opposite is taking place. Over the last three years, the austerity policies promoted by the State extinguished the local parish, closed down the daycare center for children and passed a law that hampers the devolution process of many community forests. Beyond the challenges it raises to the field of education for sustainability, the context of community forestry is a lens that allows us to experience the intersecting processes underlying the environmental problems and to imagine responses outside the box.

**Acknowledgements** The authors gratefully acknowledge the financial support of the project "SCRAM—Crises, risk management and new socio-ecological arrangements for forests: a perspective from science and technology studies", co-funded by COMPETE—Competitiveness Factors Operational Program (FEDER) and the Portuguese Foundation for Science and Technology—FCT (PTDC/CS-ECS/102041/2008; FCOMP-01-0124-FEDER-009236). Anailton Guimarães Salgado also acknowledges the FCT for financial support provided by the grant SFRH/BD/75075/2010. Special thanks go to the professionals and commoners of Vilarinho, Lousã, especially Eugénia Rodrigues and Luis Trota, for their invaluable support in all the educational activities developed.

## References


Bollier, D., & Helfrich, S. (Eds.). (2014). *The wealth of the commons: A world beyond market and state*. Levellers Press.

Brouwer, R. (1995). *Planting power: The afforestation of the commons and state formation in Portugal*. Eburon.

Bullock, R. C. & Hanna, K. S. (2012). *Community forestry: Local values, conflict and forest governance*. Cambridge University Press.

CNVTC (2010). *Programa Nacional para a Valorização dos Territórios Comunitários.*

Dardot, P. & Laval, C. (2014). *Commun: essai sur la révolution au XXIe siècle*. La Découverte.

Dewey, J. (1897). My pedagogic creed. *School Journal,* 54, 77–80 (available online http://dewey.pragmatism.org/creed.htm).

Flint, C. G., Luloff, A. E., & Finley, J. C. (2008). Where is "community" in community-based forestry? *Society and Natural Resources, 21*(6), 526–537.

Freire, P. (1996). *Pedagogia da Autonomia - Saberes Necessários à Prática Educativa* Editora Paz e Terra. Coleção Saberes.

García-Barrios, R. & Serra, R. (in press). "Reflexiones sobre la cooperación humana, el papel de la universidad pública y los derechos en la *Era de la crisis*". UNAM.

García-Barrios, R. & García-Barrios, L. (2008). The controllable society and sustainability. Chapter 10 In M Astier; O Masera & Y Galván-Miyoshi (Eds.), *Sustentabilidad: un Enfoque Dinámico y Multidimensional*. Editorial MundiPrensa. Barcelona, España (English translation).

Jeanrenaud, S. (2001). *Communities and forest management in Western Europe: A regional profile of WG-CIFM the working group on community involvement in forest management*. IUCN.

Klein, N. (2007). *The shock doctrine: The rise of disaster capitalism*. Macmillan.

Klein, R. & Stok, B. (1986). *Os baldios: een onderzoek naar de levensvatbaarheid van een socialistische wet met betrekking tot de kommunale dorpsgronden in Portugal: Een onderzoek naar de levensvatbaarheid van een socialistische wet met betrekking tot de kommunale dorpsgronden in Portugal*. Amsterdam: Universiteit.

Lopes, J., & Cristóvão, A. (2010). Restless rural spaces in the Iberian Peninsula – Montes Veciñais en Man Común in Galicia and Baldios in the north of Portugal. *Spanish Journal of Rural Development, 1*(1), 96–137.

MacIntyre, A. (1984). *After virtue* (Vol. 99). Notre Dame: University of Notre Dame Press.

Nunes, J. A., & Feijó, R. (1990). As transformações dos incultos no Noroeste (1750-1900): Uma proposta de reapreciação. *Cadernos de Ciências Sociais, 8–9*, 45–90.

Nunes, J. A., Ferreira, P., & Queirós, F. (2014). Taking part: Engaging knowledge on health in clinical encounters. *Social Science & Medicine, 123*, 194–201.

Ostrom, E. (1990). *Governing the commons: The evolution of institutions for collective action*. Cambridge University Press.

Peluso, N. L. (1992). *Rich forests, poor people: Resource control and resistance in Java*. Univ of California Press.

Santos, B. (2006). Globalizations. *Theory, Culture & Society, 23*(2–3), 393–399.

Sarre, A. (1994). What is community forestry? *Tropical Forest Update, 4*(4), 2.

Scott, J. C. (1998). Nature and space. In *Seeing like a state* (pp. 11–52, 359–369). New Haven: Yale University Press.

Tavares, A. O., Monteiro, M., Vargas, M. A., Pato, R. L., & Serra, R. (2014). Land use change and forest routing in a rural context: The relevance of the community-based management and planning framework. *Applied Geography, 52*, 153–171.

Taylor, P. J. (2010). *Unruly complexity: Ecology, interpretation, engagement*. University of Chicago Press.

Tejera-Hernández, B., & García-Barrios, R. (2008). *Agricultura y estrategias de formación de ingreso campesinas en comunidades indígenas forestales oaxaqueñas* (pp. 65–103). Instituciones y Desarrollo: Ensayos sobre la Complejidad del Campo Mexicano.
Tsing, A. (2012). Unruly edges: Mushrooms as companion species. *Environmental Humanities, 1*, 141–154.

# Part IV
# Designing Sustainable Futures

# Urban Forest Governance: FUTURE—The 100,000 Trees Project in the Porto Metropolitan Area

Marta Pinto, Conceição Almeida, Ana Maria Pereira and Margarida Silva

**Abstract**

The Porto Metropolitan Area is a region in northern Portugal with approximately 2000 km$^2$. Almost 16 % of the Portuguese population lives in the area, which is structured around the municipality of Porto with 16 other municipalities included. The region is a jigsaw puzzle of urban, agricultural, and forest areas. The Porto Metropolitan Area Environmental Strategic Plan, a broad participatory regional planning process conducted from 2003 to 2008, concluded that major challenges ahead included the improvement of the green infrastructure (forest, riverside areas, and natural corridors), the need for education and training for sustainability, as well as more and better interinstitutional coordination. In order to tackle these vital regional issues several institutions, partners within the framework of the Porto Metropolitan Area Regional Centre of Expertise on Education for Sustainable Development, decided to collaborate in order to design and implement a flagship project. The FUTURE—the 100,000 trees project in the Porto Metropolitan Area is the outcome of this process. Its purpose is to create, restore and care for native urban forests in the region with active involvement of the main stakeholders (landowners, citizens, local governments, central government, companies, non-governmental organisations, and schools, among others). Our aim is to present the case study of this project describing its context, scale, institutional framework, actors and partnerships, resources and processes, highlighting its governance model.

M. Pinto (✉) · C. Almeida · A.M. Pereira
Environmental Studies Group, Catholic University of Portugal, Porto, Portugal
e-mail: mspinto@porto.ucp.pt

M. Silva
Laboratório Associado, CBQF—Centro de Biotecnologia E Química Fina, Escola Superior de Biotecnologia, Universidade Católica Portuguesa/Porto, Apartado 2511, 4202-401 Porto, Portugal



P. Castro et al. (eds.), *Biodiversity and Education for Sustainable Development*, World Sustainability Series, DOI 10.1007/978-3-319-32318-3_12

**Keywords**
Native urban forests · Governance · Planning

# 1 Introduction

## 1.1 Urban Forests as Multifunctional Green Infrastructures

In recent years the European urban and peri-urban areas have been systematically altered by land use change, intensification and fragmentation (Antrop 2004; Naumann et al. 2011). Consequently, the spatial and functional consistency of ecosystems and landscapes, the spectrum of ecosystems services, the health and well-being of resident populations as well as overall territorial resilience have been markedly reduced (Cardinale et al. 2012; Díaz et al. 2006; Haines-Young and Potschin 2010; McKinney 2002).

Framed by this plight, green infrastructure creation is in fact an opportunity to pursue ecological, economic and social benefits through natural and multifunctional solutions (Mell 2009). This concept encompasses strategically planned networks of natural and semi-natural areas, in rural and urban zones, designed and managed with the intent of providing a wide range of ecological services (European Commission [EC] 2013a, b). Urban forests (including trees, parks, gardens, and woods) (Konijnendijk et al. 2006; Randrup et al. 2005) are key elements (EC 2013a).

The importance of urban forests has been widely studied and the services provided include at least the following ecological, economic and social dimensions:

- Air quality improvement (McDonald et al. 2007; Nowak and Heisler 2010; Tiwary et al. 2009);
- Local climate moderation, namely in the heat island effect (Kleerekoper et al. 2012; Oliveira et al. 2011);
- Climate change adaptation (Gill et al. 2007; Norton et al. 2015);
- Carbon sequestration (Caldecott et al. 2015; Rodríguez-Loinaz et al. 2013; Strohbach et al. 2012);
- Soil conservation and soil water regulation (Armson et al. 2013; European Environment Agency 2015);
- Beautification of cities, both for residents and tourists (Nowak and Dwyer 2007; Power 2005; Tyrväinen et al. 2005);
- Stimulation of memory, attention and concentration skills (Berman et al. 2008; Bratman et al. 2012);
- Stress reduction (Tyrväinen et al. 2005);
- Sleep balance (Astell-Burt et al. 2013);
- Overall sense of well-being (Haluza et al. 2014; Kaplan and Kaplan 1986; Tsunetsugu et al. 2013).

Although the Portuguese reality has been little studied, Lisbon's trees have been found to offer significant financial return. The capital's 41,247 trees provide services valued at US$8.4 million a year, a positive balance of US$4.48 dollars for each dollar invested in the trees' planting and maintenance. Each tree provides US $6.20 in energy savings, US$0.33 in carbon sink services, US$5.40 in air pollution retention, US$47.80 in reduced stormwater runoff and US$144.70 in increased real estate values (Soares et al. 2011).

## 1.2 Urban Forests as a Governance Issue

Despite its importance, urban forests and their management are still poorly studied. Understanding of the mechanisms of creation, maintenance, management and conservation has become imperative. The governance of urban forests—a dynamic that makes governments, communities, businesses and landowners, among others, interact in an iterative and co-evolutionary process—is only now just starting to grow (Bentsen et al. 2010; Jim 2011; Kronenberg 2015; Lawrence et al. 2013; Schmied and Pillmann 2003).

Among the studies on governance of urban forests, the work by Secco et al. (2011) stands out. It covered six cases in Italy: at the municipal level (Venice), regional level (Region Lombardy, Veneto Region) and also at the national level (development of the strategic framework programme for forests). From a governance point of view, none represented a genuine transition to an integral model of collective learning—and the most promising cases (involving local stakeholders and the introduction of institutional innovations based on collective learning) were developed at the local or regional levels. In addition, Buttoud et al. (2011) analysed the barriers to institutional innovations through case studies in Austria, France and Scotland, concluding that the mercantilist orientation often frustrates multifunctional forest management and respective governance styles.

Young and McPherson (2013) in the United States reviewed the major initiatives of planting trees at the metropolitan level (New York, NY, Sacramento, CA, Los Angeles, CA, Houston, TX, Denver, CO and Salt Lake County, UT) from a governance point of view and described a dominant role for the public sector in the implementation of these projects, whether in the form of the vision, planning or management. According to the authors, those projects demonstrate new types of government, although not exactly governance innovation. They also highlight that these projects are not easily institutionalized within the public authorities that lead them. One reason probably has to do with the many regulatory and organisational constraints endured by these organisations (Connolly et al. 2013). Alternatively these authors introduce the concept of "bridge-organisations" as a possible governance epicentre: autonomous structures that organize the activities of a cluster of citizen groups through a centralized interface which coordinate resources (knowledge, finance, etc.) in a balanced adaptive dance with the rest of the stakeholders, including public authorities.

Similarly, Lawrence et al. (2013) studied several case studies of governance of urban forests (in Belgium, Italy, Sweden, United Kingdom and Scotland) and designed a case presentation template easily adapted to distinctive realities (cultures, administrations, geographies and scales) and focusing on its key aspects. The analysis shows a trend towards more integrated initiatives and involvement of various parties, though with a leadership usually assumed by local, regional or central government. The "governance by government" phase has passed but was not replaced by a "governance without government" (Lawrence et al. 2013).

More recently, a comprehensive quantitative framework was published for the evaluation of urban forest governance quality through a set of aggregated indicators that overcome terminology differences, scales and models, expedite comparisons between different initiatives and significantly strengthen the evolution of this field of study (Secco et al. 2014). It does require, and builds on, qualitative characterizations as the initial systemizing step. Therefore, in this text, we aim at a first description of the FUTURE—the 100,000 trees project in the Porto Metropolitan Area as per the Lawrence et al. (2013) framework, looking ahead to paths of urban forest governance improvement in this Portuguese region.

## 2 Methods

For a description of the case study FUTURE—the 100,000 trees project in the Porto Metropolitan Area, and in order to meet the methodological requirements for case studies (George and Bennett 2004), the description and analysis of the governance model proposed by Lawrence et al. (2013) was followed. This framework structures a description of the mechanisms and processes in a project of urban forest governance in an accessible, organized and easy to apply manner. It consists of a set of key dimensions and descriptors that detail the case study both schematically and in expository form. The main dimensions include:

- Type, scale and context;
- Institutional context (policies, planning and regulations, ownership and access);
- Actors and coalitions (stakeholders, partnerships, power analysis);
- Resources (knowledge and information, funding, delivery mechanisms);
- Processes (discourses, participation and engagement, monitoring and evaluation).

The first steps of the FUTURE took place in October 2010. Since then the design, technical coordination and implementation have been the responsibility of the Catholic University of Portugal throughout. As such the team takes on the role of the participant observer (Atkinson and Hammersley 1994; Kearns 2000).

## 2.1 Case Study

### 2.1.1 Context

The Porto Metropolitan Area (PMA), a region situated in the north of Portugal, is a territory with approximately 2000 $km^2$ that comprises 17 municipalities: Arouca, Espinho, Gondomar, Maia, Matosinhos, Oliveira de Azeméis, Porto, Vila Nova de Gaia, Paredes, Santa Maria da Feira, Tirso Santo, São João da Madeira, Trofa, Vale de Cambra, Valongo, Vila do Conde, and Vila Nova de Gaia. This is the country's second most populated region: about 1.73 million inhabitants (Instituto Nacional de Estatística 2014).

Overall, it is a developed area (Carney et al. 2009) and is intensely urbanized. However, this urbanization is fairly consolidated and limited to just over 20 % of the area.

Almost 20 % of the PMA is under agricultural use (Área Metropolitana do Porto n.d.), 41 % encompasses forested areas (Direcção Nacional de Gestão Florestal 2010) and 10 % includes nature conservation expanses (Área Metropolitana do Porto 2014). However, the PMA territory, rather than spatially planned, evolved as a complex jigsaw of urban, agricultural and forest patches (Pinto and Silva 2013). The loss of forest and agricultural cover to urban expansion is ongoing. From 1991 to 2008, urban development spread at the rate of 1.8 ha per day (Grupo de Estudos Ambientais da Universidade Católica Portuguesa 2006b).

As for forestland, more than 90 % is occupied by intensive industrial monoculture of *Eucalyptus* sp. and *Pinus pinaster*. Not long ago native forest cover (*Quercus* sp., among others) dominated the region. Today, however, it covers a very fragmented 6 % of the forested territory (Direccao Nacional de Gestao Florestal 2010), enduring only along riverbanks, ridges and hedgerows (Área Metropolitana do Porto n.d.).

This brief scenario of the region must be crossed with several important challenges for urban forests (Área Metropolitana do Porto 2014), which include:

- An obvious absence or underutilization of multifunctional green spaces in the densest urban areas, as well as their discontinuation in agroforestry areas;
- Major land use fragmentation in the metropolitan area, with a mix of residential buildings, industry, and agroforestry which generate conflict and risks (such as fires: "the most forest fire occurrences in the country" [Área Metropolitana do Porto n.d.]);
- Natural areas of interest for nature conservation efforts scattered throughout the territory while invasive plants proliferate (Área Metropolitana do Porto n.d.);
- There are 33,000 trees inventoried in the Porto municipality (Câmara Municipal do Porto n.d.) but there is still a general lack of basic urban forest information in the region;
- The region also presents major climate change vulnerabilities and shows demonstrably weak adaptability (Área Metropolitana do Porto 2014). There is urgency in the application of prevention and adaptation measures (Carney et al. 2009).

### 2.1.2 Institutional Framework

From a regional governance point of view, a vast institutional fragmentation has been identified (Área Metropolitana do Porto 2014). In the case of urban forests, management has been divided among different local governments, central government services, forest landowner associations, and forest corporations.

Local governments are responsible for the management of green areas, urban trees and publicly managed forestland (amounting to just a small proportion of total forest area) as well as enforcing legislation regarding privately owned land (for example, coercing owners into regular shrub clearing). The State Forest Service, on the other hand, co-manages communally-owned areas and finances forest rangers. It should be noted that most of the region's forestland is under private management and that about 90 % of forest holdings are under five hectares (Instituto da Conservação da Natureza e das Florestas 2006).

An important basis for new environmental governance models in the region emerged with the Porto Metropolitan Area Environmental Strategic Plan (PMA-ESP), a broad and participatory regional planning process conducted between 2003 and 2008 that integrated the contributions of more than 5000 citizens and 200 organisations and brought to light a shared view of the regional environmental challenges and solutions. The need for greater collaboration and inter-municipal cooperation in education and training for sustainability and opportunities for public participation were the main conclusions identified by the study (Grupo de Estudos Ambientais da Universidade Católica Portuguesa 2006a, 2008b; Pinto and Silva 2013).

This plan also calls for actions of benefit to urban forests, in which the need for a new governance model is evident (Grupo de Estudos Ambientais da Universidade Católica Portuguesa 2008a):

- Creation of ecological corridors and areas with native vegetation as enhancing elements of biodiversity and forest fire prevention, mainly through the nurturing of natural regeneration and native species reforestation while involving the urban population;
- Increased coordination between stakeholders together with the enlisting of forest landowners and communities at large; implementation of an awareness programme involving citizens in concrete native forest promotion initiatives.

This newly found awareness and exposure triggered by the broad participatory process led to the establishment of the Porto Metropolitan Area Regional Centre of Expertise on Education for Sustainable Development (CRE.Porto) in line with the model recommended by the United Nations University-Institute of Advanced Studies (UNU-IAS 2010). The overarching objective is the creation of social networks that facilitate coordinated action and cooperation (with knowledge, trust and resources) that help move towards sustainability by overcoming the dilemmas of collective action in multiple arenas.

CRE.Porto was started with an Agreement Contract signed by 30 stakeholders—mainly local governments, non-governmental organisations, foundations, higher education institutions and local central government services—committed to making a difference in the region: do more with less; share and generate knowledge; build trust and enroll organisations and citizens in thinking and acting for change. Operations are to be based on the best knowledge, practices, cooperation, coordination, resources and participatory methodologies available.

There are two main panels working within CRE.Porto. The first is a forum representative of local and regional stakeholders—the Executive Board—whose main functions include outlining joint actions and helping in their implementation. The second panel is the Coordinating Office. This office makes things happen, prompts collaboration, manages joint projects, identifies funding sources, prepares grant applications and organises partner meetings. The Catholic University of Portugal hosts and staffs this second panel, working closely with the PMA regional government (Pinto and Silva 2013). The CRE.Porto was formally recognized by UNU-IAS in 2009.

It was at a meeting of CRE.Porto's Executive Board that the first ideas for the FUTURE project came about. Beforehand there had been no metropolitan vision or global targets for urban forests, just uncoordinated projects at various power levels. The project was designed with contributions from several partners at various stages of the planning process, which lasted for nine months. By the end of this period the methodology and key features were defined and guaranteed so as to enable the implementation on the ground. The FUTURE became a regional project for the creation and rehabilitation of native urban forests that simultaneously addresses three regional challenges:

- improving and expanding the natural capital;
- emphasizing the social capital and facilitating public participation;
- enhancing interinstitutional cooperation, so that the region evolves towards sustainability.

The goal was set: plant and care for 100,000 new native trees by 2015, in approximately 100 ha spread throughout the metropolitan territory.

Given the backdrop previously described—an intervention area with varying regimes of ownership and management—the FUTURE project focuses on urban areas (parks) and peri-urban areas (forests) managed by local governments, communally-owned areas subject to the forest regime (and co-managed by the State Forest Service), and private forestland, while privileging those of particular biodiversity value. Interventions meet specific criteria and are subject to the legal regime applied to afforestation and reforestation actions as defined by the Portuguese Decree-Law 96/2013. The National Forest Strategy, created in 2006, is the main reference in current forest planning and management.

Most interventions are carried out on land with public or communal management. However, through the forest landowners' associations, private owners have also been identified whose vision is compatible with the FUTURE's. An Agreement Contract binds CRE.Porto, the municipality and the landowner, and responsibilities are allocated. These include maintenance and site access for volunteers and staff.

### 2.1.3 Actors and Coalitions

The CRE.Porto Coordinating Office operates as an integrator and facilitator for the network of partners, experts and volunteers through project planning, implementation and evaluation. The PMA insured initial project approval and links the Coordinating Office with decision-makers. Municipalities identify areas, submit intervention plans and collaborate actively in the preparation of field activities, providing their own technical and operational teams, forest rangers or external teams. They are also responsible for post-intervention maintenance plans and interfacing with schools, among others. Although with varying degrees of involvement, all 17 municipalities of the PMA actively participate in the project.

The State Forest Service is a leading partner, providing knowledge at various project levels. It is also responsible for the intervention plans of the communally-owned land that they co-manage. In addition, they produce most of the native trees used.

Three forest landowners' associations are key partners: *ASVA—Associação de Silvicultores do Vale do Ave*, *Portucalea—Associação Florestal do Grande Porto* and *AFEDV—Associação Florestal do Entre o Douro e Vouga*. They identify landowners, draw up intervention plans, prepare land with forest rangers and guarantee regular maintenance.

In addition, dozens of other partners contribute directly to the project with knowledge, human and technical resources, materials and services. The FUTURE currently has 60 active partners—public and private, government or otherwise. A total of 164 organisations (associations, schools, companies) have participated throughout as volunteers in planting and maintenance activities.

In a network this complex it is difficult to clarify who makes every decision. Some—those determining the economic sustainability of the project—are inevitably taken by policy makers or other organisations that hold financing power. Technical decisions typically stem from consensus building with partners and volunteers.

The Coordinating Office regularly and beforehand consults key partners to elicit viewpoints on specific plans. There is also an open channel with volunteers, whose views and concerns are heard and recorded. Feedback is reviewed with partners.

### 2.1.4 Resources

Since its inception the FUTURE received financial support from the municipalities through an annual membership fee (roughly US$50,000). All other institutions contribute in kind, through services, and resources, with an estimated total annual contribution of about US$40,000. During 2014/2015 the project was partially funded by the European Regional Development Fund Program (about US $150,000), which led to a major boost in the project. These funds allowed for investment into soil preparation, seedling maintenance and the creation of a tree nursery, among other priorities.

Technical information for the tree planting interventions results from the involvement of local government staff, forest landowners associations' technical teams, State Forest Service personnel, researchers and consultants. CRE.Porto's Coordinating Office integrates contributions from the various partners and also provides know-how. Interventions proposed by any of the partners are analysed and resources identified. Each activity results from the experience, expertise and points of view of various people and institutions. Knowledge regarding plant propagation, monitoring methodology, seed collection, planting technique, plant physiology as well as ecosystem services information is sought from researchers and other experts.

From the outset, the FUTURE received native trees from state nurseries, produced and delivered free of charge under the National Native Tree Stock Initiative (www.florestacomum.org). The FUTURE's native tree nursery was set up at the Porto Municipal Nursery and benefits from the cooperation of the Porto municipal government (which provides workforce and materials) and the State Forest Service (granting management know-how). In 2014/2015 over 20,000 native plants from 13 species were produced.

In the field, operational teams (gardeners and forest rangers) perform most of the land preparation work. A total of 212 technical and operational personnel, from the various partners, have participated in the project so far. Over the four years (2011–2015) volunteer input amounted to 30,239 work hours, in a total of 9285 volunteer participations.

Volunteers are also involved in other more technical aspects of the project, such as the Geographic Information System.

### 2.1.5 Processes

The project is a niche in the sense that, through a small regional experiment, the following two dominant discourses are counteracted: (1) lack of inter-municipal coordination in creating a resilient alternative to the predominantly monoculture regime in the forestry sector and (2) generalized discredit of environmental projects, mainly due to continuity and monitoring failures.

The FUTURE gives emphasis to communication on all its fronts. People are encouraged to participate and exchange experiences. An example of this is the annual technician meeting. Two such meetings have been held (2014 and 2015) in which, in addition to reporting project results, participants can share ideas, critiques, difficulties, and successes. These meetings include training by higher education institutions and enable intervention improvements.

Over the years the concern about knowledge sharing with citizens has been taken into account. Thus training sessions have been organised, free of charge and for different communities, on plant propagation, identification methods and invasive plant control, among others. Up until July 2015, 169 training hours were organised and 460 citizens attended in total.

Specific training was given to operational teams and technical staff involved in the FUTURE, in particular on the identification, planning and control of invasive plants—a major challenge for ongoing interventions. A total of 222 people participated in the 71 h of training. Planting and maintenance activities also promote the exchange of knowledge and experiences between partners, volunteers and experts, resulting in increased mutual understanding and respect.

Actions that involve schools and students have taken place since 2013. In 2014/2015 this covered 52 schools, 130 teachers, and 2439 students (six years old and onwards). In 2015/16 the "FUTURE Schools' Network" has 46 institutions sowing new trees, enhancing school grounds and learning about their natural heritage.

The project is continuously monitored regarding activities, participants and results (Pinto and Almeida 2013). An elaborate database was created for this effect, in tandem with the Geographic Information System.

Results so far show positive growth in number of planting activities, areas intervened, trees planted, and human involvement. During the first year (2011/12) 11,409 trees were planted, whereas in 2014/15 that number rose to 29,134. Volunteer participations, in the same time frame, increased from 1858 to 3164. Volunteer hours almost doubled, from 5873 to 10,216. Throughout four consecutive winters a total of 190 ha were planted with 63,571 trees of 40 native species.

Simultaneously, some of the benefits obtained are regularly assessed and shared (Soares et al. 2011). Looking only at carbon sequestration, air pollutant retention, and soil water regulation, the trees planted annually provide US$2,990,956 in ecosystem services.

Intervened areas are monitored regularly with a biostatistical analysis methodology that determines the planting success rates. The average survival rate for trees three to six months after installation is 95 % (Grupo de Estudos Ambientais da Universidade Católica Portuguesa 2015), dropping to 60 % one to two years after installation (Grupo de Estudos Ambientais da Universidade Católica Portuguesa 2014, #72).

All reports are public and posted online, as well as sent to partners, experts and volunteers involved in the FUTURE. The website (www.100milarvores.pt) and Facebook page (www.facebook.com/100000arvores) are regularly updated. Overall the project (Table 1) is entirely transparent.

**Table 1** The FUTURE project presented as suggested by Lawrence et al. (2013)

| | |
|---|---|
| Name | FUTURE—the 100,000 trees project in the Porto Metropolitan Area, Portugal |
| Type | Project |
| Description | Regional, inter-municipal project to create, promote and maintain native urban forests |
| Scale | Porto Metropolitan Area |
| *Context* | |
| Trees, forests | Parks, natural corridors, woodlands and forestlands |
| Catchment population | 1.73 million people |

(continued)

**Table 1** (continued)

| | |
|---|---|
| Name | FUTURE—the 100,000 trees project in the Porto Metropolitan Area, Portugal |
| *Institutional framework* | |
| Policies | Porto Metropolitan Area Environmental Strategic Plan (2003–2008)<br>National Forest Strategy |
| Planning and regulations | Municipal and regional planning systems<br>Legal regime applied to afforestation and reforestation actions<br>National system of classified areas |
| Ownership | Mixed: public (state and local authorities) and private (landowners bound by agreement) |
| Access and use rights | Public access to most of the areas—in the case of private property, staff and volunteers are guaranteed access |
| *Actors and coalitions* | |
| Primary stakeholders | Porto Metropolitan Area Administration, 17 municipalities, Catholic University of Portugal, State Forest Service, forest landowner's associations, forest landowners |
| Other stakeholders | Volunteers, schools, higher education institutions, non-governmental organizations, State Agriculture and State Environment Regional Services, Lipor, Serralves Foundation, Metro do Porto |
| Partnerships | CRE.Porto conflates the partnership (Catholic University of Portugal, PMA, municipalities and other partners); also specific partnerships with private landowners and businesses |
| Power analysis | CRE.Porto leads the FUTURE, acting as a bridge-organisation between public authorities and a cluster of citizen groups as well as coordinating resources in a balanced adaptive dance |
| *Resources* | |
| Funding | Municipalities, European Commission, foundations and companies |
| Knowledge | Forest management, nature conservation, invasive species control, native plant species, nursery management, communication, community involvement |
| Delivery mechanisms | Trees produced in state nurseries and in the FUTURE's own nursery; forest rangers and gardeners provide field support; volunteers do most of the planting; the Coordinating Office brings together actors and resources |
| *Processes* | |
| Discourses | Calls for inter-municipal coordination in creating a resilient alternative to the predominant monoculture regime and a reliable community based territorial intervention |
| Participation | Governmental, non-governmental, public and private organisations actively participate in planning, implementation and monitoring; multilateral communication is promoted; citizens are involved in planting and nurturing as well as finding improvements and helping in their implementation |
| Monitoring and evaluation | Reports published annually with results on activity participation, trees planted and area covered<br>Monitoring report published annually, presenting survival rates for planting activities<br>Report on annual meeting of technical staff |

## 3 Conclusions and Future Directions

This description concludes a first exercise in systematizing a southern European regional urban forest governance initiative as per Lawrence et al. (2013). This framework demonstrated its usefulness and flexibility with the FUTURE—the 100,000 trees project in the Porto Metropolitan Area and naturally brought out the distinctive political, social and geographical puzzle that makes it unique as well as ground-breaking. In parallel, this analysis allowed for an in-depth reflection on the FUTURE's governance. The following conclusions seem noteworthy.

The FUTURE dynamics are closely linked to CRE.Porto's vitality. This broad network of public and private organisations set up a new governance model that brings together—and manages—a diverse range of resources. It embodies a more technical personality and avoids the traps of the conventional governing approach. The CRE.Porto Coordinating Office's autonomy allows it to build consensuses between citizens, local governments and other organisations. As such the Coordinating Office is a bridge-organisation (Young and McPherson 2013) which, in full compliance with existing laws and regulations, is better able to succeed where traditional public management has lagged. Scattered resources are brought together, synergies are promoted and win-win relationships are fostered among partners who traditionally evolve on parallel tracks.

The onward journey of CRE.Porto and the FUTURE depends on political will, mostly from local governments who must inject much needed funds to keep the Coordinating Office going. However, the structure, consistency and results of this network and project, against a backdrop of power dispersion in the region, guarantee a continuity that no single institution is able to ensure. Clear evidence is borne by the fact that the 2013 municipal elections, which resulted in deep political changes, elicited no threat to survival. The FUTURE being coordinated by a university probably helps insulate it further against the vagaries of political strife.

This is a clear case of "governance with government" (Lawrence et al. 2013; Secco et al. 2011; Young and McPherson 2013), where the government's role is not of leadership but of support.

This distinctive feature may be related to the institutional context in which FUTURE arose. The FUTURE and CRE.Porto are the tangible results of a long and complex participatory strategic plan carried out in the region years earlier. The broad public involvement in defining a common vision, diagnosing the challenges and identifying the priorities generated a deep knowledge base and endorsed an unprecedented social capital (of mutual knowledge, trust and shared views). People from the several organisations that regularly accompanied the development of the PMA-ESP bonded into a learning community. This community continues to grow and at the ongoing technical meetings, training sessions and field activities represent a privileged exchange forum. This important role of collective learning in the new governance models of urban forests has been already recognized by Secco et al. (2011).

Due to the concerted action of this learning community the FUTURE has developed spin-off projects that were not foreseen initially, such as the production of native plants or invasive plant control plans. It should be observed that what some may consider a partial failure—the full 100,000 native trees have not been planted by 2015—is seen by the FUTURE's community as an adjustment to the reality of the region, as the collective body recognised the need of addressing additional issues.

A new FUTURE analysis, focused on governance quality parameters in accordance with the forest framework laid out by Secco et al. (2011) is now warranted. From every end, spring ever branching beginnings.

**Acknowledgments** The authors wish to thank all citizens and organisations involved with the FUTURE—the 100,000 trees project in the Porto Metropolitan Area. Their work and enthusiasm towards this common green goal makes all efforts worthwhile.

## References

Antrop, M. (2004). Landscape change and the urbanization process in Europe. *Landscape and Urban Planning, 67*(1), 9–26.

Área Metropolitana do Porto. (2014). AMP 2020 Estratégia de Base Territorial. Porto.

Área Metropolitana do Porto. (n.d.). Prioridades de desenvolvimento para 2007–2013. Retrieved October 23, 2015, from http://tinyurl.com/h2nt8k2

Armson, D., Stringer, P., & Ennos, A. (2013). The effect of street trees and amenity grass on urban surface water runoff in Manchester, UK. *Urban Forestry & Urban Greening, 12*(3), 282–286.

Astell-Burt, T., Feng, X., & Kolt, G. (2013). Does access to neighbourhood green space promote a healthy duration of sleep? Novel findings from a cross-sectional study of 259 319 Australians. *BMJ Open, 3*(8), 1–6.

Atkinson, P., & Hammersley, M. (1994). Ethnography and participant observation. In N. Denzin & Y. Lincoln (Eds.), *Handbook of qualitative research* (pp. 248–260). London: SAGE Publications.

Bentsen, P., Lindholst, A., & Konijnendijk, C. (2010). Reviewing eight years of Urban Forestry & Urban Greening: Taking stock, looking ahead. *Urban Forestry & Urban Greening, 9*(4), 273–280.

Berman, M., Marc, G., Jonides, J., & Kaplan, S. (2008). The cognitive benefits of interacting with nature. *Psychological Science, 19*(12), 1207–1212.

Bratman, G., Hamilton, J., & Daily, G. (2012). The impacts of nature experience on human cognitive function and mental health. *Annals of the New York Academy of Sciences, 1249*(1), 118–136.

Buttoud, G., Kouplevatskaya-Buttoud, I., Slee, B., & Weiss, G. (2011). Barriers to institutional learning and innovations in the forest sector in Europe: Markets, policies and stakeholders. *Forest Policy and Economics, 13*(2), 124–131.

Caldecott, B., Lomax, G., & Workman, M. (2015). Stranded carbon assets and negative emissions technologies. *Stranded Assets Programme*. University of Oxford.

Câmara Municipal do Porto. (n.d.). A árvore no espaço urbano. Retrieved November 05, 2015, from http://tinyurl.com/glkyxbw

Cardinale, B., Duffy, J., Gonzalez, A., Hooper, D., Perrings, C., Venail, P., et al. (2012). Biodiversity loss and its impact on humanity. *Nature, 486*(7401), 59–67.

Carney, S., Green, N., Wood, R., & Read, R. (2009). Greenhouse gas emissions inventories—Brochure for individual regions. Porto: Centre for Urban and Regional Ecology of the School of Environment and Development of the University of Manchester.

Connolly, J., Svendsen, E., Fisher, D., & Campbell, L. (2013). Organizing urban ecosystem services through environmental stewardship governance in New York City. *Landscape and Urban Planning, 109*(1), 76–84.

Díaz, S., Fargione, J., Chapin, F., & Tilman, D. (2006). Biodiversity loss threatens human well-being. *PLoS Biology, 4*(8), 1300–1305.

Direcção Nacional de Gestão Florestal. (2010). Relatório Final do 5.º Inventário Florestal Nacional. Retrieved November 02, 2015, from http://tinyurl.com/ozypezy

European Commission. (2013a). *Commission Staff Working Document—Technical information on green infrastructure (GI) accompanying the document "Communication from the Commission to the European Parliament, the Council, the European Economic and Social Committee and the Committee of the Regions on Green Infrastructure (Gi)—Enhancing Europe's Natural Capital"*. (SWD (2013) 155 final). Retrieved from http://tinyurl.com/jfhyw3e

European Commission. (2013b). *Communication from the Commission to the European Parliament, the Council, the European Economic and Social Committee and the Committee of the Regions on Green Infrastructure (Gi)—Enhancing Europe's Natural Capital*. (COM (2013) 249 final). Retrieved from http://tinyurl.com/jh2ba46

European Environment Agency. (2015). Water-retention potential of Europe's forests—A European overview to support natural water-retention measures *EEA Technical report*.

George, A., & Bennett, A. (2004). *Case studies and theory development in the social sciences*. Cambridge: MIT Press.

Gill, S., Handley, J., Ennos, A., & Pauleit, S. (2007). Adapting cities for climate change: The role of the green infrastructure. *Built Environment, 33*(1), 115–133.

Grupo de Estudos Ambientais da Universidade Católica Portuguesa (Producer). (2006a). Futuro Sustentável - Diagnóstico de Ambiente do Grande Porto: Educação Ambiental/Educação para o Desenvolvimento Sustentável (Versão para consulta pública) (pp. 451): Lipor. Retrieved from http://tinyurl.com/zprjle9

Grupo de Estudos Ambientais da Universidade Católica Portuguesa (Producer). (2006b). Futuro Sustentável - Diagnóstico de Ambiente do Grande Porto: Ordenamento do Território, Espaços Verdes e Áreas Naturais (Versão para consulta pública) (pp. 151): Lipor. Retrieved from http://tinyurl.com/zprjle9

Grupo de Estudos Ambientais da Universidade Católica Portuguesa (Producer). (2008a, 2006). Futuro Sustentável - Plano de Acção (Síntese para consulta pública) (pp. 14): Lipor. Retrieved from http://tinyurl.com/hq33ue8

Grupo de Estudos Ambientais da Universidade Católica Portuguesa (Producer). (2008b, 2006). Futuro Sustentável [Fase de alargamento] - Diagnóstico de Ambiente do Grande Porto: Educação para a Sustentabilidade (Versão para consulta pública) (pp. 64): Lipor. Retrieved from http://tinyurl.com/zmblg2b

Grupo de Estudos Ambientais da Universidade Católica Portuguesa (Producer). (2015). FUTURO - Relatório de Monitorização (2015) (pp. 28): Lipor. Retrieved from http://tinyurl.com/gmxo8wv

Haines-Young, R., & Potschin, M. (2010). The links between biodiversity, ecosystem services and human well-being. In D. Raffaelli & C. Frid (Eds.), *Ecosystem Ecology: A new synthesis* (pp. 110–139). Cambridge: CUP.

Haluza, D., Schönbauer, R., & Cervinka, R. (2014). Green perspectives for public health: A narrative review on the physiological effects of experiencing outdoor nature. *International journal of environmental research and public health, 11*(5), 5445–5461.

Instituto da Conservação da Natureza e das Florestas. (2006). PROF AMPEDV - Plano Regional de Ordenamento Florestal da Área Metropolitana do Porto e Entre Douro e Vouga. Retrieved November 16, 2015, from http://tinyurl.com/ock98vt

Instituto Nacional de Estatística. (2014). Resident population by place of residence. Retrieved November 20, 2015, http://tinyurl.com/o8b3aq7

Jim, C. (2011). Holistic research agenda for sustainable management and conservation of urban woodlands. *Landscape and Urban Planning, 100*(4), 375–379.

Kaplan, R., & Kaplan, S. (1986). *The experience of nature: A psychological perspective*. Cambridge: Cambridge University Press.

Kearns, R. (2000). Being there: Research through observing and participating. In I. Hay (Ed.), *Qualitative research methods in human geography* (pp. 103–121). Oxford: Oxford University Press.

Kleerekoper, L., Esch, M., & Salcedo, T. (2012). How to make a city climate-proof, addressing the urban heat island effect. *Resources, Conservation and Recycling, 64*, 30–38.

Konijnendijk, C., Ricard, R., Kenney, A., & Randrup, T. (2006). Defining urban forestry—A comparative perspective of North America and Europe. *Urban Forestry & Urban Greening, 4*(3), 93–103.

Kronenberg, J. (2015). Why not to green a city? Institutional barriers to preserving urban ecosystem services. *Ecosystem Services, 12*, 218–227.

Lawrence, A., de Vreese, R., Johnston, M., van den Bosch, C., & Sanesi, G. (2013). Urban forest governance: Towards a framework for comparing approaches. *Urban Forestry & Urban Greening, 12*(4), 464–473.

McDonald, A., Bealey, W., Fowler, D., Dragosits, U., Skiba, U., Smith, et al. (2007). Quantifying the effect of urban tree planting on concentrations and depositions of PM 10 in two UK conurbations. *Atmospheric Environment, 41*(38), 8455–8467.

McKinney, M. (2002). Urbanization, biodiversity, and conservation the impacts of urbanization on native species are poorly studied, but educating a highly urbanized human population about these impacts can greatly improve species conservation in all ecosystems. *BioScience, 52*(10), 883–890.

Mell, I. (2009). Can green infrastructure promote urban sustainability? *Proceedings of the ICE-Engineering Sustainability, 162*(1), 23–34.

Naumann, S., Davis, M., Kaphengst, T., Pieterse, M., & Rayment, M. (2011). Design, implementation and cost elements of Green Infrastructure projects. Ecologic institute and GHK Consulting.

Norton, B., Coutts, A., Livesley, S., Harris, R., Hunter, A., & Williams, N. (2015). Planning for cooler cities: A framework to prioritise green infrastructure to mitigate high temperatures in urban landscapes. *Landscape and Urban Planning, 134*, 127–138.

Nowak, D., & Dwyer, J. (2007). Understanding the benefits and costs of urban forest ecosystems. In J. Kuser (Ed.), *Urban and community forestry in the Northeast* (pp. 25–46). Netherlands: Springer.

Nowak, D., & Heisler, G. (2010). Air quality effects of urban trees and parks. National Recreation and Park Association.

Oliveira, S., Andrade, H., & Vaz, T. (2011). The cooling effect of green spaces as a contribution to the mitigation of urban heat: A case study in Lisbon. *Building and Environment, 46*(11), 2186–2194.

Pinto, M., & Almeida, C. (2013). *FUTURE—The 100,000 trees project in Porto Metropolitan Area: 2011–2013 Report* (p. 22). Porto: Catholic University of Portugal.

Pinto, M., & Silva, M. (2013). RCE Porto: The 100,000 trees project. In Z. Fadeeva & U. Payyappallimana (Eds.), *Traditional knowledge and biodiversity: Learning for conservation and revitalisation of natural and cultural resources*. Yokohama: United Nations University Institute of Advanced Studies.

Power, T. (2005). The supply and demand for natural amenities: An overview of theory and concepts. In G. Green, S. Deller, & D. Marcouiller (Eds.), *Amenities and rural development: Theory, methods and public policy* (pp. 63–77). Northampton, MA: Edward Elgar.

Randrup, T., Konijnendijk, C., Dobbertin, M., & Prüller, R. (2005). The concept of urban forestry in Europe. In C. Konijnendijk, K. Nilsson, T. Randrup, & J. Schipperijn (Eds.), *Urban forests and trees: A reference book* (pp. 9–21). Berlin: Springer Verlag.

Rodríguez-Loinaz, G., Amezaga, I., & Onaindia, M. (2013). Use of native species to improve carbon sequestration and contribute towards solving the environmental problems of the timberlands in Biscay, Northern Spain. *Journal of Environmental Management, 120*, 18–26.

Schmied, A., & Pillmann, W. (2003). Tree protection legislation in European cities. *Urban Forestry and Urban Greening, 2*, 115–124.

Secco, L., Pettenella, D., & Gatto, P. (2011). Forestry governance and collective learning process in Italy: Likelihood or utopia? *Forest Policy and Economics, 13*(2), 104–112.

Secco, L., Re, R., Pettenella, D., & Gatto, P. (2014). Why and how to measure forest governance at local level: A set of indicators. *Forest Policy and Economics, 49*, 57–71.

Soares, A., Rego, F., McPherson, E., Simpson, J., Peper, P., & Xiao, Q. (2011). Benefits and costs of street trees in Lisbon, Portugal. *Urban Forestry & Urban Greening, 10*(2), 69–78.

Strohbach, M., Arnold, E., & Haase, D. (2012). The carbon footprint of urban green space—A life cycle approach. *Landscape and Urban Planning, 104*(2), 220–229.

Tiwary, A., Sinnett, D., Peachey, C., Chalabi, Z., Vardoulakis, S., Fletcher, T., et al. (2009). An integrated tool to assess the role of new planting in PM 10 capture and the human health benefits: A case study in London. *Environmental pollution, 157*(10), 2645–2653.

Tsunetsugu, Y., Lee, J., Park, B., Tyrväinen, L., Kagawa, T., & Miyazaki, Y. (2013). Physiological and psychological effects of viewing urban forest landscapes assessed by multiple measurements. *Landscape and Urban Planning, 113*, 90–93.

Tyrväinen, L., Pauleit, S., Seeland, K., & de Vries, S. (2005). Benefits and uses of urban forests and trees. In C. Konijnendijk, K. Nilsson, T. Randrup & J. Schipperijn (Eds.), *Urban forests and trees: A reference book* (pp. 81–87). Berlin: Springer.

UNU-IAS. (2010). *Five years of regional centres of expertise on ESD* (p. 124). UNU-IAS: Yokohama.

Young, R., & McPherson, E. (2013). Governing metropolitan green infrastructure in the United States. *Landscape and Urban Planning, 109*(1), 67–75.

# Reflexive Research and Education for Sustainable Development with Coastal Fishing Communities in the Azores Islands: A Theatre for Questions

Alison Laurie Neilson and Irina Castro

**Abstract**

Sustainable development (SD) is a controversial concept informed by conflictual narratives which reshape the way we envision the earth, the sea and the stars. Its integration in international policies and national strategy plans for development influences the ways we now know the past, our understanding of the present, and our paths to the future. It influences our lives through policies that regulate daily practices, such as the European Common Fisheries Policy which focuses its strategies for SD in trade and education. However, the problems faced by the ocean require understanding sustainable marine ecosystems through the complex interactions between ecological, social, economic, and political dimensions. Analysing the intersection of those dimensions, while respecting peoples' voices, allowed us to identify how policies and regulations for SD fail, and opened spaces for an emancipatory reflexive research on SD: responsible, accountable and transformative. This approach inevitably raised questions of environmental justice that challenged us to look critically at research and education norms for SD, as well as question how the deficit-model of research is built on the assumption that the failures of SD are due to lack of knowledge. In this chapter we bring together research experience on education and research practices, overlapping our reflexive and educational practices, with the Azores archipelago in Portugal as our background, in order to explore other possibilities. With the help of Augusto Boal's Theatre of the Oppressed, we explore the potential of multi-directional learning via aesthetic practices and action-based research to enable narrative inquiry to engage people in research, and SD policy development that are environmentally just and sustainable.

A.L. Neilson (✉) · I. Castro
Centre for Social Studies, University of Coimbra, Coimbra, Portugal
e-mail: aneilson@ces.uc.pt

 
P. Castro et al. (eds.), *Biodiversity and Education for Sustainable Development*,
World Sustainability Series, DOI 10.1007/978-3-319-32318-3_13

## 1 Introduction

This is an exploration on how research practice fits into the "world" of sustainability and education for sustainability, and how these all connect with policy and living on the sea we call home. While we explore research, education and knowledge practices as they relate to coastal fishing communities in the Azores Islands, Portugal, we invite the reader to consider tough questions about the impacts of scholarly practices and to reflect on the underlying assumptions and economic systems which profess to aim for sustainability. Multiple people have been involved in the research in the Azores of which we refer, but this analytic conversation is between only the principal researcher, Alison Neilson; and Irina Castro, not part of the Azorean projects, but who has been working on science-in-society issues and exploring practices of scientists and scientific controversies.

We started our discussions at the same point, the point of perplexity; we do this in order to spur learning (Dewey 1938). Perplexity is an inability to comprehend something or someone: difficult, yet not a frustrating dead-end, rather an opportunity to imagine what has not happened and the considerations of what is possible (Boal 2008). Being perplexed is sometimes described as an entangled condition, but this organic concept from the Brazilian theatre director, himself a perplexed man, follows the new feminism materialism in the entanglement of matter and meaning (Barad 2007) sensible to the "cuts" produced by the act of observing, and self-conscious that the cuts temporarily enacted by our presence include and exclude processes from our considerations.

For scholars of the new feminist materialism, as well as for practitioners of Theatre of the Oppressed, the goal is not to get into rigorous analysis of people's actions, narratives or representations, but rather to question how the practices of research and theatre can be more responsible and fair, while accepting that their actions of transformation are political. We consider how the metanarratives of knowledge production has colonized peoples and caused great pain, leading us to the idea that refusing some research questions and foci is "a generative, analytic practice" (Tuck and Yang 2014, p. 817) which can turn our gaze back upon the power that allows some people to construct research that begins from implicit ideas that certain other people are ignorant, immoral or savage. Embracing feminist epistemologies and Theatre of the Oppressed practices, we use our perplexity to trouble the reader into questioning the concepts of sustainability and education, as well as research practices. We also encourage researchers to take on the identity of *resident aliens* (Haraway 1997) and seek to be accountable, responsible, and transformative.

We follow the movement of our perplexity as scientists trained in ecology, as educators in environmental and science initiatives, and now as disciplinary hybrids inhabiting the territories of social sciences studying processes of informal education, environmental justice, and fishing communities; and knowledge production and governance modes. We therefore initiate a conversation on research practices and education for sustainability, defending its materialistic character (a philosophical position which acknowledges the way the real world impacts people but also that together they have the ability to change society), in order to help build a path toward a responsible sustainable development (SD). We discuss aspects of the real world and how ideas help create this world in which we are embedded. We do this while being faithful to our previous projects and research teams, to whom we owe our perplexity.

## 2 Fishing for Frameworks and Models

### 2.1 European Fishing Policy—A Beginning

We start with the EU Common Fisheries Policy [CFP] as it is perhaps at the heart of the issues of our interests, and it allows us to explore the internal contradictions of the dominant discourses of sustainable development from the most straight-forward political dimension, its governance. Policies are concrete measures created to rule and change human practices in order to promote a certain outcome, in this case sustainability/sustainable development. But as contested abstract concepts (Connelly 2007), it is in policies such as the CFP that we see what sustainability and sustainable development are in the process of actually creating, which helps in recognizing the contradictions inherent in definitions. The CFP was first established in 1970 to create equal access for all countries of the Council of the European Communities to a free trade area of the oceans, ruled by common principles. Its association with the then embryonic discourse of sustainable development came later, although today the two are inseparable. CFP is presented as a set of common rules for managing European fishing fleets and for conserving fish stocks, making it difficult for the very different practices of artisanal and industrial fishing to be regulated for both purposes, an aspect of particular interest for the Azores (Neilson et al. 2012b).

### 2.2 Difficult Match: CFP and Sustainable Development

Several authors have critiqued the dual pursuit of competition and sustainability of the ocean (Khalilian et al. 2010), while identifying the CFP, itself, as part of the problem preventing sustainability (Mansfield 2011). Approaches to sustainability based on the assumption that we need better science and technological solutions have not prevented overfishing. Nor have they been successful in providing equal access for fishing, which brings us to look for contradictions in the economic system of production in which SD is being framed. Critics have explored how the solutions presented by SD are in fact causing the problems and creating specific patterns of economic and environmental crisis by aggravating social inequality (Foster 2011; O'Connor 1998).

Although dominate discourses of SD are periodically redefined, a critique of the ways of production in capitalist economies is still mostly absent from those debates, which continues to have consequences to the processes of policy decision-making.

Perhaps this is not surprising since the concepts of diversity and biodiversity have also been constructed within the capitalist globalized economic system, resulting in a misleading understanding of the resilience of natural resources to human practices of exploitation (Jasanoff and Martello 2004; Mansfield 2011). The issue lies in the way scientists perceive the knowledge we generate as apolitical. Problems such as overfishing are not framed as a problem of globalization or as a contradiction of the capitalist system, but as inevitable tragedies caused by inherent human nature in relations with property-based conflicts, that will be overcome throughout the mechanisms of market, or common trade rules (Harding 1968; Mansfield 2011). This way of thinking affects sustainable development for oceans, forests (see Serra et al. in this book), and agriculture.

Also absent is recognition that the participatory aspects of CFP appear largely superficial (Neilson et al. 2012a) even though several models of policy making have been adopted. Research demonstrates that public participation (Brewer 2013) can create beneficial policies that elicit a high degree of compliance. However, within policy-making structures there exists a tension between participatory democracy and expert authority that has remained fundamentally unchanged for the last four decades. This involves who is included in the decision-making process, whose knowledge is valued, and the reality that participation does not always mean an effective redistribution of power (Ferreira et al. 2012).

## 2.3 Economic Systems and Sustainable Development

Defined in 1987 by the Brundtland Report as the development that meets the needs of the present without compromising the ability of future generations to meet their needs (WCED 1987), SD has become the green discursive umbrella in a globalized capitalist economy that seeks to construct a common culture and a common education. Masquerading as an apolitical concept SD serves both liberal economies (e.g., USA and UK) which are based on competitive markets, and coordinated market economies (e.g., Germany and Japan) that base their economic relations in non-market forms, by creating new fictitious commodities (Hall and Soskice 2001; Polanyi 2000), such as the Carbon Market (Brockington 2011), and patents (Jessop 2007) which promote new forms of capital accumulation and subsequently deepen social, gender and race inequalities. SD discourses under the lexicon of management proliferate worldwide through, and into institutions, organizations and decision-making procedures, becoming a moral connector amidst conflicting practices and contradictory narratives, while promoting management rationalities and market-based green governance, requiring human performances via trade and education.

The World Trade Organization [WTO] claims that international trade advances sustainable development (2011) since it is easier for everyone to access environmental goods, services and technologies and therefore reduce environmental contamination and energy losses, while stimulating the growth and basic income levels in developing nations. However, this positivist perspective regarding the role of trade in addressing environmental problems has been contested (Martinez-Alier 1993; O'Connor 1994). Furthermore, Ropke (1994) has demonstrated that international trade tends to stimulate economic growth at the costs of environmental deterioration; hence international trade is part of the problem and not of the solution.

## 2.4 Education and Sustainable Development

Sustainable development redefines education from including critical thinking to a narrower instrumental (and political) mandate to create social change (Jickling 1992). Instead of understanding and respecting different perspectives; habits and decisions are considered to be in need of becoming "more informed", privileging decision-making processes based on science and technology. Since 1987, initiatives of "Science Engaging Society" have become a priority under the compass of education for SD. As world-wide initiatives, differences between societies are expected, however, success has been limited by asymmetric notions of what people need in order to address their environmental problems (Sterling 2001), as the initiatives are being strongly influenced by the deficit-model (Dierkes 2000) that sees people to be in need of education, instead of promoting the integration of their knowledge into the processes of decision-making. *Who educated us?* (Marx 1845) is a vital question based on respecting that humans change, as well as learn from our circumstances, hence the educator herself must also be educated. One mainstream argument for education is that overfishing is a problem that impacts all people who depend directly on the ocean, but this doesn't take into account the knowledges that already exist locally. Research in historical practices suggests that humans are clearly an integral part of ecosystems (Berkes 2004) having co-evolved with interdependent and affective connections between humans and non-human nature; "depicting 'humans' as the natural enemies of wilderness is thus irrespective of the millions who have been and continue to be victims of enclosures and 'improvements', a never-ending historical process that four decades of global neo-liberal politics and trade have reinvigorated (Barca 2014).

We regularly see fisheries education with goals to "train fisheries professionals with the technical knowledge and practical skills" and to "cultivate more environmentally knowledgeable citizens" (Crook and Zint 1998), thereby telling us that the dominant explanation of problems of fishing are based on apolitical understandings, and the dominant solutions rely on the neutrality of knowledge and technologies, which are transmitted one-way from expert to ignorant fisher. This occurs in the context where the World Bank and the International Monetary Fund pressured nations to repay loans to aid their "development" (Mundy 2005), which has penalized innovative education (and environmental education cf. Bélanger 2003), and diverse models to demand compliance with a banking (of knowledge) model of education (Freire 1970). Orellana and Fauteux (2000) suggest that beginning in the mid-1980s economic concerns pushed environmental education into education for sustainable development and education for a sustainable future.

These discourses also divide researchers from society; and the role of researcher needs to be critiqued in their relationship to the commodification of scientific knowledge. The growing logic of new management and market-based knowledge production are changing the dynamics of universities and research centres to become increasingly vulnerable, and thus even more subordinated to criteria of austerity for economic recovery and neoliberal business management models (Huff 2012). Current scientific systems value knowledge that is useful to the production and distribution of other marketable commodities (Wotherspoon 2012) over those that promote sustainable just transitions (Barca 2015). Goals for engaging with and meeting the local needs of people are increasingly being relegated in research grant proposals, if accepted for inclusion at all, to secondary activity within minor work packages.

## 3 Aesthetic Research = Theory in Practice?

We now engage in a conversation to explore our practices to uncover the ways we might challenge these issues. We are in sync with the diverse economies approach of Gibson-Graham (2008), namely a "performative ontological project" designed not only to co-produce knowledge, but also to actively contribute to "novel economic performances" and to possible alternative worlds, "in which we enact and construct rather than resist (or succumb to) economic realities" (p. 615). We remind ourselves that we too, are embedded and grew up within these same systems we critique. Metrics of impact and university performance regulations pressure us to gloss over cracks to present ourselves as "good" researchers who serve up crack-free knowledge. While discussing how to reveal the cracks, R. Spannring suggests that we think about the multidimensionality and interrelatedness of the research process, asking what the process builds on (other scholars; our experiences, concepts, and beliefs; trust in colleagues to share ideas; openness to new ideas), what does it exclude (other voices, languages, concepts, life worlds) and how can the difference be productive and generative for our academic and activist work (personal communication, 24th July 2015).

**Alison**: The research that I have undertaken in the Azores was not initially focused on the fishing communities, their sustainable development or the Common Fisheries Policy. It is important to acknowledge the processes from the initial narrative inquiry to the engagement with fishing communities. In fact, all researchers, at least those dependent on external funding, undertake similar journeys in their work: an essential part of the puzzle when exploring the construction of knowledge, the connection between research, education and sustainable development and the political nature of all science scholarship, natural or social. Research publications rarely expose these details, giving the erroneous idea that question, sample sizes, and methods described were always what the researchers had planned and had been under their control from the start. This erasure of history removes the researcher and all the ways that her social, cultural, political, physical, emotional, and spiritual characteristics complicate the primacy of rational decision-making she has been taught to seek.

**Irina**: Your reflection on how the rules that govern our activity remove our organic dimensions from our daily practice of inquiry makes me think about time and space restraints. The time we are given to perform research affects the sort of decisions we make, and that time is built by epistemic reconfigurations that address the requirements of social life reproduction (Harvey 2000). Time for research isn't determined by the issues related to the inquiry, but rather with notions of time and space build in social class power relations, colonial visions and gender bias. This then pushes us to think that our knowledge and voice are more valuable than others, and that our role is to transmit that knowledge to other people.

## 3.1 A Joker in the Middle of the Ocean

**Alison**: Being in the middle of the deep ocean, the Azores is home to a myriad of resident and migrating whales supporting a great deal of cetacean research, some of it connected to the highly visible whale watching tourism, and there is a poignant history of whaling captured in literature and patrimony. Although I was provoked by my sense that the former whalers were not being remembered in a fair way, the dominant narrative seeming to be: '*our grandfathers were brave but barbarian in killing whales, thankfully we are more environmentally-friendly*'; my focus of research was to get at the ways that different people think about the ocean. I had expected there to be differences and I was particularly aware that even my framing of the research already exposed an important difference. The research project used the term Cetaceans, meaning all whales and dolphins, while the former whalers were clear in telling me that they hunted Cachalot, but not other whales. A contradiction exists to the dominant narrative as well. The intense activity around whale-watching has led some researchers to suggest that the impact on whales was far less during the period of whale hunting than during the current tourism activity (Neves 2010).

The research brought out ways that concepts such as respect were conceptualized and operationalized differently between fishers and their families, and tourist and tour operators. This part of the work also revealed strong and different emotional reactions to the photo of the tail of a whale out of the water. Former whalers trembled and sometimes shed tears as these images evoked their fear of being in the midst of a hunt on a wooden boat often smaller than the animal and the memories of brothers hurt, sometimes killed when the tail of the thrashing animal crashed their boat. For the tourists, this same photo brought smiles which led to talking about the beauty of the animal and the peaceful feelings of being on the water near them.

While interviewing and speaking casually with tourists on the street, I was shocked to notice that Azorean people, especially the fishers, were absent from the conversations (Neilson et al. 2014). Something, perhaps my fluctuating process of identifying myself with the tourists while actively resisting that identification for my preferred new identity as an Azorean, left me open to be receptive to change my underlying thinking (cf. Beausoleil 2014). The open-ended research methods that I initially chose to try to reduce the effect of my expectations were successful in that they provided an opportunity for me to hear things that provoked deeper questions then I had been asking which has drawn me and the world of research into the research problem.

**Irina**: Drowning into the research problem is the first step for a Joker in Forum Theatre sessions. This is the character of the Theatre of the Oppressed that I embody when I adopt the scientific ethos of reflexive-perplexity of accountability, responsibility, and transformation. Like the joker in a forum, I search to open the space for the possibility to re-engage in the debate of sustainable development and promote a reflexive understanding of socio-ecological problems, by bringing into my research (or forum) not just the intersection that produced social practices, but also those that bring the material world into being. The end is discovery not abstention (Boal 2006). This is the Joker's neutrality, the responsible act resulting from having made a choice.

## 3.2 Reflexive Practice

**Alison**: My education practice is based on reflexivity (Schön 1983), and the belief in the inevitability of failure when we try to deal with our own biases by simply imaging other standpoints (Stone-Mediatore 2003). As a non-native "settler" Canadian, I have been trying to listen closely to indigenous scholarship in particular. Smith (1999) who writes about the absence of indigenous approaches to research, points to the failure of critical theorists to recognize and change their own oppressive practices. While, there were never indigenous peoples on these islands, the call to transform practices which continue colonization of peoples is pertinent to the Azores.

I had begun wondering about the effects of the invisibility of the fishers on other aspects of island life. At this same time, I was beginning to interact with fishing communities beyond collecting data. For instance, I shared the stories from my research interviews via an art workshop for youth done in partnership with a woman's fisher association; we also organized a display of stories to show during a festival, and together we presented this same material for use with school curriculum, as part of a qualification upgrading workshop for teachers across the islands. Soon afterwards, I was invited to share my research at the Congress of Azorean Fishers run by fishing cooperatives and associations and later to a government and scientist forum on marine policy. It was at this point that I observed how the knowledge of fishers was undervalued and perspectives ignored. My developing collaborations with community leaders and activists in fishing communities across other islands quickly led to us to organise a 4-day forum for fishers and government and scientists (Neilson et al. 2012a). Shortly afterwards, changes to the Common Fisheries Policy were being considered, so in collaboration with researchers and others from the forum, we wrote and submitted a book about the

CFP and artisanal fishing communities for the European Commission in order to influence the outcome of the reviews (Neilson et al. 2012b).

**Irina**: A reflexive practice is also what we are doing here by letting ourselves into one another's practice. However, a reflexive sociology poses many challenges under the modern scientific model of governance. It includes a tension between being reflexive and the hostile environment of publish or perish paradigm of science, the emerging bureaucratization of scientific activities and the contradictory policy of patents. If it's true that narrative research has already made important achievements by turning visible the power relations entangled in scientific practices, I wonder if it still lacks ways to treat those voices seriously and promote spaces where they would in fact be heard.

**Alison**: These processes are also about me being quiet and leaving space, as well as me facilitating the creation of space so that other voices can be heard. But this is a worry for me as I consider having the power to create unsafe situations for some people as these can limit "our ability to access or remain reflective about our own knowledge, remain receptive to the unfamiliar, and create new responses and behaviours" (Beausoleil 2014, p. 22). In particular, researchers who are in junior or more precarious positions of employment may be put in far greater risk than I, in challenging some research paradigms. Questions of power come into play when I think about the way my reflexive actions may reflect on the other researchers involved in the Azorean research who may hold different perspectives on what is responsible research practice.

## 4 Research as Aesthetic Place

One of the aims of Theatre of the Oppressed (TO) is to use collective inquiry to find/define the common question to a certain problem. The other is to transform the reality that the question addresses. In this final section, we invite the reader to adopt the critical and aesthetic approach of a TO practitioner, and to abandon all certainties you may hold about any "scientific methods" in order to adopt the question that arises from multi-stories presented in people's narratives. In this way we are able to find the common questions that are responsible: questions that are not exhausted in their response but which shared the perplexity of all imagined answers. In this sense we will not close with certainties but open it into new possible futures by sharing our narratives.

**Irina**: It is surprising how often education is still offered as the universal solvent for problems (Campbell 2006); the scant evidence that suggests the global expansion of formal schooling as a panacea for unemployment is "one of the most romantic tales of the century" (Fagerlind and Saha 1983). In capitalist education the goal is to reproduce the labour force through education and subject people to political state ideology (Althusser 1971); while in education from a reflexive socio-ecology, the goal follows Boal's (2008) construction of an aesthetic space. Based on both Freire's Pedagogy of the Oppressed (1970) and Rancière (1991) postulate of equal intelligence, Boal's aesthetic space embraces solidarity. For a practitioner of the Theatre of the Oppressed the aesthetic space is a place where the separation between actors and spectators collapses, and where the time barrier is crossed by the coincidence of this being experienced, simultaneously in the memory of the past, and in the imagination of the future. There is no pre-existing space privileged to promote participation, but rather the possibility of an area that becomes by overlapping reality constructions with imagination produced by the desire of knowledge and the pleasure of knowing. The aesthetics of the oppressed is assumed as the place where the artistic method allows acting on the future and restoring the idea of democracy (Boal 2006). It is the exploitation of sensitive thought and symbolic thinking, understanding phenomena and the revealing of hidden forces behind a society of the spectacle (Debord 2000 [1970]) and oppression. In science, as in theatre and in politics, we are not passive beings, but active agents of transformation.

**Alison**: We live in a world of constant motion and are always in flux; seemingly with limitless potential of the imagination, yet always constrained by the limits in which our bodies will live and die along with the cycles of birth and death of all life on earth. It seems to me that we have entered some sort of parallel universe of opposites, where time and space are limitless and growing, while our imaginations are shackled in a small cell where capitalism is the only governing system, rational individualism is the only way to be and private property is the only place to make our lives. What if we listen to the narrations of the people who have experienced the phenomenon we study as if they are in a theatre workshopping the ways to express the overlapping and sometimes conflicting complexities of the multiple perspectives within the same role? What if we did that while we also workshopped the ways to express the overlapping and sometimes conflicting complexities of the multiple perspectives we can hold in the researcher role?

## 5 An Actor/Fisher in Boal's Theatre

*I could feel waves of great exhilaration* and *paralyzing fear as my small boat rides the breaking waves going out to sea. These few minutes may feel like an eternity, yet, I will only have a few hours to set my fishing lines and return to shore before night falls. If I decide to travel the 50 km on land to Ponta Delgada from my home of Ribeira Quente, I must meet with my neighbours for a ride as the cost of fuel is high to drive alone. We know the route, but the dense fog slows our progress as we watch for fallen rocks and washed out parts of the mountain roads. When we reach the meeting, we listen while the limnologist speaks for 40 min about coastal erosion harming fish habitat, then for another 40 min, another biologist talks about the status of Mero and that he would like spear-fishers to be able to take them again. Before the end, someone from the navy tells us about the limitations of the fleet for patrolling the waters to keep out foreign vessels who are illegally fishing our waters. This somewhat shorter presentation is buttressed with official words thanking the President of the Freguesia for organizing this meeting on such short notice. These public consultations are to develop the National Strategy for the Sea.*

*The Regional Director tells us that there are only a few minutes to hear from the audience before he must leave to catch the last flight to Horta. Epá! Maria was quite angry that she couldn't come with me as the baby is still sick and now she will be doubly annoyed that I didn't ask about the problem of catching Goraz which are too small to take to the Lota* (constructed from research 2008–2014).

**Alison**: It is in this world that I seek to be a researcher: someone who listens and watches and reflects on my relationship to this observing. I am conscious of limits, having the dual dis/advantage of language differences to be continually reminded of my observing and that I wear invisible blinders. Not only do I not see everything when I observe, I cannot see what I am not looking at and alone, I do not know of what I cannot see.

**Irina**: All research should be collaborative and I suggest a network metaphor to (re)conceptualize the researcher, participants and other entities within a larger and more holistic perspective of research. Otherwise, it can feel like a "world of 'too muchness', information overload, and a surfeit of meaning" (Hoskin and Stoltz 2005, p. 98). There are a limited number of pages in any publication, a limited number of minutes in any presentation, and a limited number of ideas I can consider at any one moment. How much space is left for the lived experiences of the people involved? Compartmentalizing knowledge-making from knowledge and knowers, using the techniques of empiricism (Richardson 1990), is a way in which some perspectives get excluded from policy and education.

## 5.1 An Actor/Researcher in Boal's Theatre—Scene 1

*The first words I hear in Portuguese are little more than a new music: compelling but mysterious in my lack of comprehension. My eyes begin to uncover stories, relationships, and feelings. I see welcome on faces as I walk on the cobble stones. I notice a closed hunch, apparently defending against the wind, a shiver, perhaps in response to the humid air. But my eyes too are sometimes confused and mislead by memories of other faces, other storied landscapes. There are so many clues to find, so many pieces of stories—some screaming at the top of their lungs, others whispering furtively* (research 2008—2010).

## 5.2 An Actor/Researcher in Boal's Theatre—Scene 2

*I have entered another world, legally, but which feels part invasion, part trap, as all eyes seem to question if me and my young looking assistant are lost tourists. I sit on the cold chair trying to take in every detail of the fishing Lota, where the incoming boats sell their catch. The early morning calm of Ponta Delgada infiltrates this auction where the only sounds are those of the conveyor belt, the faint clink of the bidding devices and the occasional whisper some bidders make into cell phones, checking and clicking. Beautiful colours of glistening scales on delicate creatures and strong firm bodies of others are easy to imagine alive in the nearby water only hours before. I watch as tray after tray of fish roll past the camera as numbers swirl on the monitor: name of boat, name of fish, kilograms, price of sale. My assistant and I look to each other as we simultaneously realize that the numbers are going down, not up. I will later learn that this is normal for all fish auctions, unlike auctions for cattle, milk, gold and other valued and rare items which go up in price during the sale process. But at this moment, I feel sick and in the place of the fish in the trays of ice, I see the faces of fishers who have told me of their struggles and I see the boats that I know have been recently repossessed* (research 2015).

## 6 Conclusions

**Alison**: To learn from people's narratives, I had to combine a framework of concepts from political economy and ecology, as well as historical changes, and analyse such narratives in the intersection of their personal history, history and culture-socio-economic context. This is the basis of a reflexive sociology, a practice of meaning creation that values the context, and its constant flow of transformations, of the evolving society. This wasn't however the only transformation that confronted me. Adopting a reflexive sociology has also changed me. Once you get into a structure of inquiry that recaptures you into the research, it is impossible to go back to the way you were before. Your only choice is to either ignore yourself as an agent of knowledge production and all the heterogeneous constructions in the way (Taylor 2010), or start to take yourself seriously. Taking yourself seriously is not to put yourself in the centre of your research, in self-adulation, but rather accept the responsibilities of your practices as researcher, by accepting your practice of meaning creation as a political act of transformation.

**Irina**: In being reflexive of complicity in monocultures of ideas and the problems we seek to address, we acknowledge "all our relations" by honouring all the people who have come before us as well as the other living beings with whom we share this planet (Kulnieks et al. 2013). We will remember the current and past practices which take on issues related to motivation rooted in social and cultural patterns, as well as politics of knowledge with complex histories and inequities (Sund and Öhman 2014). When we wear the title and prestige of scholar or "expert", our reflexive call to seek further learning outside of our fields, universities and ways of knowing, we create transformative possibilities beyond ourselves. We need not feel increased stress of deadlines and production by the realization that we need to engage with extensive fields of thoughts and disciplines in order to be rigorous in developing a responsible sustainable research practice. This is a step-by-step process, but only if we give up fighting for the lead, and instead follow those who perplex us, will we learn how to enjoy the dance.

**Acknowledgements** The authors gratefully acknowledge the financial support of the project "Education for active citizenship: Images of participation in marine policy" funded by the Portuguese Foundation for Science and Technology FCT—IF/01084/2013/CP1164/CT0001 and

"EDUMAR Perspectives about the sea and sea life: Cetaceans and tourism in the Azores, Portugal and Newfoundland, Canada" funded by the Azorean Regional Director of Science and Technology DRCT/FRCT—M2.1.2/F/007/2008.

## References

Althusser, L. (1971). *Lenin and philosophy and other essays*. New York & London: New Monthly Review.

Barad, K. (2007). *Meeting the universe halfway: Quantum physics and the entanglement of matter and meaning*. London: Duke University Press.

Barca, S. (2014). *Discussant for Wilderness Act forum*. An environmental history online forum. http://environmentalhistory.net/wilderness-act-forum/, October 18, 2014.

Barca, S. (2015). Greening the job: Trade unions, climate change, and the political ecology of labour. In Raymond Bryant (Ed.), *The international handbook of political ecology*. Cheltenham: Edward Elgar Publishing.

Beausoleil, E. (2014). The politics, science, and art of receptivity. *Ethics & Global Politics, 7*(1), 19–40.

Bélanger, P. (2003). Learning environments and environmental education. In L. H. Hill & D. E. Clover (Eds.), *Environmental adult education. Ecological learning, theory, and practice for socioenvironmental change* (pp. 79–88). San Francisco, CA: Jossey-Bass.

Berkes, F. (2004). Rethinking community-based conservation. *Conservation Biology, 18*(3), 621–630.

Boal, A. (2006). *The aesthetics of the oppressed* (A. Jackson, Trans.). London & New York: Routledge.

Boal, A. (2008 [1995]). *The rainbow of desire—the Boal method of theatre and therapy* (A. Jackson, Trans.). London & New York: Routledge.

Brewer, J. F. (2013). From experiential knowledge to public participation: Social learning at the community fisheries action roundtable. *Environmental Management, 52*(2), 321–334.

Brockington, D. (2011). Ecosystem services and fictitious commodities. *Environmental Conservation, 38*(4), 367–369.

Campbell, D. E. (2006). *What is education's impact on civic and social engagement? Measuring the Effects of Education On Health And Civic Engagement: Proceedings of the Copenhagen Symposium*, March 23–24, 2006.

Connelly, S. (2007). Mapping sustainable development as a contested concept. *Local Environment, 12*(3), 259–278.

Crook, A., & Zint, M. (1998). *Guide to fisheries education resources for grades K-12*. Maryland: American Fisheries Society.

Debord, G. (2000 [1970]). *Society of the spectacle*. Detroit: Black & Red.

Dewey, J. (1938). *Experience and education*. New York, NY: Collier Books.

Dierkes, M. (2000). *Between understanding and trust: The public, science and technology*. London: Routledge.

Fagerlind, I., & Saha, L. J. (1983). *Education and national development: A comparative perspective*. Oxford: Pergamon Press.

Ferreira, P. D., Coimbra, J. L., & Menezes, I. (2012). Diversity within diversity—Exploring connections between community participation and citizenship. *Journal of Social Science Education, 11*(3), 120–134.

Foster, J. B. (2011). *A ecologia de Marx: materialismo e natureza*. Rio de Janeiro: Civilização Brasileira.

Freire, P. (1970). *Pedagogy of the oppressed*. New York & London: Continuum.

Gibson-Graham, J. K. (2008). Diverse economies: Performative practices for 'other worlds'. *Progress in Human Geography, 32*(5), 613–632.

Hall, P. A., & Soskice, D. (2001). An introduction to varieties of capitalism. In P. Hall & D. Soskice (Eds.), *Varieties of capitalism: The institutional foundation of comparative advantage*. New York: Oxford University Press.

Haraway, D. (1997). *Modest_Witness @ Second_Millennium. FemanleMan© _ Meets: OncoMouseTM: Feminism and Technoscience*. New York & London: Routledge.

Harding, G. (1968). The tragedy of the commons. *Science, 162*, 1243–1248.

Harvey, D. (2000). *Condição pós-moderna: uma pesquisa sobre as origens da mudança cultural*. São Paulo: Edições Loyla.

Hoskins, M., & Stoltz, J. (2005). Fear of offending: Disclosing researcher discomfort when engaging in analysis. *Qualitative Research, 5*(1), 95–111.

Huff, T. E. (2012). The big shift: Science and the universities in crisis. In J. Dzisah & H. Etzkowitz (Eds.), *The age of knowledge. The dynamics of universities, knowledge and society*. Chicago: Haymarket Books.

Jasanoff, S., & Martello, M. L. (2004). *Earthly politics: Local and global in environmental governance*. Cambridge: MIT Press.

Jessop, B. (2007). Knowledge as a fictitious commodity: Insights and limits of a Polanyian analysis. In A. Buğra & K. Ağartan (Eds.), *Reading Karl Polanyi for the 21st century. Market economy as a political project*. Basingstoke: Palgrave.

Jickling, B. (1992). Why I don't want my children to be educated for sustainable development. *The Journal of Environmental Education, 23*(4), 5–8.

Khalilian, S., Froese, R., Proelss, A., & Requate, T. (2010). Designed to failure: A critique of the common fisheries policy of the European Union. *Marine Policy, 34*(6), 1178–1182.

Kulnieks, A., Longboat, D. R., & Young, K. (2013). *Contemporary studies in environmental and indigenous pedagogies. A curricula of stories and place*. Rotterdam: Sense Publishers.

Mansfield, B. (2011). "Modern" industrial fisheries and the crisis of overfishing. In R. Peet, P. Robbins, & M.J. Watts (Eds.), *Global political ecology* (pp. 84–99). New York: Routledge.

Martinez-Alier, J. (1993). Distribution obstacles to international environmental policy: The failures at Rio and prospects after Rio. *Environmental Values, 2*(2), 97–124.

Marx, K. (1845). *Theses on Feuerbach*. Retrived October 15, 2015 https://www.marxists.org/archive/marx/works/1845/theses/theses.htm

Mundy, K. (2005). *Discourses of development: The view from education*. Presentation at Deconstructing disciplinary discourses of development III—Centre for International Studies, Toronto, March 11, 2005.

Neilson, A. L., Bulhão Pato, C., & Sousa, L. (2012a). A short reflection on research and fishing cultures performing knowledge together. *Revista Maria Scientia., 2012*(2), 73–82.

Neilson, A. L., Cardwell, E., & Bulhão Pato, C. (2012b). Coastal fisheries in the Azores, Portugal —A question of sovereignty, sustainability and space. In K. Schriewer & T. Højrup (Eds.), *European fisheries at a tipping-point* (pp. 465–505). Murcia, Es: Cátedra Jean Monnet Universidade de Murcia.

Neilson, A. L., Gabriel, R., Arroz, A. M., & Mendonça, E. (2014). Perspectives about the sea in the Azores: Respecting narratives that sustain inshore fishing communities. In J. Urquhart, T. Acott, & M. Zhao (Eds.), *Social issues in sustainable marine fisheries management* (pp. 319–338). Dordrecht, NL: Springer.

Neves, K. (2010). Cashing in on cetourism: A critical ecological engagement with dominant E-NGO discourses on whaling, cetacean conservation, and whale watching. *Antipode, 42*(3), 719–741.

O'Connor, M. (1994). *Is capitalism sustainable? A political economy and the politics of ecology*. New York: Guildford Press.

O'Connor, J. (1998). *Natural causes: Essays in ecological marxism*. New York: Guildford Press.

Orellana, I., & Fauteux, S. (2000). Environmental education: Tracing the high points of its history. In A. Jarnet, B. Jickling, L. Sauvé, A. Wals, & P. Clarkin (Eds.), *A colloquium on the future of environmental education in a postmodern world?* (pp. 2–12). Whitehorse, Yukon: Yukon College.

Polanyi, K. (2000 [1944]). *A grande transformação: As Origens da nossa Época* (F. Wrobel, Trans., 2nd ed.). São Paulo: Editora Campus.

Rancière, J. (1991). *The ignorant schoolmaster: five lessons in intellectual emancipation*. California: Stanford University Press.

Richardson, L. (1990). Narrative and sociology. *Journal of Contemporary Ethnography, 19*(1), 116–135.

Ropke, I. (1994). Trade development and sustainability: A critical assessment of the "free trade dogma". *Ecological Economics, 9*(1), 13–22.

Schön, D. (1983). *The reflective practitioner: How professionals think in action*. London, UK: Temple Smith.

Sterling, S. (2001). *Sustainable education: Re-visioning learning and change*. Cambridge: UIT Cambridge Ltd.

Stone-Mediatore, S. (2003). *Reading across border: Storytelling and knowledges of resistance*. New York, NY: Palgrave.

Sund, L., & Öhman, J. (2014). On the need to repoliticise environmental and sustainability education: Rethinking the postpolitical consensus. *Environmental Education Research, 20*(5), 639–659.

Taylor, P. J. (2010). *Unruly complexity: Ecology, interpretation, engagement*. Chicago & London: University of Chicago Press.

Tuck, E., & Yang, K. W. (2014). Unbecoming claims: Pedagogies of refusal in qualitative research. *Qualitative Inquiry, 20*(6), 811–818.

Tuhiwai Smith, L. (1999). *Decolonizing methodologies: Research and indigenous peoples*. New York, NY: Zed Books.

WCED, World Commission on Environment and Development. (1987). *Our common future*. Oxford: Oxford University Press.

World Trade Organization. (2011). *Harnessing trade for sustainable development and a green economy*. Geneva, Switzerland: World Trade Organization. Retrieved 01 November, 2015 www.wto.org/english/res_e/publications_e/brochure_rio_20_e.pdf

Wotherspoon, T. (2012). Triple helix or triple jeopardy? Universities and the social relations of knowledge. In J. Dzisah & H. Etzkowitz (Eds.), *The age of knowledge. The dynamics of universities, knowledge and society*. Chicago: Haymarket Books.

# Public Policies and Education for Biodiversity: Brazilian Challenges in a New Global Context

Thiago Lima Klautau de Araújo

**Abstract**

This chapter aims to analyze Brazilian education and public policies for environment and biodiversity through three main points of view: (I) legal, by collecting and reviewing laws and other types of norms; (II) organizational and administrative, by considering the regulation of the subject through the creation of agencies, decisions of investments and planning to operate the public environmental policy, through an analysis using Game Theory and Marginalism; and (III) participation by the population, analyzing their contributions to the environmental field. This review allows us to understand the possibilities of environmental education (formal and informal) in the Brazilian context and the challenges that require a transformation of the present paradigm. To illustrate a good example of an integrated environmental education system, the Escola Bosque was chosen, as an internationally awarded initiative for its positive impact on education for biodiversity and the environment. Finally, new trends and some possible changes in public policies for environment and environmental education are suggested.

**Keywords**

Environmental law · Education for biodiversity · Public policies · Game theory · Brazil

T.L. Klautau de Araújo (✉)
Faculty of Economics and Centre for Social Studies,
University of Coimbra, Coimbra, Portugal
e-mail: thiagoklautau@gmail.com

 

P. Castro et al. (eds.), *Biodiversity and Education for Sustainable Development*,
World Sustainability Series, DOI 10.1007/978-3-319-32318-3_14

## 1 Introduction

With a significant number of ecosystems, Brazil is considered to be the richest country in terms of biodiversity, with 25 % of all species existing on the planet (Kageyama 2009). The Amazon, Pantanal, and Atlantic Forest are three examples among many others that contain hundreds of thousands of animal and plant species. These biomes correspond to 49.29, 1.76, and 13.06 % of the Brazilian territory, respectively (IBGE 2004). For that reason, ecologists from all over the world try to study Brazil's diversity before deforestation takes over and eliminates an incalculable number of those species and others still undiscovered (Fig. 1).

Pressures from the international community and Brazilian society forced the Brazilian State to change environmental laws and include this subject in the new Constitution. However, despite the rhetorical advances in those laws, very few—or none—have become reality. For that reason, Brazilian environmental and

**Fig. 1** Brazilian biomes (Instituto Brasileiro de Geografia e Estatística 2004)

biodiversity policy has failed and many doubts remain about the ways Brazil is dealing with environmental dilemmas and challenges, as will be seen in this chapter. Deforestation, for example, is still growing and its consequences are devastating not only locally, but also globally because it affects living ecosystems on the planet.

After the "Plano Real" (creation of a new currency and other public policies to deal with inflation, in 1994) investments in the environmental area increased, but the results have not lived up to expectations. Currently, with its economic and political crisis, Brazil is facing problems with its public budget, which lead to financial cuts, including in environmental policies. At this moment, there is thus a huge degree of uncertainty about the future.

This chapter provides an analysis based on Brazil's environmental law system and its public policies in order to establish a general overview about what has been done until the present, also identifying new trends for the future. Although the focus is on environmental and biodiversity education, an effort will be made to understand the functioning of that system as one of the decisive dimensions of public policies.

In this context, this chapter is organized into five points. The first performs an analysis of the current legal system and its peculiarities, in order to identify the changes (or lack of changes) in paradigms. The second questions the efficiency of environmental public policies in dealing with Brazilian environmental challenges, particularly deforestation and environmental degradation. The third analyzes the interactions between environmental agencies and their efficiency using the Game Theory and Marginalism. The fourth point is a reflection about environmental education, its role and the means for implementing it. Finally, the last one will highlight possible ways for adjusting those policies in the near future.

## 2 Brazilian Environmental Law: A New Perception (?)

The new environmental context that has emerged all over the world in the last four decades, especially concerning climate change and biodiversity preservation, has compelled the Brazilian legal system to adapt itself to the new trends. Law 6938/81 and the new Brazilian Constitution (CF/88), enacted on October 5th, 1988, brought a new perspective to environmental law and to biodiversity. Fauna, flora, and ecosystems were raised to another level of protection, and from that moment on they were considered as important to all society and as a common right. Before these two legal milestones, nature was protected in most cases merely as a thing, as property of the government. Thus, the protection goal was to save economic interests and not to conserve nature (Chiuvite 2010). Especially with the CF/88, popular participation in the process of drafting public policies, laws, surveillance against pollution (of all kinds), and deforestation began to be taken into account. From a legal point of view and according to CF/88, art. 225, society, along with the government, began to be holder of such rights and duties. This constitutional article, by the way, is the main environmental law reference in the Constitution. However, its assumptions were conditioned by the need to create supplementary enabling regulations, and that is one of the biggest problems with the Brazilian legal system. The Constitution brought a significant innovation to environmental preservation by introducing the need for new relations between governments (Federal, States, and Cities), society, and nature. Even with this new approach, constitutional provisions could not be applied instantaneously. Just to mention the case of this article (CF/88, art. 225), its complementary regulation came almost twelve years later, with law 9985/2000. A few years later came MP 2186-16/2001 (substituted by law 13123/2015) and law 11105/2005. Those norms represent the main regulatory system for the environmental law provided in the Constitution. As it can be seen, there was a vacuum between the first steps and the possibility of enforcing constitutional rules. Nevertheless, deforestation and environmental degradation have not been halted during all those years since the new Constitution, and even after this complex system was implemented, the results in this area reveal that biodiversity in Brazil is still highly endangered.

Despite all the legislative efforts, the higher number of laws has not been translated into a more eco-friendly situation in the Brazilian dynamic. As said at the beginning of this point, since the 1980s the law system has brought a new and different perception about the environment. But the complexity of the instruments for conserving fauna, flora, ecosystems, and natural resources is so high that the results are totally the opposite of the original law's purposes. Participation by the community is limited because although citizens can contribute to building a better way to manage nature in their home areas, in most cases, they lack the technical knowledge for interacting with the government or for taking legal or out-of-court action (Klautau de Araújo 2014).

It is important to highlight the fact that the regions in Brazil that own the largest share of biodiversity are among the poorest and most disconnected from large urban areas. In those places, populations suffer from lack of schools, health centers, electricity, roads, piped water, sewerage, sanitation and also internet, television, phones and cell phones. It is not possible that environmental preservation will be a priority for populations in this context where basic needs are still not satisfied.

More than the government's desire to be modern, laws must be adjusted to the social and economic situation, in order to have the needed connection with the "real" world, not to mention the necessity of being feasible. Otherwise, the law will be simply a wish list, without any effectiveness. Those two *sine qua non* conditions do not always receive the proper attention from the authorities. It is time for an important reflection: is Brazilian legislation really evolving its perception in order to solve environmental problems, or is it merely an outward change to ease the pressure made by environmentalists and other sectors of society?

## 3 Public Policies for Biodiversity: Is Brazil Really Doing All that Is Needed?

The answer to the last question of the previous topic is not found in the laws, but arises when one analyzes public policies. Very few actions have been taken to turn legal provisions into concrete results. This reveals an unexpected side to those laws: the government's evasion of its responsibilities. As awkward as the last phrase may sound, the different levels of government in Brazil are receiving considerable pressure from all parts of society and from the international community, both worried about increasing deforestation, losses of genetic assets, climate change, etc. As a response to those critics, the government did create laws, but did not execute them properly, or did not create instruments to make their enforcement possible. Equipped with a huge and strong bureaucracy, the different levels of government in Brazil do not display a disposition to turn legal provisions into reality. For example, the National Solid Waste Policy—law 12305/2010—in its art. 54 set a deadline for extinguishing solid waste dumps. This deadline expired in August 2014, at which time 60 % of the cities in Brazil had still not complied with the law (Torres 2015). Examples like this are numerous. Brazil still has a legal culture that views laws as a priority and the solution for all institutional problems. Things are indeed changing, but this conception that laws in themselves can bring about change remains very present today.

The law itself is not a solution, especially when it is related with the environment. Without surveillance, prevention, repression, and other investments such as education, it is impossible to stop environmental degradation. The law is merely a legislative framework to be combined with duly planned and appropriately adjusted public policies. Perpetual investments are needed for protecting Brazilian biomes, and a considerable portion of the criticisms leveled at the governments are directed exactly to the issue of financial resources applied to nature conservation. Despite that, the federal government has recently announced a 72 % reduction to the budget against deforestation in the Amazon Forest (Leite 2015), even after a 282 % increase in deforestation seen in February 2015 compared with the same month in 2014 (de Castro 2015). The future is really uncertain, as the political and economical crisis advances in Brazil in 2015, and other cuts in the budget for 2016 and the following years have already been announced.

The main problem with public policies in Brazil—in general, and not only related with environmental issues—against deforestation, the results of one policy can appear quickly, but can is that they are not considered as a State Policy, but as the Policy of a specific Government. That means a lack of planning and vision in the long term. The way Brazilian Governments, at different levels, treat some problems is often by giving them "solutions" that can be managed in four or eight years (the time for one term of office, or two if the president, governor or mayor is reelected). But it is incorrect to think that this way of acting will solve the structural problems we are facing concerning environment and biodiversity. In some subjects such as protection only be maintained if there is a long-term planning to keep that policy working. In other cases, the results only appear many years later, as it is the case of reforestation. However, in almost all situations related to environmental and biodiversity law, investments and planning must emphasize long-term strategies and actions to achieve, but above all, to maintain results, since it is easier to prevent damage than to recover from it because environmental recovery is not always possible.

This instability is not the only guilty scenario for the failure in Brazil's environmental public policies. The segregation between related areas contributes to results to be even weaker. What is the point of having separately a National Environmental Policy (law 6938/81), Law of Access to Brazilian Genetic Assets ("Biodiversity Legal Framework", Law 13123/2015), National Policy for Environmental Education (Law 9795/99), and Climate Change National Policy (Law 12187/2009), among other related public policies/laws? All of them approach different aspects of the same subject: environmental law and nature conservation. If one of those actions is taken without consulting the other parts, its effects will probably not be as accurate as they should. If the different government sectors act in different ways, their public policies may nullify each other.

When analyzing the issue more in depth, we can see there is an almost total overlap of responsibilities between the components of SISNAMA (National Environmental System), established by art. 3 of Decree 99274/90. For example, ICMBio (Chico Mendes Institute for Biodiversity Conservation), whose purpose is listed in article 1 of Law 11516/07 is almost identical to the duties of IBAMA (Brazilian Institute of Environment and Renewable Natural Resources), set out in article 2 of Law 7735/89. There are also: the SBF (Brazilian Forest Service), with functions similar to those of the previous institutions, but restricted to forest issues (art. 55 of Law 11284/2006) and the Environmental Education Governing Body (arts. 2 and 3 of Decree 4281/2002). In addition to these, there is CONAMA (National Environment Council), a consultative and deliberative body, which has its tasks set out in art. 7 of Decree 99274/90. There are many other agencies that play very similar roles to those, but with more specific and limited competences.

This is not to say that the specifics of these organisms in the environmental problem approach should be forgotten. On the contrary, they must be promoted in an integrated, articulated, multidimensional system that meets the specific characteristics and their interdependencies without losing sight of the whole.

For those reasons, what are the advantages of having so many public agencies with almost the same functions? For example, IBAMA and ICMBio have police powers and duties that involve monitoring, protection, preservation, and promoting environmental education. With the exception of police power, the SBF has the same functions for forestry. Assignments that distinguish these three agencies are perfectly amenable of being performed by only one of them. The segregation of duties brings huge financial losses, lack of efficiency and harm to the public, and above all, to nature. The financial losses are due to the maintenance of multiple separate institutions that perform similar functions involving additional infrastructural expenses as well as commissioned positions, and other administrative costs that do not contribute to improve the quality of the public service. The loss of efficiency due to this separation is evident by the fact that specialization, in some cases, is so great that the lack of communication between the agencies can cause damage and raise barriers to public policies implemented by the same government. Also, agencies that have the same competence when working separately, as demonstrated by the Brazilian experience, not only do not improve performance, but also decrease it. The damage to the population can be checked through the increase in deforested areas; the enormous amount of time and bureaucracy that is needed to license environmental projects, lack of environmental law enforcement, and others. This is loss for local populations and a global loss for humanity.

The lack of tools that enable law enforcement brings severe consequences to the effectiveness of the proposed measures. Even when there is enforcement (which does not always happen), the flexibility of Brazilian law ends up generating a high rate of impunity for polluters and those who harm biodiversity. Between 2011 and 2014, only 8.7 % of the total amount of fines imposed by IBAMA were paid (Geraque and Mena 2015), which shows that offenders benefit from system's slowness and extreme bureaucracy, which drags out payment of fines for years. However, even if those fines were paid promptly, that would not be enough to curb avoidable environmental damage because their amounts are very low. Some data show that it is much cheaper to pay fines than to take measures to prevent environmental damage as we may see in the recent example of the environmental disaster that occurred in November 2015, in the city of Mariana, Minas Gerais. A catastrophe resulting from a dam holding toxic waste was breached, and the resulting mud advanced along the Rio Doce and reached the sea; Samarco (the mining company responsible for the barrage) was fined in R$362 million (approximately US$100 million), while the insurance would have cost US$1 billion (Oliveira 2015).

The Mariana case is emblematic for several reasons. The damage to the environment and populations is incalculable, as many experts believe that the life of Rio Doce was decimated by heavy metals found in the river water (Redação G1 2015). Water cannot be used for irrigation and drinking, and the mud spot already extends for several kilometers along the ocean shore, also threatening marine biodiversity. The first information conveyed by the media reported that the breach occurred due to the negligence of the company with dam conservation and lack of control by the competent state bodies. This should be investigated in more depth, but all indications suggest that these assumptions are real. Brazil currently has 16 high-risk dams, which, if breached, can affect the basins of the Paraguay River (in the Pantanal), the Amazon River (in the Amazon Forest) and the São Francisco River (Sanches 2015). If there is an incident, we will probably witness an environmental apocalypse in Brazil.

So, while the environmental legislation does not fit Brazil's reality and ways for applying it are not created, situations like the Samarco disaster will occur and the damage will be irreversible, with harm to future generations. Especially in what concerns biodiversity, investments in environmental education will be useless if the government does not act in environmental preservation. Concerned citizens are fundamental for the improvement of environmental conditions, but they cannot replace the role of the state.

In a new global context, where societies demand more efficiency from governments (do more, spending less), maintaining different committees, consultants and executive authorities, and hundreds of public employees working in related roles in an uncoordinated way is not only unnecessary but also counterproductive. It would be much more effective and cheaper if the government creates a special agency to manage environmental issues in a coordinated, planned, and unified manner, joining forces to implement responsible public policies that show awareness of environmental problems and all their implications for the communities involved. Of course, it is necessary to ensure two aspects in this conception: the focus in planning and execution must combine all dimensions of environment, promoting coordinated actions in order to deal with current challenges; there is also the need to decentralize the decision process, providing a real opportunity to involve local communities in participating in the establishment of public policies, that must be rooted in the specific contexts where they will be performed. This last point is essential in Brazil's case, as the huge territory has its specificities that cannot, under any circumstances, be disregarded at the risk of causing incalculable harm. This has been the case with the Amazon Forest in the past (Klautau de Araújo 1995). In Brazil, at this moment we may observe the opposite: planning and execution are dispersed, and the decision-making processes are centralized.

Nature and biodiversity must not be seen as a duty or a heavy weight for all society to carry, but as an opportunity to grow economically and provide equality in social issues, while preserving nature. These are the three aspects of sustainable development. Destroying natural resources is the shortest route to the bottom in all aspects. Nowadays, Brazil is exploiting its forests, minerals, and oil without creating a productive chain that will make it possible to create jobs, adding more value to the products and providing financial resources to the government. Only this sustainable strategy that integrates the population's needs with responsible natural resources management and their preservation can help to reduce the need for forest devastation.

## 4 Game Theory, Marginalism and the Inefficiency of Environmental Agencies in Brazil

The Brazilian Constitution of 1988 raised Environmental Law to another level. One way to promote the environment preservation was to assign supervisory tasks and responsibilities to various federal agencies, states, and municipalities, as well as to the citizens and the Public Prosecution Service. In other words, the environmental issue has been expanded in order to all of society be able to participate in their conservation, preservation and restoration. Federal environmental agencies end up having their obligations and powers coinciding. In the view of the legislator, all of these shared responsibilities could actually provide ways for improving environmental surveillance and enforcement.

**Table 1** The umbrella game

| | B—with umbrella | B—without umbrella |
|---|---|---|
| A—with umbrella | 0, 0 | 0, 1 |
| A—without umbrella | 1, 0 | −1, −1 |

Nearly thirty years after promulgation of the new Constitution, there have been many advances in environmental issues, but still far from what was expected. Environmental agencies are extremely criticized for their inefficiency by environmentalists and other sectors of society. But, how can the expansion of supervisory responsibility to many other agents reduce efficiency? Game theory and Marginalism can help in understanding this problem.

One can characterize the existence of numerous agents with similar functions in solving the same issues as a cooperative game, given that the performance of any of them can be influenced by the others. In short, this is similar to the Battle of Sexes Game (McAdams 2008) or a variety of the Umbrella Game (Fudenberg and Tirole 1991). Mixing elements of these two games, let us suppose a situation where two people in the same building will meet and go out, the possibility of rain is uncertain, and beyond the fact that they can get wet, there is also the inconvenience of carrying the umbrella. If there is no communication about the subject between the players, the game would go this way, on the assumption that, in fact, it will rain (Table 1).

Taking the umbrella implies getting less wet but carrying it is a disadvantage, so it is valued as 0. If neither of them brings the object, both players will get wet so the value is −1. But if one of them takes the umbrella and the other does not, the one who does not carry achieves a better result, because he will still be relatively dry, but did not have the effort of taking the object. For that reason, the best possible result for A or B only occurs for one of them if there is no action by one and the other simultaneously acts. Therefore, the most likely outcome of the interaction is that both get wet, as the best outcome involves the omission of conduct. This is the total opposite of the legislators' intention when the Constitution and laws were made. Instead of providing an increase in enforcement, there has been a lack of accountability for the environmental agencies.

Analyzing the performance and overlapped competences imposed on different environmental agencies it is possible to draw a parallel between this situation and the Umbrella Game. In matters in which agencies A and B have concurrent jurisdiction (a choice made in the legislation so that both can monitor and act), the most likely outcome is that neither of them will fulfill their role. This is because action by the agencies against environmental degradation generates budgetary costs and institutional and political friction. Furthermore, the text of the legislation is vague, the population is unaware of the legal system in most cases, and there is a plethora of agencies with environmental responsibilities in Brazil (besides the federal level, there are 26 states, the Federal District, and 5570 municipalities). So, it is quite easy for them to avoid accusations of inaction or incompetence by dodging their responsibilities and passing them on to other agencies.

On the other hand, the Marginalism approach could also provide a basis for understanding the inefficiency of numerous environmental agencies, whose numbers have actually increased over the years. In short, one of the assumptions with Marginalism is that the addition of a factor, at a certain point, begins to reduce its efficiency until it loses its usefulness and begins having a negative contribution to the system/individual/production (Marshall 1996). In an analogy with the Brazilian environmental system, the dispersion of investments among different players, instead of strengthening a limited number of environmental agencies, may be a hindering factor and significantly reducing the efficiency of public policy in the sector.

## 5 Education for the Environment and Biodiversity: The Way Beyond the Classrooms

Disrupting educational paradigms is a complex task. When one adds the fact that the system is immersed in a scenario of unstable planning and disjointed (or even contradictory) public policies, those paradigms become almost insurmountable barriers. Brazil is not able to meet its goals for improving education indicators and is also failing to provide an acceptable education to the population. Many schools in the countryside have lack of teachers, school meals, and basic structure in classrooms, not to mention the frequent lack of transportation to the schools. How can a child have a good school performance if she or he has to face hours getting to school, braving mud, rain, sun, and even dangers along the way? In such a chaotic scenario, it is unthinkable to have an education with quality for biodiversity and environmental protection.

Only with an educational system that works properly it is possible to establish an efficient environmental education for biodiversity. The classroom is the first step in order to prepare future ecological citizens in a context where basic needs are guaranteed. That is not the case in Brazil. There are many things to be done, many involving investments in several areas mentioned earlier in this chapter: conservation, forest management, biodiversity research, offering economic alternatives for local people, labor qualification, environmental monitoring, improving life conditions, integrated vision of public policies, etc. It is necessary to create conditions for environmental education to flourish and these conditions at present are far from being achieved.

The law for the promotion of environmental education was created in 1999 (Law 9795/99) and has been harshly criticized by some authors, namely Antunes (2010). Since that law was passed, very little has been done with its new guidelines in order to integrate them into formal and informal Brazilian education. Another situation that deserves attention and reflection is that there was no provision (in the law) for creating a school curriculum in "Environmental Education" (Machado 2014).

Brazil is too large to have standardized policies for all regions, especially in environmental education. Consequently, it is rather common to teach other regions contents and specificities. However, it is very important to wake students' interest with issues closest to their reality, namely achieving an appreciation of the country's biodiversity. It turns out that this centralization of decisions ends up transforming environmental education into something very distant and unattractive for children and young people. In the long run, there is even disunity between regions. The North, Northeast and Central West have various topics about the South and Southeast in their content, but the reverse is not true. So those three more marginal regions are truly unknown to the South-Southeast, the wealthiest part of the country.

The main challenge in implementing environmental education is, therefore, the practical definition of what should be done to make it happen. The legislation is extremely vague, both in concepts and in establishing policies, and does not point the agents to their specific purposes. Instead of it, the law makes all actors responsible for everything. In fact, the wording of the legislation exempts agencies from responsibility, so that they do not act on the assumption that the other co-guardians will act. However, even if the agencies were responsible for specific parts of the implementation of an environmental education policy, the situation would hardly change, since in practice this policy is nothing more than a letter of intent that has no substance to be transformed into reality, due to the law's imprecision and lack of enabling instruments that can materialize it.

Within this context there are only a few individual examples of success. Among the experiences in environmental education and biodiversity in Brazil, the "Escola Bosque" deserves to be highlighted. Initially created to be installed in the city of Belém, Pará, the concept was also taken to the archipelago of Bailique, Amapá in the Amazon River estuary. The latter effort was more successful in maintaining longer the social-environmental method developed by sociologist Mariano Klautau (Klautau de Araújo and Lima 1997a). The Escola Bosque won numerous national and international awards for innovations adopted, from planning and execution until the formation of the first classes. The population has always participated in decision-making, which was an important factor for the community to embrace the project and its guidelines were laid down privileging the location of the town and the population involved (Klautau de Araújo and Lima 1997b). Later, the school was built based on six points, four of which are highlighted: (1) integration of the buildings architecture in the landscape, using local materials and improving local construction techniques; (2) local recruitment of workers; (3) disciplines focused on learning aspects of the locality in addition to conventional educational curriculum; and (4) implement training plans for community teachers (Klautau de Araújo and Lima 1997c). These features, added to the fact that the school was sustainable, taking advantage of the sunlight, rainwater and its area to plant foods, transformed the school into an attractive hub for the whole community. Families participated in numerous activities, sharing and building knowledge with students, and some courses were geared to families, such as new planting techniques and crops, use of leftover food, composting, recycling, etc.

As a result, the project made possible the reduction of environmental degradation and child and families malnutrition, combined with an improvement in school indicators and health indices. Another aspect to be noted is that young people were no longer forced to move to the state capital if they wanted to continue their studies (Lima 2013). They could stay in Bailique and obtain professional qualification in addition to regular education, in areas such as Flora or Fauna Management, Eco-tourism, Mineralogy, Archeology, Traditional Fishing, and others (Klautau de Araújo and Lima 1997a). The dynamic generated by this educational project for the environment was an example of sustainable development.

The Escola Bosque was a revolutionary project, but it was implemented as a Government Policy and at the end of the governor's term in Amapá the project was closed with regard to its primary philosophy, and the social-environmental method was discontinued. The Hotel Escola Bosque, for eco-tourism and hotel management training was also abandoned. The building was ready and equipped, only awaiting the formal opening. Today it lies in ruins in the jungle (Lima 2013).

This is the fate of all public policies that are not assimilated as State Policies. It has to be understood that Brazil is larger than any political dispute and that every day that we lack a coherent and sustainable public policy for biodiversity we lose an invaluable amount of species and natural resources that will be missed in the future.

## 6 Possible New Paths for Preserving Brazil's Biodiversity

As seen in the previous topics of this chapter, there may be numerous alternatives for the preservation of biodiversity, but almost all of them require effective public participation. More than listening to people or winning their adherence to the projects and actions to be taken, it is very important to involve them in all the stages of the process in order to promote their understanding of public policies that are designed to protect the interest of all society. When citizens engage they become the main protectors and inspectors of nature, supporting the state in vigilance against environmental damage. In addition to this fact, there is also the important aspect that changes in individual behavior and interaction with nature that on a larger scale are crucial for achieving the status of an ecological society.

However, for such targets to be met, all levels of government in Brazil need to organize their public policies and join efforts in carrying out public policies and achieving paradigm shifts. After observing the current policies and system failures, some ideas are provided here to highlight suggestions for possible future paths.

The first measure that can be suggested is the creation of a unified law broadly covering the environmental issue. As there is a Civil Code, Criminal Code, etc., it is necessary to create an Environmental Code. Milaré (2014) states that codification of environmental law in Brazil will be a necessary step for providing legal certainty to this area. Considering this context, the mentioned need is highlighted by the fact that the sparse legislation, in its current state, causes two major problems: disarticulation between the different environmental policies, because they are treated as entirely separate issues for each one of the laws; and the great difficulty with handling such legislation, because it requires expertise that is not readily accessible to the population. The law itself does not solve environmental issues, but bringing in a more streamlined legislation can democratize access to it. There are several cases where people commit environmental crimes simply because they do not have information about the law, and it is quite common for people not to know whom to seek out when they become aware of any irregularity.

An advanced technical knowledge is required, and the cost of hiring a lawyer with experience in environmental law is high, which may make it impossible for some people to do so (Klautau de Araújo 2014). There are cases where political organizations fund the hiring of such professionals, but merely as a means of political bargaining and to influence those populations. This politicization of environmental issues causes damages in the implementation of environmental law and creates prejudice against the needed awareness among some part of the population and law professionals. Thus, the creation of codified legislation could facilitate people's access to rules and authorities, untangling the procedures and making the mechanisms of state action more transparent.

The second suggestion is the creation of a government body that concentrates the State's actions in the environmental and biodiversity areas: e.g. preventive and educational actions, enforcement and surveillance, application of fines, regulations, etc. From planning to execution and evaluation of public policies, this body would concentrate both advisory and executive roles. This concentration of competencies is crucial to correct the present lack of coordination of public policies.

In other words, this institution would require the fusion of the existing ones, as CONAMA, IBAMA, ICMBio, SBF, as well as other environmental agencies, consultative and deliberative, with the aim of developing public policies and join forces towards maximizing the efficiency of State action. Currently, the budget for environmental policies is so dispersed among several government bodies that their performances are increasingly restricted. By incorporating all institutions into a single one, there will be savings with costs, maintenance, and commissioned positions, decreasing administrative expenses, enabling investment to achieve its primary function: preservation of the environment, biodiversity, and promotion of environmental education.

This is essential in the current situation in which Brazil finds itself. The government is going through great difficulties to control its spending and avoiding finishing 2016 with a deficit of more than R$30 billion (Martello and Passarinho 2015). One of the areas that suffered a budget cut, as mentioned earlier in this chapter, was the fight against deforestation. Reducing investments in environmental protection while maintaining administrative spending is an almost surefire recipe for a future environmental disaster. Instead, the costs could be optimized with this alternative unification of environmental agencies.

The third suggestion is to create a public company for the administration of all products and ecosystem services obtained from Brazil's biodiversity and biodiversity research. This is essential to achieve nature conservation. The exploitation of these resources, if not managed by the State, will be handled by private companies, irregularly and without adequate financial return for the country. The rich Brazilian biodiversity is the passport to a future of development, quality of life, and good public services. Thus, much more than an environmental issue, the management of genetic resources is essential to national sovereignty.

Law 13123/2015 in its art. 20 provides that the payment of royalties in return for the use of genetic resources, will be up to 1 % of gross revenue, but can be reduced to 0.1 %. In the current legal scenario, this insignificant amount goes through an intricate system—that will never work in practice—and the country will give away its rich genetic heritage almost for free.

Once Petrobras was created it contributed decisively to the scientific and economic development of the country, a public company that manages the genetic heritage can earn billions of dollars, and assist in environmental conservation and economic development of the regions that hold biodiversity. One has to understand that decisions of this scale cannot be postponed: either Brazil takes on a new position and begins profiting from biodiversity resources as a way to preserve nature, or Brazilian biodiversity will be lost without leaving any concrete benefit for the country.

It is not possible to predict the future, but the current indicators are very disturbing. The decrease of forests in Brazil has begun to have consequences even for large cities, as seen in the water crisis faced by the Southeast. Studies show that deforestation in the Amazon Forest lowers the flow of moisture from the air inside Brazilian territory, which reduces rainfall in the Southeast (Carvalho 2014). The paths followed so far have been seriously distorted from their original purposes. Brazil continues failing to accomplish environmental education; the forests degradation is increasing; the population still does not get involved; time keeps ticking and the situation only gets worse. The presented suggestions highlight a new Environmental Law that supports an integrated and multidimensional intervention as an alternative to a system with serious operational limitations that needs to be simplified urgently, in order to become closer to reality.

## 7 Conclusions

Brazil has an advanced legislation for the environment. However, the complexity of terms and the high numbers of laws, decrees, and other regulations, turn Brazil's environmental law system into a truly inaccessible instrument to enable society to cooperate with the state and cope with environmental challenges. More than that, the governments have not been able to implement legislation.

One of the reasons why this happened was the lack of coordination between different environmental agencies. Erroneously it was thought in the 1980s that more institutions taking care of the environment would represent more efficiency in nature conservation. Almost thirty years after the New Constitution, the results have proved that this conception is very expensive and has not achieved expectations.

All of the structure, complexity, and inefficiency of Brazilian environmental law and public policies systems are disabling implementation of environmental education. For that reason, a deeper reflection is needed to involve society in this process and to build a more ecological society, where biodiversity and the environment are important to the educational system.

There are many possible paths that Brazil may find to solve these problems. This chapter has pointed to three feasible options, but the debate is always open in order to refine the environmental scenario in Brazil. A better and more sustainable country can be a reality in a near future if society wants that. Many reforms have to be made. Brazil is a reference in terms of its natural wealth and can become an example of nature conservation, ecological economic growth, and satisfactory living conditions.

## References

Antunes, P. B. (2010). *Direito Ambiental* (12th ed.). Rio de Janeiro: Lumen Juris.

Brasil. (1981). Lei nº 6938, de 31 de agosto de 1981. Retrieved from: http://www.planalto.gov.br/ccivil_03/Leis/L6938.htm

Brasil. (1988). Constituição da República Federativa do Brasil de 1988, promulgada em 05 de outubro de 1988. Retrieved from: http://www.planalto.gov.br/ccivil_03/Constituicao/Constituicao.htm

Brasil. (1989). Lei nº 7735, de 22 de fevereiro de 1989. Retrieved from: http://www.planalto.gov.br/ccivil_03/leis/l7735.htm

Brasil. (1990). Decreto nº 99274, de 06 de junho de 1990. Retrieved from: http://www.planalto.gov.br/ccivil_03/decreto/antigos/d99274.htm

Brasil. (1999). Lei nº 9795, de 27 de abril de 1999. Retrieved from: http://www.planalto.gov.br/ccivil_03/leis/l9795.htm

Brasil. (2000). Lei nº 9985, de 18 de julho de 2000. Retrieved from: http://www.planalto.gov.br/ccivil_03/Leis/L9985.htm

Brasil. (2001). Medida Provisória nº 2.186-16, de 23 de agosto de 2001. Retrieved from: http://www.planalto.gov.br/ccivil_03/MPV/2186-16.htm

Brasil. (2005). Lei nº 11105, de 24 de março de 2005. Retrieved from: http://www.planalto.gov.br/ccivil_03/_Ato2004-2006/2005/Lei/L11105.htm

Brasil. (2006). Lei nº 11284, de 02 de março de 2006. Retrieved from: http://www.planalto.gov.br/ccivil_03/_ato2004-2006/2006/lei/l11284.htm

Brasil. (2007). Lei nº 11516, de 28 de agosto de 2007. Retrieved from: http://www.planalto.gov.br/ccivil_03/_ato2007-2010/2007/lei/l11516.htm

Brasil. (2009). Lei nº 12187, de 29 de dezembro de 2009. Retrieved from: http://www.planalto.gov.br/ccivil_03/_ato2007-2010/2009/lei/l12187.htm

Brasil. (2010). Lei nº 12305, de 02 de agosto de 2010. Retrieved from: http://www.planalto.gov.br/ccivil_03/_ato2007-2010/2010/lei/l12305.htm

Brasil. (2015). Lei nº 13123, de 20 de maio de 2015. Retrieved from: http://www.planalto.gov.br/ccivil_03/_Ato2015-2018/2015/Lei/L13123.htm

Carvalho, E. (2014, October 30). Novo estudo liga desmatamento da Amazônia à seca no país. Retrieved from: http://g1.globo.com/natureza/noticia/2014/10/novo-estudo-liga-desmatamento-da-amazonia-seca-no-pais.html

Castro, F. (2015, March 21). Desmatamento aumentou 282 % na Amazônia Legal em Fevereiro. Retrieved from: http://www.estadao.com.br/noticias/geral,desmatamento-aumentou-282-na-amazonia-legal-em-fevereiro,1655430

Chiuvite, T. B. S. (2010). *Direito Ambiental*. São Paulo: Barros, Fischer & Associados.

Fudenberg, D., & Tirole, J. (1991). *Game theory*. Cambridge, MA: MIT Press.

Geraque, E., & Mena, F. (2015, November 22). Infratores ambientais pagam só 8,7 % das multas aplicadas pelo Ibama. Retrieved from: http://www1.folha.uol.com.br/cotidiano/2015/11/1709460-infratores-ambientais-pagam-so-87-das-multas-aplicadas-pelo-ibama.shtml

Instituto Brasileiro de Geografia e Estatística. (2004, May 21). IBGE lança o Mapa de Biomas do Brasil e o Mapa de Vegetação do Brasil, em comemoração ao Dia Mundial da Biodiversidade. Retrieved from: http://www.ibge.gov.br/home/presidencia/noticias/21052004biomashtml.shtm

Kageyama, P. Y. (2009). Biodiversidade e Biopirataria: contradição entre a biodiversidade e a pobreza no mundo. In A. Aleixo, C. Azevedo-Ramos, E. Camargo, P. Y. Kageyama, M. C. Maio, D. M. Nascimento, & N. S. Oliveira (Eds.), *Amazônia e Desenvolvimento Sustentável*. Rio de Janeiro: Fundação Konrad Adenauer.

Klautau de Araújo, J. M. (1995). *Caligrafias de Belém – vol. I: a dimensão insular*. Belém: Imprensa Oficial do Estado do Pará.

Klautau de Araújo, J. M., & Lima, D. M. B. (1997a). Projeto Escola Bosque do Amapá – Centro de Referência em Educação Ambiental da Ilha de Santana. Macapá: Governo do Estado do Amapá.

Klautau de Araújo, J. M. K., & Lima, D. M. B. (1997b). Projeto Escola Bosque do Amapá – O método sócio-ambiental. Macapá: Governo do Estado do Amapá.

Klautau de Araújo, J. M., & Lima, D. M. B. (1997c). Projeto Escola Bosque do Amapá – Projeto de Socialização. Macapá: Governo do Estado do Amapá.

Klautau de Araújo, T. L. (2014). Environmental law, public policies, and climate change: A social-legal analysis in the brazilian context. In Leal Filho, W. (Eds.), *Handbook of climate change adaptation* (pp. 973–982). Berlin: Springer. doi:10.1007/978-3-642-40455-9_115-1; ISBN: 978-3-642-40455-9

Leite, M. (2015, March 31). Dilma Corta 72 % da verba contra desmatamento na Amazônia. Retrieved from: http://www1.folha.uol.com.br/ambiente/2015/03/1610479-dilma-corta-72-da-verba-contra-desmatamento-na-amazonia.shtml

Lima, D. M. B. (2013). The Escola Bosque project: Building ways to an ecological society. In *Proceedings 7th World Environmental Education Congress, Marrakech, Morocco*, June 9–14, 2013.

Machado, P. A. L. (2014). *Direito Ambiental Brasileiro* (22th ed.). São Paulo: Malheiros Editores.

Marshall, A. (1996). *Princípios de Economia: Tratado Introdutório* – (Vol. I). São Paulo: Nova Cultural.

Martello, A., & Passarinho, N. (2015, September 14). Governo anuncia pacote de corte de gastos e volta da CPMF. Retrieved from: http://g1.globo.com/economia/noticia/2015/09/governo-anuncia-bloqueio-de-gastos-no-orcamento-de-2016.html

McAdams, R. H. (2008). Beyond the prisoner's dilemma: Coordination, game theory and the law. Retrieved from: http://www.law.uchicago.edu/files/files/LE437.pdf

Milaré, É. (2014). *Direito Ambiental* (12th ed.). São Paulo: Revista dos Tribunais.

Oliveira, M. (2015, November 24). Pagar multa custa menos que prevenir dano ambiental. Retrieved from: http://economia.estadao.com.br/noticias/geral,pagar-multa-custa-menos-que-prevenir-dano-ambiental,10000002788

Redação G1. (2015, November 13). É oficial: o Rio Doce está completamente morto. Retrieved from: http://revistagalileu.globo.com/Ciencia/Meio-Ambiente/noticia/2015/11/e-oficial-o-rio-doce-esta-completamente-morto.html

Sanches, M. (2015, November 22). Barragens de alto risco ameaçam 540 mil pessoas. Retrieved from: http://oglobo.globo.com/brasil/barragens-de-alto-risco-ameacam-540-mil-pessoas-18111236

Torres, A. (2015, August 6). Maioria dos Municípios Brasileiros não cumpre a Lei que proíbe os lixões. Retrieved from: http://g1.globo.com/jornal-nacional/noticia/2015/08/maioria-dos-municipios-brasileiros-nao-cumpre-lei-que-proibe-os-lixoes.html

# Education for Sustainable Development in Brazil: Challenges for Inclusive, Differentiated and Multicultural Education

Andréia Faraoni Freitas Setti and Ulisses Miranda Azeiteiro

**Abstract**

Economic progress has been noteworthy in almost all fields in the last 70 years. However, the unsustainable use of non renewable resources, the destruction of biological diversity, and greenhouse gas emissions accelerated the environmental crisis and highlighted social inequalities. The accountability over this civilizational crisis is diffuse, but environmental education in Brazil is mostly focused on those who cause the least environmental damage. This chapter analyses the interactions between the determinants of the current environmental crisis and the contribution of education to sustainable development, critically considering the sustainability of the current means of production and consumption and as a strategy for the promotion of autonomy and equity in the Brazilian reality. A qualitative methodology was used to understand a movement in environmental education based on the critical analysis of the sustainability of the means of production and consumption. For such, documents, laws, and texts were analyzed to approach three important aspects, not intending to exhaust the issue. The first aspect is the original, foundational, and practical constitution of environmental education. The second is about understanding the challenges of institutionalizing environmental education in the Brazilian context. The third has

A.F.F. Setti (✉)
Center for the Studies, Research and Documentation on Healthy Cities, University of São Paulo, School of Public Health, São Paulo, Brazil
e-mail: andreiasetti@gmail.com

A.F.F. Setti
Centre for Functional Ecology, Department of Life Sciences, University of Coimbra, Coimbra, Portugal

U.M. Azeiteiro
Department of Science and Technology, Universidade Aberta and Centre for Functional Ecology, University of Coimbra, Coimbra, Portugal

 
P. Castro et al. (eds.), *Biodiversity and Education for Sustainable Development*, World Sustainability Series, DOI 10.1007/978-3-319-32318-3_15

to do with the paths of political action to be carried out to attain the Sustainable Development Goals (SDGs).

**Keywords**
Environmental education · Sustainable development · Traditional communities · Differentiated education · Sustainable Development Goals

# 1 Introduction

## 1.1 The Challenge of Paradigm Shift

Sustainable development should designate a socially-fair, economically-inclusive and environmentally-responsible process of production of goods and services, but integrating these three pillars has been challenging. The more the hegemonic model of production and consumption advances, the greatest is the global risk of environmental collapse—related not only to the extinction of animals, global warming, or low oil reserves, but also to the consequences of social inequality: poverty, food, and nutritional insecurity, housing shortage, lack of access to potable water and basic sanitation, among others (Martine and Alves 2015).

Our risk society produces growing negativities. According to United Nations [UN] report "World Population Prospects: The 2015 Revision," the world population should reach 9.6 billion people by 2050. Including all these people in the average standard of consumption implies a huge demand for natural resources and the negative impact on the environment may be irreversible. However, if most people remain socially excluded, the fight between peoples to meet their basic needs will intensify (UN 2015a).

Given its reach, omnipresence, level of acceleration and impact, the environmental crisis should be considered Humanity's central and most essential problem, a priority in the development of public policies and governance tools. This implies a change in the development paradigm based on economic growth, in which the well-being of societies is measured by the growth rate of the Gross Domestic Product (GDP) and determined by the culture of consumption, the "culture-merchandise" and the "culture-value" mentioned by Guattari and Rolnik (2000).

A new form of conduct based on the development and the implementation of values based on the respect of life, values that consider complex issues and are developed for the long term, that stimulate a new perspective on the globalized world, on nature, on society and knowledge, with the goal of implementing a healthy, solidary and sustainable model of development is necessary. Therefore, environmental education seeks to provide solutions for a new balance that allows the effective education of critical, reflective and participative citizens that are capable of making decisions fitting of more emancipatory and solidary initiatives dedicated to consolidating true and inclusive democracies (Medina and Santos 2000; Setti and Gallo 2013).

A qualitative methodology was used in this chapter to understand a movement in environmental education based on the critical analysis of the sustainability of the means of production and consumption. For such, documents, laws and texts were analyzed to approach three important aspects, not intending to exhaust the issue. The first aspect is the original, foundational and practical constitution of environmental education. The second is about understanding the challenges of institutionalizing environmental education in the Brazilian context. The third has to do with the paths of political action to be carried out to attain the Sustainable Development Goals (SDGs).

Next sections intend to contextualize this debate that points to diversity and multiculturality as strategies for sustainable societies in the reality and context of Brazil.

## 1.2 Interdisciplinarity in Education

The environmental approach requires interdisciplinary knowledge in order for the complex political challenges and the challenges of environmental management to be understood and addressed. Studies of biodiversity, for instance, must consider not only the perspective of Botanics and Zoology, but also the dimensions of social diversity, productive activities, the climate, the soil, water resources and other elements that affect and are affected by the biodiversity in question (Bursztyn and Bursztyn 2012; Alves et al. 2013).

Interdisciplinary approaches must focus on the environment as within the disciplines, but furthermore must also establish a methodological communication between them, unifying the process as a whole in schools through integrative programs that account for the complexity and the interconnection of the various components of the global ecosystem (Vincent 2011).

The need for integrated knowledge arises from the realization of tensions between plural, scientific knowledge and local, traditional knowledge, which implies inherent issues to science, technology, economy, sociology, politics, and culture (Alves et al. 2013). With multidisciplinary and interdisciplinary interventions, environmental education approaches environmental, educational, political, and social issues in an articulation that makes use both of the objective parlance of science and of resources that bring about ethics and a form of politics. Therefore, the educational process must include different forms of knowledge (science and common sense) stemming from many sources, whether produced by scientists or by ethnic groups and social and popular groups movements.

## 1.3 Dialogue Between Different Forms of Knowledge

The ideology behind the movement for environmental education relates to understanding beings within knowledge, incorporating ethical principles that emerge from various senses that break from unidimensional thinking, from the scientific and instrumental rationality. Therefore, social knowledge emerges from the dialogue between knowledges and is translated into a pedagogical guideline that recognizes the need for a plurality of heterogeneous (often contrary) forms of knowledge in the construction of a sustainable future without risking a loss of autonomy (Santos 2007; Leff 2001).

The complexity of the environment inaugurates a new pedagogy of a dialogic and subjective reappropriation of knowledge, moved by values, interests and utopias, a confluence of different forms of knowledge in permanent diversification and differentiation. The environmental rationality includes various forms of cultural rationality, opening different worlds, articulating realities and complex thinking in the reconstruction of identities. Therefore, the construction of environmental knowledge implies the deconstruction of the simplified and unitary knowledge trapped into disciplines (Leff 2001).

> This is the greatest challenge of education today: accountability—the task of cooperating with this reconstruction, of educating so that the new men and women of the world are capable of standing the burden of this civilizational crisis and converting it into the meaning of their existence, so that life brings a new enchantment and that the world is reconstructed. These paths have been open by the environmental rationality, and these are the veins through which the blood of environment education runs in Latin America (Leff 2001).

## 1.4 Fundamental Humanitarian Principles and Brazilian Environmental Law

The ecological crisis is also a crisis of human values and an ethical crisis, in all dimensions. Together with legal norms, social norms also regulate the conduct of people in relation to one another and may fit into the category of "Morality"; the discipline dedicated to studying it can be called "Ethics" (Kelsen 2000). The fact that human beings are capable of minimally agreeing among themselves on principles such as justice, equal rights, the dignity of the human person, full citizenship, solidarity, etc., creates a chance for these principles to be put into practice, but does not assure their fulfillment.

Nations have come to an agreement around many ethical principles, such as, for instance, in the Universal Declaration of Human Rights (UN 1948). However, Ethics is not sufficient as a theory—or a general principle agreed between nations, peoples, religions, etc.—or the fact that the Constitutions (like the Brazilian Constitution, in 1988) mirrored these principles.

Brazilian Environmental Law includes the principles of protecting all forms of life to promote sustainable development, considering the saturation caused by the economic system in the ecological system because of the disproportionate and continuous use of natural resources to maintain the productive system above nature's capacity to replace them.

The principles of Environmental Law are the bases of the Constitutional system of environmental protection, since they inform, in all aspects (administrative, civil and criminal, in the terms of Art. 225, §3 of the Federal Constitution), the legal system that protects the environment (DOU 1988).

This chapter highlights the structuring principles that assure rights, such as the principle of an ecologically-balanced environment—related to the fundamental right to a healthy life (both regarding physical existence, the health of individuals and the dignity of existence)—and the principle of participation, education and information, which assure people the right to be educated and informed in order to actively participate in the management of the environment (Machado 2000).

Also highlighted here are the principles of repressive and preventive order, which guide educational actions, such as the polluter pays principle, which seeks to make polluters accountable for actions or omissions that negatively impact the environment in order to recover what was degraded. The precautionary principle is focused on the need to make decisions in an environment of scientific certainty concerning the potential future damage of a given activity before such environmental damage takes place. And the preventive principle, a classic principle of environmental law that makes polluters accountable whenever there's actual or potential damage and a causal relationship between such damage and an action or omission (Milaré 2001).

Such legal principles not only influence the development of the legal system, but also guide the actions of the State concerning the protection of the environment.

The historical and determinant relationship between legal norms and democracy —via its institutional representations, which demand weighing the interests involved and common economic, legal and political sense in guiding the valuation of the Constitution, especially its fundamental principles—should be highlighted (Leal 2010). This process takes place by means of conscious actions, based on ethical values, such as equity, solidarity, and social justice and is often opposed to the hegemonic model of economic power. Therefore, environmental education is a strategic process dedicated to shaping the necessary values, skills and abilities that will guide the transition towards sustainability.

## 1.5 Environmental Education: A Matter of Citizenry

With the Military Dictatorship in the 1960s, mass education in Brazil became technical, focusing on efficiency, productivity and scientific neutrality, aimed at providing workers for industrial and technical jobs. This liberal/technical pedagogical trend was also responsible for the fragmentation of knowledge and the "banking concept of education," of depositing information on students, the Cartesian form of education (Brüger 1999; Stahlschmidt 2012).

Further analyzing this instrumentalization of science as an ideology and as a means of domination (operated from the 1970s onwards), environmental education became a priority for the government, to break from the parameters of Cartesian education, of a fragmented, compartmentalized, disjunctive, reductionist intelligence that divides problems, separates what is tied together and makes unidimensional what is multidimensional (Morin 2001).

Environmental education was recognized at the United Nations Conference on the Human Environment (1972), in Stockholm, as a strategy to fight the environmental problems of the time. At the time, the Special Secretariat for the Environment was created in Brazil to implement an integrated form of management.

From then on, Brazil played an increasingly important role in the field, having hosted two international conferences on sustainability: The United Nations Conference on Environment and Development (the Earth Summit) and the United Nations Conference on Sustainable Development (the Rio+20), which expanded the participation of the civil society, of the scientific and academic communities and of the private sector, strengthening the legislation, even though its enforcement remained deficient.

The National Environmental Education Policy, instituted by Law 9795 (from April 27, 1999), has been contributing to the institutionalization of Environmental Education in Brazil, which is being implemented through a collaboration between government spheres and the strengthening of the National Environment System (DOU 1999).

However, environmental actions and plans are still being implemented in different secretariats and councils, their guidelines are developed and implemented separately, mirroring university projects originated from isolated initiatives, not emanated from a broad, strategic approach (Shiel et al. 2015).

Environmental problems are mostly linked to the climate change (global warming), solid waste and the availability of natural resources (water, soil, forests), which demonstrates a perspective of nature detached from people, opposed to the complexity suggested by Morin (2001).

For Brüger (1999), a large part of environmental courses operate philosophically within an instrumental worldview, which is demonstrated by the use of utilitarian nouns such as "management" and "administration" in natural/ecological issues, which undermines the effectiveness of their application in the promotion of sustainable development. The more waste is recycled, the more waste is produced and the more people live in subhuman standards, depending on landfills, which shows that recycling, separating trash and the sustainable use are insufficient in tackling the problem, which should also encompass issues such as equity, human rights and empowerment (Brüger, 1999). Therefore, environmental education in Brazil needs to focus on the population that mostly causes environmental problems, which, in turn, must be understood in all its complexity.

The Brazilian Constitution, in its Art. 205, states that education (in general) and environmental education (especially) should be promoted and incentivized with the collaboration of the society, aiming at human development (DOU 1988). Incentivizing permanent and accountable individual and collective participation in the preservation and balance of the environment is understood as inseparable from the exercise of citizenry (DOU 1999).

The systems of social pressure towards State agents and instances of power—such as the mechanisms that assure citizenry, the institutions and instruments of collective participation, the forms of association and organization, etc., which integrate citizens and the public sphere—should be known and appropriated as to assure and defend individual and collective rights. Therefore, environment-related information cannot be the privilege of public administrators, technicians or scientists; it should be broadly disseminated and is crucial for democracy. It is fundamental that environmental education is incorporated to the teaching/learning process, making sustainability a moral and ethical imperative, respecting cultural diversity and traditional knowledge.

## 2 Differentiated Education and Future Perspectives

### 2.1 Ethnic, Racial, and Gender Identities and Movements for Differentiated Education

Given the issues concerning environmental education in Brazil, new social movements emerged to reaffirm ethnic, racial and gender identities, linked to the protection of territories and the traditional use of natural resources.

Most indigenous and quilombola lands are inalienable goods of the Federation. Traditional peoples have the permanent ownership and the exclusive right of use of their existing natural resources, and few traditional peoples have proprietary rights over the property (RRI 2015).

The cultural dimension of diversity is associated to social movements, especially those working with identity and articulated around the protection of the so-called “policies of difference” (Hall 2003). In Brazilian education, diversity emerges in the demands of ethnic groups for differentiated schools. In the legal realm, both Law 10.639/03—which makes it mandatory for African-Brazilian history and culture to be taught in all schools in the country—and Law 11.645/08—which includes the history and culture of indigenous peoples in school curriculums—were approved (DOU 2003, 2008).

Diversity has to do with a demand for recognition at the public and political spheres of the so-called “minority” groups and certain varieties of feminism. In Brazil, such claim emerged in a few sectors of the feminist and black movements in the 1980s and intensified in the following years, when it also involved the indigenous and disability rights movements (Moehlecke 2009).

Understanding the history of the movement for differentiated education implies comprehending the contradictions and tensions concerning the access to knowledge in society and problematizing the dominant knowledge and the epistemological hierarchization of the society, which delegitimizes the knowledge produced by part of its population.

Rural education emerged during the struggle of peasants, put forth because transnational companies began controlling the agricultural production, exacerbating the capital and its expansionist rationale. In Brazil, different and associated movements kicked (and continue to kick) workers out of the fields while promising to include them in the technological modernity of agribusiness, subduing all, in one way or another, to the technological model that has been called "industrial agriculture," which maintain awful labor conditions (Caldart 2010).

## 2.2 Indigenous Ethnicities and Indigenous School Education: Future Perspectives

There is a whole array of indigenous ethnicities and cultures in Brazil (305 ethnicities and 274 languages), and 17.5 % of indigenous individuals do not speak Portuguese (IBGE 2010). Indigenous school education is becoming differentiated, bilingual, and intercultural, but the model has not been implemented equally throughout the country. The challenge of indigenous school education is attaining a high-quality and differentiated education system that meets the specificities of a different people in the national society, respecting their culture, language and teaching/learning methodologies (Cohn 2005).

Schools are often instances of social, economic, and cultural reproduction, which, at best, offer limited mobility to socially excluded groups. How to create schools that are directly related to real life, that are the syntheses of multiple relationships, determinations, diversities, and concretenesses that real life imposes? How to develop a curriculum that includes different educational dimensions and that articulates pedagogical work at the knowledge level with practical applications concerning work, culture, and social struggle? This demands breaking from a structure of values, from a worldview that follows the rationale of capital, politicizing, therefore, the struggle for the right to established forms of education and for public policies that assure the access to public, differentiated schooling (at the basic and higher levels), coupled to the access to land, to dignified labour, to culture, to political participation and the protection of the environment (Caldart 2010).

As to traditional communities, the goal is contributing to a model of equitable development that integrates the protection of the environment and natural resources, economic growth and the promotion of a better quality of life (Setti et al. 2016a). Affirmative action can contribute to traditional communities if they affirm them as producers of their own knowledge, education and culture and actors of the decisions that affect their lives, opening spaces for participation, based on their organization and on the history of their struggle. Therefore, since they have identified the need to include and legitimize both traditional knowledges and pedagogies—which are crucial for the maintenance of their culture and way of life—, this represents a resistance to the hegemonic model of education. Hence, they criticize practices that are unrelated to the local reality and to their struggle for land legalization, the valorization of their culture, the fight against prejudice and racism, to the nascent public policies that recognize, repair and assure their rights to housing, labor, education and health, adapted to their specificities (Setti et al. 2016b).

Inclusive, equitable, high-quality education that is accessible to all children, that respects cultural and traditional identities was highlighted among the targets to reach Sustainable Development Goal #04. It was established that content related to the contributions of African descendants and indigenous people to the development of nations should be included in school curricula (UN 2015b). The complex and unbreakable relationship between the different education targets and their contribution to the other SDGs were highlighted in the 2030 Agenda and included development, employment, inequity, population dynamics, and health (UN 2015b).

The culture and ways of life of traditional communities were recognized as a driver of sustainable development, especially regarding the various forms of knowledge, technologies, traditional cultural practices and their actions in the territories, valorizing diversity in all its dimensions. Therefore, targets dedicated to protecting and assuring the ways of creating, doing and living of traditional peoples and communities were established: preserving historical sites, mapping, cataloguing and researching the memory, rituals and celebrations of traditional communities; preserving linguistic diversity, forms of expression, artistic expressions, and cultural practices of various ethnicities; promoting the culture of diversity, of solidarity, of equality, and inclusion in the media, with the goal of suppressing situations of discrimination, racism, homophobia, sexism, and any other form of exclusion; promoting the strengthening of peoples and cultures for the planning of climate change (UN 2015b).

The global cooperation for the implementation, monitoring and evaluation of programs, actions and guidelines for the work of the United Nations and its member countries (in the 2030 Agenda) toward sustainable development between 2016 and 2030 is being promoted. Such cooperation must be based in the principles of equity, autonomy, sustainability, solidarity, and respect to humanity and shared responsibilities depending on each person's capacities.

Governance actions must mobilize intersectoral government policies, including shared goals and joint planning, at the national and global levels, by means of negotiations within governments and the participation of different segments of the civil society (Setti et al. 2016a).

## 3 Closing Remarks

The current model of development is based on the premise that nature has a limitless capacity to provide resources and absorb waste, both of which are necessary in the productive process. For centuries, the environment was explored without any form of protection, which led to a considerable loss in biodiversity and the ensuing ecological unbalance. The change in attitudes of these societies will depend on the understanding of the problem that nature is going through, of the environmental crisis, their direct and indirect causes (such as the health-disease process, poverty, violence, unemployment), all of which are based in an economic model of superproduction and superconsumption for a small group and underconsumption and a lack of proper conditions for the vast majority.

History shows that there are two types of political acts: those that preserve reality, maintain the status quo, and those that change, transform reality. In both cases, it could be said that political acts (in the perspective of cultural relativism) can both promote and degrade people and their environments. The practical implementation of sustainable development depends on political acts capable of transforming the current reality, promoting equity and autonomy and reincluding people into nature, considering the interdependence between the social, the socio-economic and the cultural spheres.

Using the law as an instrument of social transformation (or even conservation) necessarily confronts politics and power. Having rights assured in a text isn't enough, rights need to be effective. Moreover, States (institutions dedicated to the protection of the common good, promoters and safe-keepers of human dignity) play a contradictory role in maintaining the balance between capital and the labor force, being the keeper of the economic system. In this context, environmental knowledge emerges, characterized by a complex rationality, bringing together and clashing theory and practice, promoting the dialogue between scientific and popular forms of knowledge to repurpose the social and political territory in a process of interaction, discussion, and critical evaluation of environmental problems.

The contributions of environmental education for a more participatory and democratic political management is based on the idea of empowering individual and collective agents, promoting autonomy and making politics emancipatory. The goal of this chapter was to ponder how specific educational policies have been considered and developed and whether they promote dialogue, contention or agreement with specific situations experienced by traditional communities, which play a fundamental role in the preservation and conservation of ecosystems, which has been recognized in the Convention on Biological Diversity. Therefore, differentiated education is a human right, assured by law, and should be articulated with a territorial development project that connects work, culture and the social function of traditional communities via the recognition of their diversity and the respect to their difference.

Differentiated education has also been recognized in the new universal agenda, the 2030 Agenda, in the Declaration of the 17 Sustainable Development Goals (SDGs) and in its 169 targets. Although the SDGs are applicable globally and universally, they also dialogue with policies and actions aimed at the regional and local levels.

Tackling the asymmetries of humanity and making sustainability a moral and ethical imperative—in which cultural diversity and traditional knowledge need to be respected—is fundamental for the fulfillment of the SDGs. This idea strengthens the perspective and the practices of a popular, participatory, critical, transformative and emancipatory environmental education in the development of socio-environmental public policies in Brazil.

## References

Alves, F., Filho, W. L., Araújo, M. J., & Azeiteiro, U. M. (2013). Crossing borders and linking plural knowledge: biodiversity conservation, ecosystem services and human well-being. *International Journal of Innovation and Sustainable Development* (7), 111–125. 10.1504/IJISD.2013.053323. Accessed June 3, 2015.

Brüger, P. (1999). *Educação ou adestramento Ambiental?*. Florianópolis: Editora Letras Contemporâneas.

Bursztyn, M., & Bursztyn, M. A. (2012). Fundamentos de política e gestão ambiental: os caminhos do desenvolvimento sustentável. Rio de Janeiro, Garamond. ISBN: 978-85-7617-290-1.

Caldart, R. S. (2010). Educação do Campo: notas para análise de percurso. In: Hilário E. Educação do Campo: Semiárido, Agroecologia, Trabalho e Projeto Político Pedagógico. Cadernos Temáticos, Prefeitura Municipal de Santa Maria da Boa Vista.

Cohn, C. (2005). Educação escolar indígena: para uma discussão de cultura, criança e cidadania ativa. *Perspectiva, Florianópolis, 23*(02), 485–515. http://www.ced.ufsc.br/nucleos/nup/perspectiva.html

Guattari, F., & Rolnik, S. (2000). Micropolítica. Cartografias do desejo. Petrópolis. Rio de Janeiro, Vozes.

Hall, S. (2003). Da diáspora: identidades e mediações culturais. Belo Horizonte: Ed. UFMG, Brasília, Unesco.

IBGE. (2010). Os indígenas no Censo Demográfico 2010: primeiras considerações com base no quesito cor ou raça. http://www.ibge.gov.br/indigenas/indigena_censo2010.pdf. Accessed November 25, 2015.

Kelsen, H. (2000). *Teoria pura do direito*. São Paulo: Martins Fontes.

Leal, R. G. (2010) Impactos econômicos e sociais das decisões judiciais: aspectos introdutórios. Brasília, Escola Nacional de Formação e Aperfeiçoamento de Magistrados – ENFAM. http://www.enfam.jus.br/wp-content/uploads/2014/04/Impactos-Economicos_site.pdf. Accessed December 3, 2015.

Leff, E. (2001). Saber ambiental: sustentabilidade, racionalidade, complexidade, poder. Trad. Lúcia Mathilde Endlich Orth. Petrópolis, RJ: Vozes.

Machado, P. A. L. (2000). *Direito ambiental brasileiro* (8th ed.). São Paulo: Malheiros. ISBN 9788539202799

Martine, G., & Alves, J. E. D. (2015). Economia, sociedade e meio ambiente no século 21: triplé ou trilema da sustentabilidade? Revista Brasileira de Estudos de População, Rio de Janeiro. doi:10.1590/S0102-3098201500000027P

Medina, N. M., & Santos, E. C. (2000). *Educação ambiental: uma metodologia participativa de formação*. Petrópolis: Vozes. ISBN 9788532622792.

Milaré, E. (2001). *Direito do ambiente: doutrina, prática, jurisprudência, glossário*. São Paulo: Revista dos Tribunais.

Moehlecke, S. (2009). As políticas de diversidade na educação no governo Lula. Cadernos de Pesquisa, *39*, 137. http://www.scielo.br/pdf/cp/v39n137/v39n137a08. Accessed November 25, 2015.

Morin, E. (2001). *Os sete Saberes Necessários à Educação do Futuro* (3ª ed., p. 118p). São Paulo: Editora Cortez.

Rights and Resources Initiative. (2015). *Who owns the world's land? A global baseline offormally recognized indigenous and community land rights*. Washington, DC: RRI. 1http://www.rightsandresources.org/wp-1content/uploads/GlobalBaseline_web1.pdf. Accessed November 25, 2015.

Santos, B. S. (2007). Para além do pensamento abissal: das linhas globais a uma ecologia de saberes. *Novos estudos CEBRAP, 79*, 71–94. doi:10.1590/S0101-33002007000300004.

Setti, A. F. F., & Gallo, E. (2013). Educação e Meio Ambiente: princípios e desafios para produção da autonomia. In: Mandarino ACS, Gallo E, Gomberg E (Org.) Informar e Educar em Saúde: análises e experiências, Salvador (1)103–120.

Setti, A. F. F., Ribeiro, H., Azeiteiro, U. M., & Gallo, E. (2016a). Governance and the promotion of sustainable and healthy territories: the experience of Bocaina, Brazil. Revista da Gestão Costeira Integrada. http://www.aprh.pt/rgci/rgci612.html.

Setti, A. F. F., Ribeiro, H., Gallo, E., Alves, F., & Azeiteiro, U. M. (2016b). Climate Change and health: Governance mechanisms in traditional communities of Mosaico Bocaina/Brazil. In W. L. Filho, U. M. Azeiteiro, & F. Alves. (Org.), *Climate change and health: Improving resilience and reducing risks* (1 ed.). Berlin: Springer International Publishing.

Shiel, C., Filho, W. L., Paço, A., & Brandli, L. (2015). Evaluating the engagement of universities in capacity building for sustainable development in local communities. *Evaluation and Program Planning*.

Stahlschmidt, A. P. M. (2012). Integralidade, construção e socialização de conhecimentos no contexto da educação permanente e atuação de profissionais da área da saúde. Interface Botucatu, *16*(42). doi:10.1590/S1414-32832012005000038

United Nations. (1948). Universal declaration of human rights. http://www.un.org/en/ga/search/view_doc.asp?symbol=a/res/217(III). Accessed December 15, 2015.

United Nations, Department of Economic and Social Affairs, Population Division (2015a) World Population Prospects: The 2015 Revision, Key Findings and Advance Tables. Working Paper No. ESA/P/WP.241.http://esa.un.org/unpd/wpp/publications/files/key_findings_wpp_2015.pdf . Accessed November 25, 2015.

United Nations. (2015b). Transforming our world: The 2030 Agenda for Sustainable Development. https://sustainabledevelopment.un.org/post2015/transformingourworld. Accessed December 15, 2015.

Vincent, S. (2011). Interdisciplinary environmental education: Elements of field identity and curriculum design. *Journal of Environmental Studies and Sciences,*. doi:10.1007/s13412-011-0007-2. (Springer).

## Legislation

DOU. (1988). Constituição da República Federativa do Brasil. Diário Oficial da União – Seção 1 - 5/10/1988, p.1. Disponível em https://www.planalto.gov.br/ccivil_03/constituicao/constituicao.htm

DOU. (1999). Política Nacional de Educação Ambiental. Lei n. 9.795 de 27 de abril de 1999. Diário Oficial, Brasília (DF).

DOU. (2003). Lei n. 10.639, Altera a Lei no 9.394/1996, que estabelece as diretrizes e bases da educação nacional, para incluir no currículo oficial da Rede de Ensino a obrigatoriedade da temática "História e Cultura Afro-Brasileira", e dá outras providências. Brasília. http://www.planalto.gov.br/ccivil_03/leis/2003/L10.639.htm. Accessed December 15, 2015.

DOU. (2008). Lei n. 11.645, Altera a Lei n° 9.394/1996, modificada pela Lei n° 10.639/2003, que estabelece as diretrizes e bases da educação nacional, para incluir no currículo oficial da rede de ensino a obrigatoriedade da temática "História e Cultura Afro-Brasileira e Indígena". Brasília. http://www.planalto.gov.br/ccivil_03/_ato2007-2010/2008/lei/l11645.htm. Accessed December 15, 2015.

# Protecting Soil Biodiversity and Soil Functions: Current Status and Future Challenges

Sara Mendes, Anabela Marisa Azul, Paula Castro, Jörg Römbke and José Paulo Sousa

**Abstract**

Living soils are fundamental for human life as we know it. The top layer of Earth's crust, essentially composed of minerals, water and air, harbours an immense variety of organisms, from plants to microorganisms, which qualifies it as a living system. Soil biodiversity is the main actor underlying the provision of services that are essential for regulating, providing and supporting human life. The increasing level of human activity has been subjecting soil to multiple pressures, resulting in soil degradation and biodiversity decline, hence deterioration in the system's capability to render those ecosystem services. Growing concern on this resource's misuse has led to a series of conventions and strategies targeting its conservation (such as the Thematic Strategy for Soil Protection in the European Union (EC) 2006) and advocating for specific protection measures that can ensure a sustainable use of soil. These measures mainly focus on extending our knowledge on how soil functions but also on developing monitoring programmes that can detect trends and changes in soil biodiversity. This demand boosted research on soil ecology over the last decades, with significant increases of scientific knowledge on its structural and functioning complexity. However, there are still some gaps and needs to be

S. Mendes (✉) · P. Castro · J.P. Sousa
Centre for Functional Ecology, Department of Life Sciences,
University of Coimbra, Coimbra, Portugal
e-mail: sara.mendes@gmail.com

A.M. Azul
Center for Neuroscience and Cell Biology, University of Coimbra, Coimbra, Portugal

J. Römbke
ECT Oekotoxikologie GmbH, Flörsheim, Germany

 
P. Castro et al. (eds.), *Biodiversity and Education for Sustainable Development*,
World Sustainability Series, DOI 10.1007/978-3-319-32318-3_16

addressed in order to design adequate measures for soil protection. In this chapter we review the main advances in soil ecological research and monitoring and further discuss the status of current strategies towards soil protection and sustainability. Moreover, we present here a strategy, consisting of three action lines, for effectively contributing to soil protection. It is based on monitoring and mapping, experimentation and raising awareness towards soil issues, which hopefully can change the way we perceive and use soil, this very dynamic but non-renewable resource at the human life time scale.

## 1 Soil—The Hidden Living World

As fundamental as it might be, soil is frequently overlooked when protection and conservation plans are designed (Dominati et al. 2010). In public opinion, soil lacks the attractiveness and charisma of other natural elements and, although it is fundamental in linking the atmosphere with subsurface and aquatic compartments (Faber et al. 2013), it is usually perceived as less important than air or water (Beck et al. 2005). The functions and services soil provides, ranging from supporting human housing and a plethora of activities, to providing food or filtering water, are essential for human life. Soil, as defined by the International Organisation for Standardisation (ISO) (2015), is the upper layer of the Earth's crust transformed by weathering and physical/chemical and biological processes, composed of mineral particles, organic matter, water, air, and living organisms organized in genetic soil horizons (ISO 2015). Therefore, soil is not only the outcome of a series of abiotic components, but it is, in fact, a living system, whose life is usually hidden but of paramount importance when it comes to delivering ecosystem functions and services. Soil biological diversity is immense and considerably higher than that found aboveground (Gardi et al. 2009; Pulleman et al. 2012), with at least ¼ of the described species living in the soil (Decaëns et al. 2006).

## 2 Soil Biodiversity, Soil Functions and the Provision of Ecosystem Services

Soil biodiversity, as outlined in Turbé et al. (2010) following the definition of biodiversity from the 1992 Rio de Janeiro Convention, is "the variation in soil life, from genes to communities, and the variation in soil habitats, from micro-aggregates to entire landscapes". The diversity and interactions between and within soil organisms are responsible for the processes that ultimately make soil such an important resource (Breure and Römbke 2005). Soil organisms can array from plants to animals, fungi or microorganisms and have traditionally been classified according to their taxonomic position and body size (Jeffery et al. 2010; Pulleman et al. 2012) into microfauna/flora (size range 1–100 μm), mesofauna (size range 100 μm–2 mm) and macro/megafauna (size range >2 mm). Taxonomical classification, however, has proven to be a very difficult task for soil organisms, not only because the vast majority of soil species remains to be identified but also due to the lack of taxonomic specialists in soil science (Pulleman et al. 2012; Wurst et al. 2012). To overcome this difficulty, soil organisms can broadly be classified into functional groups based on the main roles they perform in soil (Pulleman et al. 2012; Turbé et al. 2010):

- Chemical engineers, also labelled transformers or decomposers, are mainly microorganisms (bacteria, protozoa, fungi) involved in the decomposition of plant organic matter, responsible for carbon transformation as well as making nutrients (such as nitrogen and phosphorus) readily available for plants or animals;
- Biological regulators or biocontrollers encompass a wide variety of small invertebrates, such as nematodes, mites, springtails, and pot worms, that regulate the dynamics of other organisms in soil (plants, invertebrates, microorganisms) through predation, grazing or parasitism;
- Ecosystem engineers, such as earthworms, termites or ants, maintain and modify the structure of soil by aggregating and transporting soil particles, forming pore networks and bio-structures, creating habitats for other soil organisms.

This classification generally reflects most of the functions the soil community is responsible for, from supporting plant productivity, nutrient and water cycling and soil formation, to regulating soil erosion and water purification, or simply providing a pool of biodiversity (European Commission [EC] 2006). Soil functions are of utmost importance and fundamental for ecosystem services provision, with most of the benefits human beings gain from natural systems relying on soil functioning, being it supporting, provisioning, regulating or cultural services (Brussaard 2012; Millennium Ecosystem Assessment [MEA] 2005). Soil quality, defined as the capacity of soil to function, within natural or managed ecosystem boundaries, to sustain plant and animal production, maintain or enhance water and air quality, and support human health and habitation (Karlen et al. 1997), further highlights the importance and value of soil biodiversity and its functions (Creamer et al. 2015) whose environmental and economical (direct and indirect) benefits can to an extent be assessed and valued (Pascual et al. 2015).

## 3 Soil Resources Under Threat

External drivers associated to increasing levels of human activity have led to biodiversity loss, with the disappearance of several life forms and a decrease in the abundance of many species due to a variety of pressures (European Environment Agency [EEA] 2007). Soil biodiversity is no exception, and with its decline soil functions and services have been deteriorating. Such reduction, which may result from different inappropriate management practices including soil overexploitation, makes soils more vulnerable to other degradation processes (Orgiazzi et al. 2016). These have been listed separately in the TSSP-Thematic Strategy for Soil Protection (Bispo et al. 2009; Commission of the European Communities 2006; EC 2006), although occurring simultaneously. Specifically, the following threats are of utmost importance:

- Erosion—This is a natural process occurring over geological time, resulting on the removal of soil particles by physical forces (such as water or wind), leading to a progressive exposition of underlying rocks. Although erosion might be triggered by a combination of non-anthropogenic factors like climate, geology or topography, human activities (namely deforestation, inappropriate land use, deep tillage or mineral fertilization) can and cause significantly accelerated erosion rates. Soil erosion leads not only to a decline in soil functions (and consequent loss of agricultural productivity and degradation of water courses and reservoirs, for example) but ultimately can result in losing soil itself.
- Decline in organic matter—Soil organic matter (SOM) is constantly built-up and decomposed, as it incorporates organic material, living organisms, and humus (resulting from organic material decomposition in soil). SOM is crucial for several soil functions, securing the binding and buffering capacity of soils, which, in turn, limit pollution and diffusion of contaminants. Carbon is one of the main components of SOM, which is one of the reasons why soil is so important in the carbon cycle (hence in climate change): carbon is released in the atmosphere as $CO_2$ and recaptured though photosynthesis, but this sequestration is occurring at much slower rates than those of current SOM depletion. Farming and forestry malpractices (such as conversion of semi-natural ecosystems to agricultural lands, use of monocultures, deep ploughing, deforestation, overgrazing or excessive irrigation) accelerate soil erosion and leaching, causing a rapid depletion of the soil carbon pool.
- Soil contamination—Contaminants make their way to the soil coming from a wide range of sources (mining, industry, waste landfills, traffic, agriculture), generally resulting in damage or loss of soil functions. Depending on the type of contaminant and the extent of contamination, different problems arise: e.g., mining and smelting facilities may be sources of local contamination where risk is mainly associated with storage or disposal of metal contaminated tailings and other type of residues; the latter, together with other industrial activities may also originate diffuse contamination with the generation of airborne contaminants and the consequent land deposition far from the source; intensive farming activities, with the overuse of pesticides, fertilizers, and antibiotics may cause contamination of soil, surface and groundwater, and also the increase of resistance of microorganisms and the increase of potentially pathogenic organisms (like viruses and bacteria) in soil. Soil contamination raises issues on soil and water safety, as the chemical imbalance raised by contaminants frequently decreases the buffering capacity of soils (with consequent massive releases and leaching of contaminants that can enter groundwater and/or surface water).

Ultimately it raises concerns at human health level since increasing contaminant load in soil may cause not only a direct exposure via contact with contaminated soil particles, but also exposure via the food chain.

- Sealing—Soil sealing is commonly referred as the covering of soil with an impermeable layer. Even though it can happen naturally, most of soil sealing nowadays is of anthropogenic origin and directly linked to urban expansion and human population growth. Inadequate urban planning has allowed the establishment of structures (housing, roads or land developments) on fertile soils (and valuable for agricultural production), in a process that is almost irreversible. Soil sealing blocks exchanges between above and belowground communities, reducing the area for soil organisms to carry out their functions and even driving most soil organisms to a slow death due to resource exhaustion and subsequent changes in the soil community composition and behaviour. Soil sealing also generates great impacts indirectly on surrounding soils, as the process creates a waterproof surface that changes flow patterns for water, nutrients, and organisms.
- Compaction—Intensive agricultural practices, such as the use of heavy machinery or overgrazing, exert mechanical pressure on soil that can lead soil compaction. This modification in soil structure is very difficult to reverse and has immediate impacts in reducing soil porosity and, therefore, decreasing the habitat availability for soil organisms, especially for those living in upper soil layers and within interstitial spaces in soil. The space between soil particles gets reduced and so compaction also affects the absorptive and water storage capacity of soils, restricting root growth and ultimately having consequences in fertility, stability and biological activity in soils. For the same reason, changes in water flowing arise and because water infiltration rates can be severely altered, in face of heavy rainfall events, water that can't infiltrate soil will runoff, increasing erosion risks and causing floods.
- Salinization—Soluble salts of calcium, magnesium, and sodium can accumulate in soils naturally but this accumulation may also be human-induced. Generally, salinization is associated with irrigation, more specifically with bad irrigation practices in which saline water ends up being used and/or soil characteristics inhibiting salt washing (high evapotranspiration rates, soil textural characteristics) further add to the problem. Overexploitation of groundwater can also be the cause behind salinization, especially in coastal areas where the lowering of the water table makes way for the intrusion of marine water. Soil salinity is a key factor influencing soil organisms, controlling their metabolism and limiting their distribution (due to different sensitivities in soil organisms to salinity concentrations), which is why irrigation with high salt content waters can have drastic effects in soil communities. Usually, salinization tends to cause a decrease in plant growth and, hence, significantly altering crop productivity, being commonly considered one of the main causes for desertification.
- Floods and landslides—A slightly different type of threat emerges from floods and landslides, which in spite of being natural hazards are closely related to soil and land management. Whether they happen because soil water cycle is

hampered (namely due to changes caused by compaction or sealing) or because land management has favoured erosion (deforestation, land abandonment), these mass movements of soil can end-up causing erosion, pollution, damage to infrastructures, loss of agricultural land and loss of soil resources, with considerable impacts on human activities and lives.

The list of threats presented may seem generalized or reduced, but it's a starting point to raise awareness and in effect work towards soil protection and sustainable use of soil (Orgiazzi et al. 2016), i.e., a use of soil that can meet the needs of the present without compromising the ability of future generations to meet their own needs (Beck et al. 2005). Soil protection and soil issues have been receiving greater attention from policy makers (EC 2011; United Nations [UN] 2014) and the general public (Banwart 2011; Hartemink 2008; Wall et al. 2015), but still there's little implementation of effective protection measures towards soil. Current rates of soil degradation are too fast for rates of soil formation (or recovery) to cope with (Pulleman et al. 2012), which is one of the reasons why soil protection strategies and frameworks have been sought over the last decades. Over the next section we present some of these and elaborate on a three vector strategy to efficiently protect soils.

## 4 Soil Protection and Conservation Strategies

The international community has been trying to address biodiversity conservation and a lot of debate has been put on strategies and frameworks to halt biodiversity decline. Emerging from the 1992 United Nations Conference on Environment and Development (also known as the Rio Earth Summit), the Convention on Biological Diversity (CBD) (UN 1992) marked one of the first steps towards sustainable development. The legally binding document laid down the general principles on an environmental agreement addressing the conservation of biological diversity, the sustainable use of its components, and the fair and equitable sharing of benefits arising from the use of genetic resources (Jeffery et al. 2010; UN 1992), recognizing the importance of biodiversity for ecosystem functioning and the provisioning of soil services to mankind (Rutgers et al. 2009). Biannual meetings of the Conference of the Parties for the CBD assessed and discussed further details for policies and guidance targeting these objectives, whilst also adopting work programmes to deal with specific goals or management approaches. While developing the work programme on agricultural biodiversity, soil biodiversity became visible as an area requiring particular attention, earning a spotlight that eventually led to the creation of the International Initiative for the Conservation and Sustainable Use of Soil Biodiversity (Jeffery et al. 2010). Implementing CBD in light of the role of biodiversity in the maintenance of ecological functions in the soil, policy makers and researchers felt that new tools had to be developed or implemented and that more data had to be collected in order to assess the quality and resilience of soil ecosystem services (Rutgers et al. 2009). For that reason, several European Union

[EU] initiatives arose, aiming to develop tools and sets of indicators to monitor biodiversity trends (Sousa et al. 2009).

It was not long before an articulate approach to soil protection reached the European political agenda: in 2006 the TSSP (EC 2006) was proposed and adopted, compelling EU states to embrace measures to halt and reverse soil degradation whilst stating the essential and irreplaceable role of soil and the functions depending on it. The objective of TSSP was to delineate common approaches oriented for soil protection and preservation but, at the same time, to also restore and prevent further degradation of soils. The strategy was based on four pillars:

- Framework legislation [with the proposal of a Soil Framework Directive (SFD) (Commission of the European Communities 2006)];
- Integration of soil protection in national and Community policies;
- Increased research on soils to assist for policies;
- Raising public awareness.

Even though the SFD failed to meet ratification and was withdrawn in 2014, research conducted under this strategy set the basis for protecting and monitoring soil organism communities and functions at EU and national levels (Römbke et al. 2016).

Stemming from that same Rio Earth Summit, the United Nations Convention to Combat Desertification in Those Countries Experiencing Serious Drought and/or Desertification, Particularly in Africa (UNCCD), adopted in 1994 and put into force in 1996 (UN 1994), aimed to combat desertification and mitigate drought effects. Soil protection falls in the scope of this convention, namely because its objectives include preventing and reducing land degradation, restoring partly degraded land and reclaiming desertified land (Turbé et al. 2010). Similarly to what happened with the implementation of CBD, a series of recommendations were made looking for deeper assessment and knowledge of soil information as to provide tools for policy makers and help design strategies to promote sustainable land uses.

Soil research (specifically soil biodiversity) studies are thus critical for coping with the paucity of knowledge on this subject (Gardi et al. 2009) as there's yet not enough information on soil biodiversity distribution (or which drivers/pressures affect this distribution) (Orgiazzi et al. 2016) and our understanding of relationship between soil biodiversity and soil or ecosystem functions is still incomplete (Bardgett and van der Putten 2014). Also, there's a pressing need for monitoring soil biodiversity, which could not only allow the detection of biodiversity hot spots and areas subject to change but also be useful in the implementation of ecosystem management successfully. The accessibility to all this information and the communication between soil researchers, policy makers and public in general will ultimately allow for effective soil protection programmes and operational tools to be designed and implemented. A good strategy for soil protection could then be delineated using three action lines: monitoring, experimentation and raising awareness. Current status and opportunities for each of these lines is presented below.

## 4.1 Action Line 1: Monitoring

Soil monitoring in general can be perceived as a systematic determination of soil variables in order to register their temporal and spatial variation (Jeffery et al. 2010). It is essential to record data and to detect possible changes and/or decline in soil quality (Morvan et al. 2008). Monitoring aims to provide guidelines for decision makers, enabling the design and implementation of policy measures to support the provision of ecosystem goods and services by protecting and maintaining the sustainable use of soil (or reverse undesired trends) (Gardi et al. 2009; Jeffery et al. 2010; Niemelä 2000). In order to develop soil quality measures, policy makers will then need to know the current status and establish baseline and threshold values (normal operating range—NOR) for soil biodiversity to which new policies can be applied to (Orgiazzi et al. 2016; Stone et al. 2016). Increasing pressures in soil biodiversity have inspired the adoption of soil biodiversity monitoring programmes (Turbé et al. 2010) for which a management framework that allows for systematic monitoring is necessary (Beck et al. 2005; Orgiazzi et al. 2016). Even though a consensus on a single monitoring scheme that fits all scenarios, scopes and goals can be hard to find (Rutgers et al. 2009; Stone et al. 2016), some common principles should be kept in mind when designing monitoring programmes. It is fundamental that prior to establishing a monitoring programme matters like the goal, data collection and processing are well analysed and defined (Niemelä 2000). Site selection for monitoring programmes should be made according to its goal and follow an adequate sampling design, both in terms of the spatial arrangement of the sampling points within each site, and of the sampling effort necessary to have a robust estimation of the indicators to be assessed (this is particularly relevant when dealing with biodiversity indicators—see Reis et al. (2016). Evaluating and assessing the sustainable use of soils should follow a holistic approach, encompassing the characterization of physical, chemical and biological variables of soil (Turbé et al. 2010). For monitoring programmes to be legitimate, all of these measurements should be made following standardized, quantitative, and repeatable protocols (such as those published by ISO), allowing for results to be comparable among sites and time (Jeffery et al. 2010; Stone et al. 2016). Some of these measurements are too complex to assess (for example, identification of soil organisms can prove to be a hard task due to both the lack of taxonomists and the current hampered knowledge on soil species) and even more difficult to communicate to managers and policy makers, which is a vital step in any monitoring activity (Niemelä 2000). A way to bypass this complexity is to use indicators, measurable surrogates for environmental endpoints too complex to assess or communicate (Pulleman et al. 2012). Selecting and developing indicators should take into consideration several criteria, like their measurability, power to monitor at multiple spatial and temporal scales, relevance, and ability to be interpreted and communicated (Pulleman et al. 2012). Since it comprises so many different and intricate aspects, no single indicator will suffice to assess soil biodiversity (Römbke et al. 2016), which is why large scale monitoring relies on sets of indicators. The information and metrics provided by several indicators can, afterwards, be quantified in terms of synthetic numerical indexes, usually derived by using multivariate analysis on physical, chemical, and biological indicators measured (Ruiz et al. 2011; Velasquez et al. 2007) or based only on biodiversity data using a trait-based approach like the eco-morphological QBS index (Parisi et al. 2005).

Selecting the best set of indicators is not an easy task, but it's an important step to reduce operational costs and, most of all, provide data fit to the monitoring purpose (Stone et al. 2016). Looking to contribute for the harmonization of monitoring schemes in Europe, EU FP6 project ENVASSO (Environmental Assessment of Soil for Monitoring) set out to design a single, integrated, operational set of EU-wide criteria and indicators for soil biodiversity (Bispo et al. 2009). Selecting indicators that were complementary, have standardized sampling methodologies and were easy to interpret at both scientific and policy levels, the project came to a set of 3 key indicators as a minimum to be used as surrogate for overall biodiversity decline: diversity of earthworms (in acid soils, naturally without earthworms: enchytraeids), diversity of Collembola, and soil microbial respiration. These indicators were validated and proven to be effective and sensitive to detect changes in land use across Europe (Bispo et al. 2009). The project also defined extended sets of indicators for a second and third more intensive levels of monitoring (depending on relevance to specific issues or for when resources are available) and suggested further systematic sampling across EU relating to land use and soil type categories to derive baseline and threshold values for soil biodiversity (Bispo et al. 2009). Following the work of ENVASSO, the EU FP7 project EcoFINDERS (Ecological Function and Biodiversity Indicators in European Soils) set out to standardize methodologies for the assessment of biological soil indicators, delineate NOR for soil biodiversity according to climatic zones, soil and land use types, investigate associations between biological diversity and land use management types in Europe and analyse how the relationship between organisms within and across taxonomic trophic levels relates to key ecosystem services in soil (Creamer et al. 2015; Pulleman et al. 2012). Looking for potential indicators of biodiversity and ecosystem functioning suitable for the European scale, a first list of soil biodiversity indicators was compiled from literature, after which a "logical sieve" approach was applied to establish key indicators to be used (Faber et al. 2013). The final indicator shortlist thus obtained included process measurements, organism-based assays and biodiversity (organism-based assays as well as developing molecular biology tools), which were evaluated at EcoFINDERS sites (Stone et al. 2016).

Significant progress has been recorded in the selection and definition of indicators for European areas, but this is only a step in designing a monitoring programme (Faber et al. 2013) and while there are monitoring networks implemented for soil abiotic properties, very rarely these include indicators related to the decline of soil biodiversity (Stone et al. 2016). There are some exceptions, like Australia's Earthworms Downunder programme (Baker et al. 1997) and, in Europe, France's Réseau multi-institutionnel de Mesures de la Qualité des Sols (RMQS) (Arrouays et al. 2002), UK's Countryside Survey (Black et al. 2003) or the most noteworthy German biological soil classification scheme (BBSK) (Ruf et al. 2003) and Dutch

Biological Indicator of Soil Quality (BISQ) (Rutgers et al. 2009). Both BBSK and BISQ use ecological classification and assessment of soils based on reference data for soil organisms, using community approach methodologies and previously established monitoring networks (Ruf et al. 2003; Sousa et al. 2009). BISQ uses the Dutch Soil Quality Network (DSQN), which includes 200 sites on 10 soil type/land use combinations and was established to obtain policy relevant information on soil status and trends (Gardi et al. 2009). Following the ratification of the CBD by the Dutch government, the role of biodiversity in maintaining ecological functions in the soil was one of the subjects policy makers felt needed more attention (Rutgers et al. 2009). BISQ was designed to answer this need: composed of 25 indicators comprising both abiotic, biotic (nematodes, earthworms, enchytraeids, soil microarthropods) and functional (microbial biomass, respiration, structural and functional diversity, and C and N-cycling) parameters, it compares indicator values measured in a site with the reference values taken from reference sites, enabling an assessment of bacterial diversity under different management regimes/soil structure and an estimation of the impact of land use and human activities. As for BBSK, it relies on the simple assumptions that soil fauna community composition is mainly determined by abiotic parameters (therefore sites with similar soils should have similar communities and reference communities can be defined) and that it's possible to find the fundamental parameters that influence soil fauna (Ruf et al. 2003). BBSK accounts for multiple biotic factors (like life history traits, feeding guilds or diversity and abundance of meso- and macrofauna); however, contrary to BISQ, the network in which BBSK is applied is not centrally coordinated and biodiversity monitoring endpoints vary by the individual German states according to the interest of management responsible (Gardi et al. 2009; Ruf et al. 2003). However, at least a central database (called Edaphobase) containing all German soil biological data has been set up (Burkhardt et al. 2014).

Even though no consensus has been reached so far on which is the best approach as to monitor soil biodiversity, the fact is that almost a decade after the SFD was presented the status of soil biodiversity has certainly improved (Römbke et al. 2016). There's still work to be done on establishing long-term systematic sampling schemes (Bispo et al. 2009), using standardized methods (Römbke et al. 2005), finding baselines/reference values, integrating new promising techniques (such as "omics" tools) (Faber et al. 2013) and above all, understand and describe the relationships, links and processes connecting soil biodiversity and ecosystem service provision, which is why experimentation remains a key issue for soil protection.

## 4.2 Action Line 2: Experimentation

Whilst our knowledge on soil and the importance of its biodiversity is expanding rapidly, the relationship between soil organisms, soil processes and ecosystem services is not yet fully described. This task is being undertaken with the help of recently developed technologies and approaches—and is expected to continue to be pursued on more studies and experimentation led by soil researchers.

Empirical studies have led to a growing consensus that quantifying functional traits in biotic communities might be the most meaningful way to assess ecosystem services (de Bello et al. 2010). A functional trait is defined as "a characteristic of an organism, which has demonstrable links to the organism's function" (de Bello et al. 2010), being it a morphological, physiological, behavioural or life-history attribute of an organism (Pulleman et al. 2012; Vandewalle et al. 2010). Classifying soil organisms according to their functional traits may help for a better understanding of the relationships that ultimately enable the provision of ecosystem services, with the advantage of being eco-region independence (Pulleman et al. 2012). Relationships between ecosystem services and traits have been documented for several organisms (de Bello et al. 2010), but more experimental-based studies are necessary to disentangle these relationships in order to fully understand the role of soil biodiversity in ecosystem functioning.

A promising tool to attain these goals is to use the so called "semi-field" methods, namely mesoscoms, such as the terrestrial model ecosystems (TME). TMEs consist of intact soil cores, extracted from the field and meant to be used to under controlled laboratory conditions. These types of methods ally the realism of the field environment by working with intact soil cores and the natural soil community, with the ability to control for some abiotic parameters potentially acting as confounding factors in the experiment, thus focusing on particular structure-function relationship questions. The use of these semi-field experiments in ecological studies proliferated over recent years and has so far allowed for the understanding of some mechanisms by which soil organisms mediate nutrient cycling in soil (Pulleman et al. 2012) and further linkages between biodiversity and ecosystem functions are expected to be unearthed by running such experiments. Actually, this method has been proposed as a standard approach in high-tier ecotoxicological assessments (Schaeffer et al. 2010) and are also being used to assess effects of climate change on shifting the risk posed by toxicants (Bandow et al. 2016).

Advances in technology, bioinformatics and data management also unveil an auspicious future for soil research. Management of large databases is becoming a conceivable task, whilst the evolution of statistical analysis and novel conceptual designs (like network based approaches) have allowed for further obtaining relevant information and comprehension on the interactions between organisms, trophic levels and how these shape the natural communities in soil (Burkhardt et al. 2014; Creamer et al. 2015).

## 4.3 Action Line 3: Raising Awareness

The ultimate goal of soil protection is to maintain soil quality in order to secure its functioning for current and future generations. For that reason, involving the general public in soil protection is not only advisable but essential. The need to raise awareness has been emphasized at global scale and efforts have been made in order to increase everyone's knowledge about soil biodiversity and its importance (Jeffery et al. 2010). These efforts run through research and education at every level (from pre-school children to university students, but also considering key participants in soil use and management such as farmers, land-owners or policy makers), trusting that the more people learn about the role soil organisms play in soil functions and, therefore, sustaining the environment, the more they will be likely to protect it (Jeffery et al. 2010). Soil has reached the global agenda and the threats on it have raised enough concern to convince the UN General Assembly of the importance of a sustainable use of soil and, thus, adopting a resolution to designate December 5 as the World Soil Day and declare 2015 as the International Years of Soils (IYS) (UN 2014). The work under the TSSP, as well as the activities prepared during the 2015 IYS and the importance and funding that government agencies have been giving to soil matters (namely the LIFE programme in the EU (EC 2014), have considerably helped raising awareness towards the importance of soil and a number of educational activities (such as lesson plans, games or films) have been and are still being developed to intensify this message. Soil is becoming a current theme on general public perception, which hopefully will pave the way for a better and more sustainable use of this living, fundamental and dynamic, yet, at the human life scale, non-renewable resource.

## References

Arrouays, D., Jolivet, C., Boulonne, L., Bodineau, G., Saby, N., & Grolleau, E. (2002). Une initiative nouvelle en France: La mise en place d'un réseau institutionnel de mesure de la qualité des sols (RMQS). *Comptes Rendus de l'Académie d'Agriculture de France, 88*(5), 93–103.

Baker, G. H., Thumlert, T. A., Meisel, L. S., Carter, P. J., & Kilpin, G. P. (1997). "Earthworms Downunder": A survey of the earthworm fauna of urban and agricultural soils in Australia. *Soil Biology and Biochemistry, 29*(3–4), 589–597. doi:10.1016/S0038-0717(96)00184-8

Bandow, C., Ng, E. L., Schmelz, R. M., Sousa, J. P., & Römbke, J. (2016). A TME study with the fungicide pyrimethanil combined with different moisture regimes: Effects on enchytraeids. *Ecotoxicology, 25*(1), 213–224. doi:10.1007/s10646-015-1581-y

Banwart, S. (2011). Save our soils. *Nature, 474*(7350), 151–152. doi:10.1038/474151a

Bardgett, R. D., & van der Putten, W. H. (2014). Belowground biodiversity and ecosystem functioning. *Nature, 515*(7528), 505–511. doi:10.1038/nature13855

Beck, L., Römbke, J., Breure, A. M., & Mulder, C. (2005). Considerations for the use of soil ecological classification and assessment concepts in soil protection. *Ecotoxicology and Environmental Safety, 62*(2 SPEC. ISS.), 189–200. doi:10.1016/j.ecoenv.2005.03.024

Bispo, A., Cluzeau, D., Creamer, R., Dombos, M., Graefe, U., Krogh, P., et al. (2009). Indicators for monitoring soil biodiversity. *Integrated Environmental Assessment and Management, 5*(4), 717. doi:10.1897/IEAM-2009-064.1

Black, H. I., Parekh, N., Chaplow, J., Monson, F., Watkins, J., Creamer, R., et al. (2003). Assessing soil biodiversity across Great Britain: National trends in the occurrence of heterotrophic bacteria and invertebrates in soil. *Journal of Environmental Management, 67*(3), 255–266. doi:10.1016/S0301-4797(02)00178-0

Breure, A. M., & Römbke, J. (2005). The ecological classification and assessment of soils. *Ecotoxicology and Environmental Safety, 62*(2), 185–186. doi:10.1016/j.ecoenv.2005.03.022

Brussaard, L. (2012). Ecosystem Services Provided by the Soil Biota. In D. H. Wall, R. D. Bardgett, V. Behan-Pelletier, J. E. Herrick, T. H. Jones, K. Ritz, et al. (Eds.), *Soil Ecology and Ecosystem Services* (pp. 45–58). Oxford, United Kingdom: Oxford University Press.

Burkhardt, U., Russell, D. J., Decker, P., Döhler, M., Höfer, H., Lesch, S., et al. (2014). The Edaphobase project of GBIF-Germany—A new online soil-zoological data warehouse. *Applied Soil Ecology, 83*, 3–12. doi:10.1016/j.apsoil.2014.03.021

Commission of the European Communities. (2006). Proposal for a Directive of the European Parliament and of the Council establishing a framework for the protection of soil and amending Directive 2004/35/EC. COM(2006) 232 final. Brussels.

Creamer, R. E., Hannula, S. E., Leeuwen, J. P. V., Stone, D., Rutgers, M., Schmelz, R. M., et al. (2015). Ecological network analysis reveals the inter-connection between soil biodiversity and ecosystem function as affected by land use across Europe. *Applied Soil Ecology, 97*, 112–124. doi:10.1016/j.apsoil.2015.08.006

de Bello, F., Lavorel, S., Díaz, S., Harrington, R., Cornelissen, J. H. C., Bardgett, R. D., et al. (2010). Towards an assessment of multiple ecosystem processes and services via functional traits. *Biodiversity and Conservation, 19*(10), 2873–2893. doi:10.1007/s10531-010-9850-9

Decaëns, T., Jiménez, J. J., Gioia, C., Measey, G. J., & Lavelle, P. (2006). The values of soil animals for conservation biology. *European Journal of Soil Biology, 42*(suppl. 1), S23–S38. doi:10.1016/j.ejsobi.2006.07.001

Dominati, E., Patterson, M., & Mackay, A. (2010). A framework for classifying and quantifying the natural capital and ecosystem services of soils. *Ecological Economics, 69*(9), 1858–1868. doi:10.1016/j.ecolecon.2010.05.002

European Commission. (2006). Communication from the Commission to the Council, the European Parliament, the Economic and Social Committee and the Committee of the Regions. Thematic Strategy for Soil Protection. COM(2006)231 final. Brussels.

European Commission. (2011). Communication from the Commission to the Council, the European Parliament, the Economic and Social Committee and the Committee of the Regions. Roadmap to a Resource Efficient Europe. COM(2011) 571 final. Brussels.

European Commission. (2014). *LIFE and soil protection*. Luxembourg: Publications Office of the European Union. doi:10.2779/64447

European Environment Agency. (2007). *Halting the loss of biodiversity by 2010: Proposal for a first set of indicators to monitor progress in Europe*. EEA Technical Report 11/2007. Copenhagen, Denmark. Retrieved from http://www.eea.europa, http://reports.eea.europa.eu/technical_report_2007_11/en/

Faber, J. H., Creamer, R. E., Mulder, C., Römbke, J., Rutgers, M., Sousa, J. P., et al. (2013). The practicalities and pitfalls of establishing a policy-relevant and cost-effective soil biological monitoring scheme. *Integrated Environmental Assessment and Management, 9*(2), 276–284. doi:10.1002/ieam.1398

Gardi, C., Montanarella, L., Arrouays, D., Bispo, A., Lemanceau, P., Jolivet, C., et al. (2009). Soil biodiversity monitoring in Europe: Ongoing activities and challenges. *European Journal of Soil Science, 60*(5), 807–819. doi:10.1111/j.1365-2389.2009.01177.x

Hartemink, A. E. (2008). Soils are back on the global agenda. *Soil Use and Management, 24*(4), 327–330. doi:10.1111/j.1475-2743.2008.00187.x

ISO (International Organization for Standardization). (2015). Soil quality—vocabulary. ISO 11074:2015. Geneva, Switzerland.

Jeffery, S., Gardi, C., Jones, A., Montanarella, L., Marmo, L., Miko, L., et al. (2010). *European Atlas of Soil Biodiversity*. Luxembourg: Publications Office of the European Union. doi:10.2788/94222

Karlen, D. L., Mausbach, M. J., Doran, J. W., Cline, R. G., Harris, R. F., & Schuman, G. E. (1997). Soil quality: A Concept, Definition, and Framework for Evaluation (A Guest Editorial). *Soil Science Society of America Journal, 61*(1), 4. doi:10.2136/sssaj1997.03615995006100010001x

Millennium Ecosystem Assessment. (2005). *Ecosystems and Human Well-being: Biodiversity Synthesis*. Washington, DC: World Resources Institute.

Morvan, X., Saby, N. P. A., Arrouays, D., Le Bas, C., Jones, R. J. A., Verheijen, F. G. A., et al. (2008). Soil monitoring in Europe: A review of existing systems and requirements for harmonisation. *Science of The Total Environment*, *391*(1), 1–12. doi:10.1016/j.scitotenv.2007.10.046

Niemelä, J. (2000). Biodiversity monitoring for decision-making. *Annales Zoologici Fennici*, *37* (December), 307–317.

Orgiazzi, A., Panagos, P., Yigini, Y., Dunbar, M. B., Gardi, C., Montanarella, L., et al. (2016). A knowledge-based approach to estimating the magnitude and spatial patterns of potential threats to soil biodiversity. *Science of The Total Environment*, *545–546*, 11–20. doi:10.1016/j.scitotenv.2015.12.092

Parisi, V., Menta, C., Gardi, C., Jacomini, C., & Mozzanica, E. (2005). Microarthropod communities as a tool to assess soil quality and biodiversity: A new approach in Italy. *Agriculture, Ecosystems & Environment*, *105*(1–2), 323–333. doi:10.1016/j.agee.2004.02.002

Pascual, U., Termansen, M., Hedlund, K., Brussaard, L., Faber, J. H., Foudi, S., et al. (2015). On the value of soil biodiversity and ecosystem services. *Ecosystem Services*, *15*, 11–18. doi:10.1016/j.ecoser.2015.06.002

Pulleman, M., Creamer, R., Hamer, U., Helder, J., Pelosi, C., Pérès, G., et al. (2012). Soil biodiversity, biological indicators and soil ecosystem services—An overview of European approaches. *Current Opinion in Environmental Sustainability*, *4*(5), 529–538. doi:10.1016/j.cosust.2012.10.009

Reis, F., Carvalho, F., Martins da Silva, P., Mendes, S., Santos, S. A. P., & Sousa, J. P. (2016). The use of a functional approach as surrogate of Collembola species richness in European perennial crops and forests. *Ecological Indicators*, *61*, 676–682. doi:10.1016/j.ecolind.2015.10.019

Römbke, J., Breure, A. M., Mulder, C., & Rutgers, M. (2005). Legislation and ecological quality assessment of soil: Implementation of ecological indication systems in Europe. *Ecotoxicology and Environmental Safety*, *62*(2), 201–210. doi:10.1016/j.ecoenv.2005.03.023

Römbke, J., Gardi, C., Creamer, R., & Miko, L. (2016). Soil biodiversity data: Actual and potential use in European and national legislation. *Applied Soil Ecology*, *97*, 125–133. doi:10.1016/j.apsoil.2015.07.003

Ruf, A., Beck, L., Dreher, P., Hund-Rinke, K., Römbke, J., & Spelda, J. (2003). A biological classification concept for the assessment of soil quality: "biological soil classification scheme" (BBSK). *Agriculture, Ecosystems & Environment*, *98*(1–3), 263–271. doi:10.1016/S0167-8809(03)00086-0

Ruiz, N., Jérôme, M., Léonide, C., Christine, R., Hommay, G., Etienne, I., et al. (2011). IBQS: A synthetic index of soil quality based on soil macro-invertebrate communities. *Soil Biology and Biochemistry*, *43*(10), 2032–2045. doi:10.1016/j.soilbio.2011.05.019

Rutgers, M., Schouten, A. J., Bloem, J., van Eekeren, N., de Goede, R. G. M., Jagersop Akkerhuis, G. A. J. M., et al. (2009). Biological measurements in a nationwide soil monitoring network. *European Journal of Soil Science*, *60*(5), 820–832. doi:10.1111/j.1365-2389.2009.01163.x

Schaeffer, A., van den Brink, P. J., Heimbach, F., Hoy, S. P., de Jong, F. M. W., Römbke, J., et al. (Eds.). (2010). *Guidance from the SETAC Europe Workshop: Semi-field methods for the environmental risk assessment of pesticides in soil (PERAS)*. Boca Raton, FL: CRC Press.

Sousa, J. P., da Silva, P. M., Feld, C. K., de Bello, F., Feio, M. J., Grandin, U., et al. (2009). *Web report comparing indicator systems targeting deviation from undisturbed areas and disturbance gradients*. Retrieved from http://www.rubicode.net/rubicode/RUBICODE_Report_on_Indicators_of_Disturbance.pdf

Stone, D., Blomkvist, P., Hendriksen, N. B., Bonkowski, M., Jørgensen, H. B., Carvalho, F., et al. (2016). A method of establishing a transect for biodiversity and ecosystem function monitoring across Europe. *Applied Soil Ecology*, *97*, 3–11. doi:10.1016/j.apsoil.2015.06.017

Stone, D., Ritz, K., Griffiths, B. G., Orgiazzi, A., & Creamer, R. E. (2016). Selection of biological indicators appropriate for European soil monitoring. *Applied Soil Ecology*, *97*, 12–22. doi:10.1016/j.apsoil.2015.08.005

Turbé, A., De Toni, A., Benito, P., Lavelle, P., Lavelle, P., Lavelle, P., Ruiz, N., et al. (2010). *Soil biodiversity: Functions, threats and tools for policy makers. Bio Intelligence Service, IRD, and NIOO, Report for European Commission (DG Environment)*. Retrieved from http://ec.europa.eu/environment/archives/soil/pdf/biodiversity_report.pdf

United Nations. (1992). *Convention on biological diversity*. Retrieved from http://www.cbd.int/doc/legal/cbd-en.pdf

United Nations. (1994). *United Nations Convention to Combat Desertification in those Countries Experiencing Serious Drought and/or Desertification, Particularly in Africa*. Retrieved from http://www.unccd.int/Lists/SiteDocumentLibrary/Publications/UNCCD_Convention_ENG.pdf

United Nations. (2014). *General Assembly Resolution 68/232. World Soil Day and International Year of Soils, A/RES/68/232 (20 December 2013)*. Retrieved from undocs.org/A/RES/68/232

Vandewalle, M., de Bello, F., Berg, M. P., Bolger, T., Dolédec, S., Dubs, F., et al. (2010). Functional traits as indicators of biodiversity response to land use changes across ecosystems and organisms. *Biodiversity and Conservation*, *19*(10), 2921–2947. doi:10.1007/s10531-010-9798-9

Velasquez, E., Lavelle, P., & Andrade, M. (2007). GISQ, a multifunctional indicator of soil quality. *Soil Biology and Biochemistry*, *39*(12), 3066–3080. doi:10.1016/j.soilbio.2007.06.013

Wall, D. H., Nielsen, U. N., & Six, J. (2015). Soil biodiversity and human health. *Nature*, 1–8. doi:10.1038/nature15744

Wurst, S., De Deyn, G. B., & Orwin, K. (2012). Soil Biodiversity and Functions. In D. H. Wall, R. D. Bardgett, V. Behan-Pelletier, J. E. Herrick, T. H. Jones, K. Ritz, et al. (Eds.), *Soil Ecology and Ecosystem Services* (pp. 28–44). Oxford, United Kingdom: Oxford University Press.

# Interdisciplinary and Participatory Research at Early Childhood to Biodiversity Education and Sustainable Development

João Miranda, Raquel Maricato, Joana Vila Nova, Joana Margarida Baptista, João Lourenço Monteiro, Nuno Freitas, Odete Gonçalves, Vera Vale and Anabela Marisa Azul

**Abstract**

The biodiversity is the ultimate concept of life on Earth and includes all living organisms but remains a serious challenge at global scale, principally as result of human activities. The perception of biodiversity and their linkages to the ecosystems functioning and human well-being may have significant impacts in terms of Biodiversity Education and Sustainable Development (BESD). Interdisciplinary and participatory research, have increasingly strong evidence in the biodiversity conservation awareness. In early childhood, however, certain challenges come around in the understanding of basic concepts of biology and ecology, and their combination with the areas of proximal development of

J. Miranda
Centre for 20th Century Interdisciplinary Studies, Faculty of Arts and Humanities, University of Coimbra, Coimbra, Portugal

R. Maricato · J.V. Nova · J.M. Baptista · N. Freitas · O. Gonçalves
Kindergarten JISASUC–Jardim de Infância dos Serviços de Ação Social da Universidade de Coimbra, Av. Dias da Silva, 7, 3030-136 Coimbra, Portugal

J.L. Monteiro
CIUHCT- Interuniversity Center for the History of Science and Technology, Department of Applied Social Sciences, Faculty of Science and Technology, Universidade NOVA de Lisboa, Lisbon, Portugal

V. Vale
College of Education of Coimbra (ESEC-Escola Superior de Educação de Coimbra), Rua Dom João III – Solum, 3030-329 Coimbra, Portugal

A.M. Azul (✉)
CNC-Centre for Neuroscience and Cell Biology, University of Coimbra, Pólo I – Edifício da FMUC, Rua Larga, 3004-517 Coimbra, Portugal
e-mail: amjrazul@ci.uc.pt



P. Castro et al. (eds.), *Biodiversity and Education for Sustainable Development*,
World Sustainability Series, DOI 10.1007/978-3-319-32318-3_17

children. This chapter explores the effectiveness of a participatory research at early childhood through a discovery process focused on biodiversity experience. We intended to explore how biological understanding of the biodiversity and ecological processes at early childhood may contribute to BESD awareness. The participatory research undertook a constructive programme, with the active collaboration of researchers from life sciences, humanities, science education, kindergarten teachers, children, and artists. The perceptions, evaluation, and validation of the approach are emphasised in the ateliers progressively designed, in the drawings by the children, in the documentation by the kindergarten teachers, and in the interviews to the children. The findings focused attention on the understanding of biological and ecological interactions, the adaptations to climate, the food and the products of the Mediterranean forests, and the biodiversity legacy in Mediterranean region. We highlight the construction of a conceptual design from the child's perspective that includes the children as actors/authors of knowledge, which resulted from mutual learning and active collaboration. The participatory research linked to real-life of children and local/regional context clearly contributed the extension to families and society. Moreover, the science-art collaborations did engage the children expressively. The paper concludes with remarks addressing the participatory research in early childhood to rise awarenes in BESD context, with attitude gains and lifelong outcomes.

## 1 Introduction

Biodiversity is the ultimate concept of life on earth and includes all living organisms found on land in water, and in the atmosphere. Biodiversity supports the ecological networks that underpin the ecosystems functioning and the human existence (World Health Organization [WHO] and Secretariat of the Convention on Biological Diversity 2015). Despite the magnitude of biodiversity to human health and well-being, the loss of biodiversity remains a serious challenge at global scale, mainly as result of human activities (WHO 2015; European Environment Agency [EEA] 2015).

Mediterranean region is recognized as a global biodiversity hotspot. However, over the last decades, the risk of biodiversity loss is increasing (EEA 2015) due to the over-exploitation of natural resources, habitat loss, pollution, and growing impacts of invasive alien species. About 60 % of protected species assessments and 77 % of habitat assessments recorded an unfavourable conservation status in Mediterranean region (EEA 2015). Also, climate change is projected to increase water shortages, with direct impacts on water resources and indirect effects on food production, ecosystem functions, services and health, and people health (IPCC 2014). Some impacts of biodiversity and habitat loss can be estimated—nearly 5 % of the original vegetation in the Mediterranean region remains relatively intact (FAO 2013)—but other effects in ecosystem such as soil degradation—the microorganisms loss and biogeochemical changes—are very difficult to estimate. The soil biodiversity is increasingly recognized to deliver existential benefits—in terms of food production clean air, water and nutrients—with the ability to promote human health (Wall et al. 2015). The increasingly global sustainable and health challenges are intimately tied to the interactions between ecosystems processes and people behaviour.

There is a compelling need for understanding the identity of biodiversity and ecological interactions and networks at local and regional levels, but also to promote the affective dimension in terms of the sensations of own experiences and identity from nature (Hinds and Sparks 2008). The Mediterranean diet, acclaimed as Intangible Cultural Heritage of Humanity by UNESCO in 2013, brings the opportunity to reflexions at multi-layered contexts, from nutritional to socio-economic (Bach-Faig et al. 2011), and with the "biodiversity" featuring a healthy and sustainable dietary pattern (Dernini and Berry 2015).

Environmental communication and education programmes have been dominated by approaches with end-users as receivers of simplified and overgeneralized information. Such assumptions too often create barriers in the communication and misconceptions in the appropriation of knowledge. This ultimately translates in the ineffective environmental awareness and the incapacity to remodel individual and collective behaviors in terms of natural resources sustainability and human well-being.

Education for Sustainable Development (ESD) promotes efforts to rethink educational programmes using participatory research and recognizes the pluralistic and the continuous mutual learning processes throughout lifetime (UNESCO 2007, 2008, 2015). However, there is a recognized gap between early childhood educational programmes and systems—methodologies and topics—in perspective of the child's (UNESCO 2008).

Knowledge of biodiversity in their extension to ecological interactions and how human well-being is interconnected and influence their existence in future will have a profound impact on conservation education in terms of Biodiversity and Education for Sustainable Development (BESD). Understanding the biological mechanisms and ecological processes that underlie linkages between nature and people, and the effects of interdisciplinary research, mutual learning environment, emotion, and age on experiencing nature could driver educational strategies to design BESD programs that promote awareness and attitude gains from individual to collective and from local to global scale.

Interdisciplinary and participatory research can now offer some understanding of how linkages between nature and people can be strengthened for improved explorations at early childhood and processes the experiences and knowledge throughout life.

We explored how discovery process and biodiversity experience at early childhood underpin attitudes and behaviours at local and regional levels for improving BESD at global scale. We specially focused on the understanding of biological and ecological principles—abiotic and biotic interactions—and the biodiversity legacy in the Mediterranean region.

Interdisciplinary and participatory research was previously explored within pre-schoolers (Wells 2013; Wells et al. 2015), primary scholars (Azul et al. 2009), secondary scholars (Azul 2009), teachers (Azul et al. 2008), and forest producers (Azul et al. 2010; Azul 2011).

## 2 Perception of Biodiversity and Their Linkages to the Ecosystems Functioning and Human Well-Being

### 2.1 Approach: Mediterranean Forest Experience

The main objective of the project «Exploring and Experiencing Mediterranean» was to link the biological understanding of biodiversity in Mediterranean region as it inhere the network of ecological interactions, and to promote sustainability awareness connected to natural resources conservation and human well being.

We intended to report the extraordinary biodiversity—plants, animals, and mushrooms—in Mediterranean forests, particularly in the oak woodlands, traditionally called «*Montado*» in Portugal and «*Dehesas*» in Spain. These oak woodlands correspond to man-made ecosystems and are recognized as a successful example of sustainable land use in Europe. We also meant to explore basic concepts of biology and ecology—abiotic and biotic interactions—and the connections between human and nature in Mediterranean region—food, materials, and history (see the glossary, Table 1).

The «Exploring and Experiencing Mediterranean» project used an interdisciplinary and participatory research, with the active collaboration of researchers from life sciences, humanities, science education, kindergarten teachers, and artists. We intended to explore how biological understanding of the biodiversity and ecological processes may contribute to biological conservation awareness and attitude gains in terms of sustainability.

The JISASUC–Jardim de Infância dos Serviços de Ação Social da Universidade de Coimbra, is a kindergarten from the University of Coimbra, with four classes with children 3 to 6 years old. The JISASUC have wide experience in participatory research with researchers of the Coimbra Education School and the University of Coimbra. The approaches at the JISASUC have being inspired by several constructivists authors, such as Dewey (1989), Piaget (1926), Vygotsky (1967), Freire (1996), and Loris Malaguzzi (Smidt 2013), and is increasingly demonstrated interconnected regarding the cognitive, emotional, and physical well-being. The observation, the reflection, and the documentation, are the three main pillars that JISASUC explore to support children in their areas of proximal development.

**Table 1** Ateliers accomplished along with the participatory research

| Term | Definition |
|---|---|
| Biodiversity | *The variability among living organisms from all sources including, among others, terrestrial, marine and other aquatic ecosystems and the ecological complexes of which they are part; this includes diversity within species, between species and of ecosystems* (Convention on Biological Diversity, Article 2; 1992) |
| Biome | Regions on Earth with similar climatic conditions, especially temperature and rainfall, characterized by large communities of plants and animals as result of adaptation to climate |
| Climate | Climate represents the average of weather in a place over many years (the weather include temperature, precipitation, cloudiness, and wind speed, at a certain place; the weather may change according to regular events, such as the seasons). The climate is associated to gradual chances over large periods of time |
| Ecosystem | Community of living organisms, combining the relationships they establish with each other, and in conjunction with surrounding environment—air, water, and mineral matrix of soil |
| Food chain | The food chain represents the fluxes of nutrients and energy between the organisms. The food chain has at the basis the organisms that produce their own food, such as the plants, and then the organisms that feed from other organisms. Some animals eat plants others eat animals or fungi. The fungi feed themselves by absorption |
| Fungi | The kingdom of Fungi is composed by organisms, which the structural unit is the hyphae. The fungi do not move by themselves, and do not produce their own food—get feed by absorption |
| Lichen | The lichen corresponds to a symbiotic relationship that arises from junction of a fungus and algae or cyanobacteria (or both). The lichens have many colors, sizes, and forms |
| Mediterranean | The Mediterranean biome is characterized by temperate climate with hot and dry summers, and precipitation restricted to a few months in a relatively warm winter and spring |
| Montado | Savannah-like landscape dominated by open oak formations—*Quercus suber* L., *Quercus rotundifolia* L., and *Quercus ilex* L.—but also some dispersed pines and other tree species, that results from an extensive agro-silvo-pastoral exploitation. Known by *Montados* in Portugal and *Dehesas* in Spain, have accompanied human history in Mediterranean basin and represent a classic example of sustainable land use in Europe by combining biological conservation and socio-economical value. The multifunctional use of *Montados* is associated to multiple products, e.g., cork, wood, meat, fruits, apiculture, medicinal plants, mushrooms, habitat for fauna, tourism, and ecosystems services. The *Montados* dominate the landscape in the Alentejo region (southern Portugal) |
| Mycorrhizas | Mycorrhizas (from Greek mykes = fungi and rhiza = root) are mutualisms between fungi and plant roots. The fungi translocate water and nutrients to plant roots; plants translocate carbon to the fungus. The mycorrhizas are common associations in nature that promote the ecosystems functioning below- and aboveground |

(continued)

**Table 1** (continued)

| Term | Definition |
|---|---|
| Mushrooms | The mushrooms are associated to the sexual fruiting bodies of higher fungi, which produce the spores. The fruiting bodies growing above-ground are called mushrooms, the fruiting bodies growing below-ground are called truffles |
| Soil | Soil encompasses mineral particles, organic materials in various stages of decomposition, living soil organisms. The living soil matrix includes fungi, bacteria, insects, and earthworms |
| *Rhizobium* | *Rhizobia* are bacteria; the *Rhizobium*-legume symbioses promote the biological fixation of $N_2$, and can play a significant role in improving the fertility and productivity of soils |
| Symbiosis | In ecology, symbiosis is associated to long-term interaction between two different biological species. The mutualism is a symbiosis in which both benefit from the association |
| Sustainability | In ecology, sustainability is associated to the biological and ecological processes and functioning, as well as to the use of the resources without compromising the balance in nature |

# 3 Materials and Methods

## 3.1 Sample

The approach comprised all children at the kindergarten JISASUC; in total, 53 children, between 3 and 6 years: 3 year—7 children; 4 year—18 children; 5 year—19 children; and 6 year—9 children (52 % boys and 48 % girls). Informed consent from the parents and kindergarten teachers was obtained through written collaboration agreement in order to discuss the progress of the research and to disseminate the results.

## 3.2 Participatory Research Approach

The participatory research experienced three phases: (1) conceptual design, (2) realization of the ateliers, and (3) perceptions, evaluation, and validation of the approach. The approach encompassed a constructivist programme over a 9 months period that accompanied the project from the beginning.

Phase 1: *Conceptual design*

The conceptual design was primarily planned taking into account basic concepts of biology and ecology (see glossary, Table 1) to improve the understanding of the biological processes associated to biodiversity in Mediterranean at early childhood. The conceptual design followed a discovery process by children and kindergarten teachers, and support of the researchers and the multiple actors. The constructivist programme engaged children as active actors/authors. The discovery-based process followed a holistic exploration of the biodiversity through a flexible and spontaneous programme, within a combination of basic concepts of biology and ecology and creativity.

The participatory research team—researchers and kindergarten teachers—met regularly to discuss the approach and the tools to be explored in the ateliers—books, website, newspapers, videos, experiences, games, and artistic forms. The regular meetings were crucial to accurate the programme in terms of the biodiversity related concepts the development and interests of children. The regular meetings contributed to strength the active collaboration between the research team and the invited actors that participated in the project (researchers and artists).

Phase 2: *Realization of the ateliers*

The discovery-based process resulted in sixteen ateliers (Table 2). The ateliers specially focused on the biodiversity in Mediterranean forests (A1, A4–9, A13, A15–16), adaptations to climate in Mediterranean region (A1, A6, A9, A16), traits of plants (A4–6, A9, A16), traits of fungi (A10–11, A15), Mediterranean diet (A1, A4, A6, A9, A15, A16), products of the Mediterranean forests (A1–2, A9, A13, A15–16), ecological interactions (A6, A10–11, A16), efficient use of water (A5–6, A16), connections between biodiversity and the real-life of children (A2, A9, A15, A16), history of the Mediterranean region (A3, A12), science-art collaborations (A3, A8, A12, A13), and participative actions (A3, A7, A9). The ateliers were mostly conducted by the kindergarten teachers and the actors invited during the project. The collaborative actions attempt science from the children's perspective with children as actors and authors in the appropriation of knowledge.

Phase 3: *Perceptions, evaluation, and validation of the approach*

The perceptions, evaluation and validation of the approach include the active collaboration in the conceptual design of the ateliers/activities/actions, the perceptions from the children's perspective, and the consultation process through an interview/questionnaire to the children.

The kindergarten teachers documented the perceptions and interests of the children during the ateliers/activities/actions, and the questions and initiatives prior to the ateliers. The documentation and perceptions from the children's perspective were registered during the ateliers and discussed on the regular meetings. The documentation included drawings by the children, photos, and videos during the ateliers and actions, and documentation by the kindergarten teachers (Edwards et al. 2011).

In order to ascertain the appropriateness of the approach, a consultation process was conducted with the children at the end of the 9-month experience. The children were interviewed based on a questionnaire with 27 closed questions. The consultation process involved a selection of topics, ateliers, and actions presented randomly. The questionnaire was firstly presented as a pre-test. The researchers and kindergarten teachers optimized collectively the consultation process in terms of the specific scientific contents and concepts regarding. The consultation process/interview was primarily oriented to the perceptions of the biodiversity in the

**Table 2** Ateliers accomplished along with the participatory research

| Atelier | Action | Aspiration/brief description |
|---|---|---|
| A1 *Where acorns come from?* | Research hands-on | *Research about trees with acorns*/exploration of oak trees—leaves, bark, roots—that produce acorns, and their habitats in the Mediterranean. Children used multiples tools (books, web) with their families and at the kindergarten. Discussion groups promoted dialogues about inhabitants, habitats and products of oaks and oak woodlands (traditionally called *Montados*) |
| A2 *Discovering the acorn in the kitchen!* | Inquiry cooking | *Confection of bread with acorn flour*/children participated actively in the confection of bread with acorn flour |
| A3 *Dinosaurs in the prehistoric forests* | Research painting drama | *Drama about dinosaurs and their forests*/the dinosaurs—a topic chosen by children—was used to introduce the history of Mediterranean forest. Children discovered a dinosaur—*Plesiossauro*—that inhabited the Portuguese territory, and giant mushrooms in prehistory forests. Afterwards, they participated actively in the construction of the dino-masks, scenario, and drama |
| A4 *Seeds of Mediterranean plants* | Research exploration | *Exploration and observation of seeds of Mediterranean plants*/children explored seeds with their families. At the kindergarten, children observed the seeds by using magnifying glasses and overhead projectors. Children compared colours, sizes, and textures, and discussed the seeds present in the Mediterranean dietary |
| A5 *Vertical garden* | Research hands-on | *Cultivation of plants in small areas*/children and educators explored edible plants that may be cultivated in balconies and then constructed a vertical structure with recycled materials. Children sowed selected seeds (from A4), and accompanied the life cycle of the plants |
| A6 *Eat the garden you sow!* | Inquiry hands-on gardening | *Calendar of edible plants*/children and educators inquired legumes and vegetables of Mediterranean that are sown during winter and spring. Children cultivated a small garden outside. The gardening involved the techniques to use water efficiently. Besides, children followed the growth of plants and explored particular symbiosis between legumes and the bacteria *Rhizobium* |

(continued)

**Table 2** (continued)

| Atelier | Action | Aspiration/brief description |
|---|---|---|
| A7 *Oaks in the city!* | Participative action | *Plantation of a cork oak at a garden in the city*/children, educators, and researchers, made a walk to a garden of the city to plant the cork oak seedlings. Children identified oaks in different areas of the city |
| A8 *Dance of Mediterranean forest* | Creative dance | *Performance of sounds and movements of Mediterranean*/children experienced sounds and movements associated to abiotic—sun, wind, rain, sea, river, soil—and biotic—animals, plants—components of ecosystems, in particular the Mediterranean forests. The creative dance involved the active collaboration between children and a professional dancer |
| A9 *Aromatics and teas* | Investigation hands-on | *Identification and preparation of aromatics and teas*/children identified and catalogued aromatic plants of the Mediterranean region. Children selected: (i) aromatic herbs commonly used for tea, (ii) aromatics to make aromatic bags; (iii) aromatics to flavour salt, olive oil and vinegar; and (iv) aromatic herbs to ice cubes. Children dried some plants at the class and then made tea and fresh drinks at the kindergarten for their families. Children took the aromatic bags to their homes |
| A10 *What is a mycorrhiza?* | Survey hands-on | *Observation of cork oak mycorrhizas*/children imagined what could be a mycorrhiza. Children observed mycorrhizas in a block of soil with roots of cork oak; they observed the mycorrhizas by using magnifying glasses |
| A11 *There is an internet growing my under by feet!* | Role playing | *Simulation of development of mycelium in soil*/through a role playing, children, educators, and researchers, explored the mutual symbiosis between plant and fungi—*mycorrhizas*—and the fructification of mushrooms and truffles. Children also made a panel of a forest with mycelia (1 spaghetti = 1 hyphae) |
| A12 *Soil and houses from Mediterranean* | Expedition to Roman ruins visual arts | *Construction of a mosaic with puzzle pieces of clay*/children visited Roman ruins (Conímbriga) and explored the soil materials used to build the city. Children created puzzle pieces with clay and then constructed a mosaic. Children discussed about the products derived from Mediterranean soil |

(continued)

**Table 2** (continued)

| Atelier | Action | Aspiration/brief description |
|---|---|---|
| A13 *Painting with Nature* | Research expedition to botanic garden painting | *Painting with natural ink*/children prepared ink with acorns and other fruits, legumes, spices, and mushrooms. The performance was conducted in two different moments: (i) individual at the Botanic Garden of the University of Coimbra (JBUC); and (ii) collective at the kindergarten. Children explored the biodiversity and applications from biodiversity of Mediterranean. Children experienced *painting with nature in Nature*. The painting at the JBUC included a visit to the glasshouse and a drama performed by the educators |
| A14 *How much did it rain?* | Experiment hands-on | *Use water resources efficiently*/children constructed a pluviometer, registered the evolution of water according to the rainfall, and construct a strategy to irrigate efficiently. Children discussed about the importance of water |
| A15 *Risotto of wild mushrooms* | Exploration culinary | *Exploration of mushrooms diversity and confection of risotto*/children participated actively in the risotto confection that counted with a collaboration of a researcher from social sciences as "chef". Children were able to discuss the diversity of mushrooms and the products from forest that contribute to dietary |
| A16 *Mediterranean garden* | Exploration hands-on gardening | *Construction of a Mediterranean garden*/researchers selected wild plants, children replanted the wild plants over the perimeter of the kindergarten, after exploration the main characteristics and names. Children also replanted some plantlets from A5 and A6. Once the garden established, children also explored the trees that were previously present and another symbiosis: lichens |

The detailed report can be consulted at the web page www.montadomedia.com

Mediterranean region, the insights about the ecological interactions, the engagement and attitude gains regarding food and products of Mediterranean forests, the impact of science-art collaborations, and the preference for sources and actions. The consultation process comprised all children at the kindergarten JISASUC (53 children). The data was analysed using the computer program Statistical Package for the Social Sciences (SPSS) version 20.0.

# 4 Results and Discussion

## 4.1 Participatory Research from the Children's Perspective

The children feel exceptionally comfortable with the new questions and the initiatives associated to the research at the kindergarten and at home with families. To the question, "*How can we research*?" children clearly answered "On your computer, on the internet"; "And we can make drawings"; "We can also go to the Google"; "And [consult] in the books"; "And we could also go a museum"; "And go to the library where I often go".

## 4.2 Biodiversity in Mediterranean Region

Table 3 reports the list of plants, animals, and mushrooms, after the ateliers/activities/actions done by the children. The Mediterranean forest experience resulted in well-known species, particularly plants and mushrooms. The exploration of biodiversity in Mediterranean forests associated to traits and adaptations to climate, food, and products, clearly contributed to depict main characteristics of species but also to debate the importance of biodiversity in terms of conservation and sustainable development. The exploration and experiencing of biodiversity stimulated the construction of the vertical garden (Table 2 A5), "the garden to eat" (Table 2 A5), and the Mediterranean garden (Table 2 A13). Likewise, the exploration of biodiversity revealed to be a good practice to introduce interactions between species and the symbiotic associations, such as mycorrhizas (Fig. 1x, xi; i, j), *Rhizobium* (Fig. 1xiii; l), and lichens (Fig. 1xvi; m).

## 4.3 Perceptions from the Children's Perspective: Documentation and Drawings

To investigate the perceptions from the children's perspective we combined the documentation and drawings throughout the participatory approach. The kindergarten teachers documented the perceptions, questions, interests, and suggestions of the children during the ateliers, group discussions, and sharing experiences. Figure 1 exemplifies the documentation during the research, explorations, experiences, and creative and participative actions by the children.

The drawings were explored before, right through, and after the ateliers (Fig. 1). The drawings produced by the children may expressively contribute to better comprehend the appropriation of knowledge regarding the topics explored (Fig. 1b–k) and experienced (Figs. 1j and 2), but also the interests and engagement of the children to perform in multiple circumstances, such as the discovery from the unknown (Fig. 1a, i, l, m).

**Table 3** Biodiversity in Mediterranean region explored and experienced during the participatory research

| Group | |
|---|---|
| Plants | Trees: azinheira (holm oak, *Quercus rotundifolia*), carrasco (kermes oak, *Quercus coccifera*), sobreiro (cork oak, *Quercus suber*), azereiro (Portuguese laurel cherry, *Olea oleaster*), loureiro (laurel, *Laurus nobilis*), pinheiro-manso (stone pine, *Pinus pinea*), medronheiro (strawberry tree, *Arbutus unedo*), limoeiro (lemon, *Citrus limon*), oliveira (olive tree, *Olea europaea*)<br>Shrubs: carqueja (*Pterospartum tridentatum*), giesta (broom, *Cytisus* spp), gilbardeira (butcher's-broom, *Ruscus aculeatus*), esteva (rockrose, *Cistus ladanifer*), sargaço (sage-leaved rock rose, *Cistus salvifolius*), tojo (gorse, *Ulex* sp.; Figure 1k), alecrim (rosemary, *Rosmarinus officinalis*), rosmaninho (lavender, *Lavandula* spp), rosmaninho-verde (green lavender, *Lavandula viridis*), erica arbórea (tree heath, *Erica arborea*), urze (heath, *Erica* spp.)<br>Herbs: borragem (borage, *Borago officinalis*; Fig. 1 documentation xii), cardo (thistle, *Cynara cardunculus*), dedaleira (foxglove, *Digitalis purpurea*), erva-de-S-Roberto (herb-of-S. Roberto, *Geranium robertianum*), ervilhaca (vetch, *Vicia sativa*), espargo-silvestre (asparagus, *Asparagus aphyllus*), hipericão (Perforate St John's-wort, *Hypericum perforatum*), malmequer (chop-suey greens, *Chrysanthemum coronarium*), morango-silvestre (wild strawberry, *Fragaria vesca*) |
| Animals | Domestic mammals: black pig, cow (Fig. 1e), donkey, goat, horse and sheep<br>Domestic birds: chicken, duck and turkey<br>Wild mammals: coelho-bravo (wild rabbit, *Oryctolagus cuniculus*), corço (buck, *Capreolus capreolus*), javali (boar, *Sus scrofa*; Fig. 1d), texugo (badger, *Meles meles*), gineta (genet, *Genetta genetta*), lebre (hare, *Lepus granatensis*), lince (Iberian lynx), lontra (*Lutra lutra*), morcego (bat, *Tadarida* spp.), ouriço-cacheiro (hedgehog, *Erinaceus europaeus*), toupeira (mole, *Talpa occidentalis*), raposa (fox, *Vulpes vulpes*; Fig. 1f), rato-do-campo (wood mouse, *Apodemus sylvaticus*), saca-rabos (ichneumon, *Herpestes ichneumon*)<br>Wild birds: águia-imperial (imperial eagle, *Aquila adalberti*), poupa (hoopoe, *Upupa epops*)<br>Insects: beetles, flies, spiders, ants, bees, butterflies, grasshopper, cricket, wood louse, centipedes, earthworms |
| Mushrooms | Wild mushrooms: Crista-de-galo (Chanterelle, *Cantarellus cibarius*; Fig. 1g), *Astraeus hygrometricus, Amanita muscaria, Boletus chrysenteron, Boletus edulis, Boletus subtomentosus, Bovista* spp, *Laccaria laccata, Lactarius delicious, Pisolithus* (*Pisolithus tinctorius* Fig. 1h) *Russula cyanoxantha, Russula sororia, Schizophyllum commune, Scleroderma* spp. and *Trametes versicolor*<br>Cultivated mushrooms: repolga (oyster mushroom, *Pleurotus ostreatus*) and portobello mushroom (*Agaricus bisporus*) |

The documentation (Fig. 1; Table 4) contributed to ascertain the curiosity, the engagement, and the appropriation of knowledge in understanding the biological and ecological principles—abiotic and biotic interactions—the attitude gains, and their combination with the areas of proximal development of children.

The documentation to perception, evaluation, and validation of the approach attempted to be diverse. The ateliers were realized through a discovery-based process, in which children explored/experienced individually and/or collectively.

| Documentation | Drawings |
|---|---|
| i) *Where acorns came from?* (3, 4 years old) Before research: "The acorns come from the soil, because they are nuts..." "... and because pigs eat them. Pigs do not climb the trees!" "No! Acorns come from trees..." "The tree is called «boloteira»." (Fig. 1 a) "The acorns are green..." "...and then become brown and fall down to the soil." "When the acorns fall down, no longer return to the trees... we can not stick acorns on trees!"<br>ii) After research (3, 4 years old): "No! The tree that produces acorns is called cork oak!" "The cork oak belongs to the family of oaks." (Fig. 1 b, left to right) "This is a leaf of cork oak. This is a leaf of oak... of course! ... They have little waves"<br>iii) After research (4-5years old): *What gives us the cork oak?* "Give us acorns! It is a fruit. Pigs and wild boar eat acorns. And we too!" " "Give us cork stoppers... and many things... walls, bags, shoes, umbrellas, coats and dresses..." "Give us wood..." "...Forests, biodiversity..." "It also gives shadow. The shadow is nice not to take sun on the head." "Oxygen... it's a lot!" "To make flour of acorn, we take off its hat?"<br><br>iv) "When the acorn grows, comes out a seedling from the soil... This little plant grows and then turns into a cork oak." (Fig. 1 c, 5 years old)<br><br>v) *Seeds of Mediterranean plants* (3-4 years old) "We've been watching seeds with a magnifying glass that lights." "My seed was of lettuce and I thought it was yellow, and then it was not. It is very little."<br>vi) *Garden to eat* (3, 4, 5, 6 years old): "We sow the seeds in the soft soil and then we water them." "We sow the seeds, water the seeds and come to see as they are." "The carrot seeds will germinate and will give carrots" "...then we eat." "I am working in our garden!"<br><br>vii) Some of the animals of Mediterranean forests (*Montados*) (3, 4, 5, 6 years old): boar (Fig. 1 d), wild rabbit, Iberian lynx, genet, bat, hedgehog, pig, sheep, goat, cow (Fig. 1 e), horse, chicken, duck, turkey, donkey, fox (Fig. 1 f), and imperial eagle.<br><br>viii) Discussion group about foxes: "The foxes live in a labyrinth" "It's a den!" "The animals live in that dens. Then, when they leave, the foxes go there." "The foxes have a good smell and good hearing" "The foxes are also mammals." "[They] have orange hair and white legs and tail" "Foxes run very fast" "They eat animals, are carnivorous...and eat eggs" "Foxes eat animals that are inside the eggs. And also eat chicks of chickens that are there [in the eggs]." "The foxes hunt... and hunt rabbits. They also hunt at night" "Because they take the opportunity to go out at night."<br><br>ix) Some of the Mediterranean forests (A13, A15; 5, 6 years old) Chanterelle (*Cantarellus cibarius*; Fig. 1 g); *Pisolithus* (*Pisolithus tinctorius*; Fig. 1 h): "There are two stones or bones of dinosaurs" "The stones have no sand!" "It is not a stone because it broke on the stairs, I think it is clay!" "These mushrooms are strange. I've never seen mushrooms like these!" | **(a)** 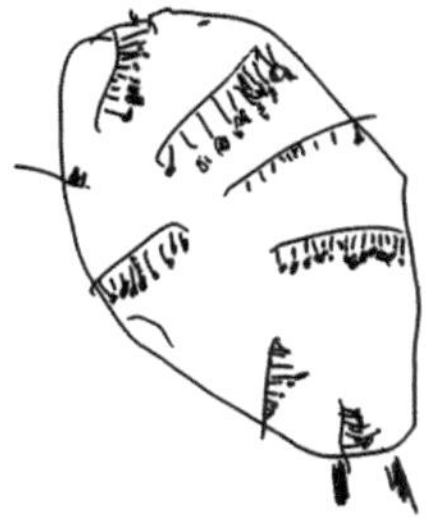<br>**(b)** <br>**(c)** <br>**(d)** <br>**(e)** <br>**(f)** <br>**(g)** <br>**(h)**  |

**Fig. 1** Documentation of perceptions and interests of the children during the approach and actions

x) *What is a mycorrhiza?* (A10; 5, 6 years old) Before the research: "It's a scary person. When you are watching TV, the mycorrhiza hides behind the sofa and... it scares you!" (Fig. 1 i) "It's a thing that sews clothes. When the clothes are shabby we put them on the mycorrhiza and it sews them, so we don't get cold." "It's a house." "It's a name for stones. And everybody loves the mycorrhiza because it has a funny name." "It's a microbe because it could be a microbe name." "It's a machine. You put damaged things in it... and you get new things! This machine is in the stores."

xi) After the research, during the role-playing (A11; 5, 6 years old): "The mycelium is what goes below-ground. It seems the internet. When it finds a tree, the mycelium forms a mycorrhiza" (Fig. 1 j). "I have an Internet under my feet".

xii) Plants of Mediterranean (A16; 3, 6 years old)
Borage (*Borago officinalis*): "Herb with blue star-shaped flowers and larger leaves. [Leave] It is hairy and tastes like cucumber with salt!"
Carqueja (*Pterospartum tridentatum*): "Shrub with yellow flowers. It has green branches. The leaves are below."
Gorse (*Ulex* sp.): "The leaves are thorns and are everywhere and it has yellow flowers." (Fig. 1 k)
Heather (*Calluna vulgaris*): "Shrub with tiny leaves. The flowers seem balls with fringes and smell like honey".
Rosemary (*Rosmarinus officinalis*): "Bush with brown and green branches, narrow leaves, and purple flowers. It smells great!" "With the magnifying glass, it's rather easy to see the purple flowers."

xiii) What is a *Rhizobium*? Before the exploration of legumes in the garden (A6; 4, 5 years old): "... This arm is a hook and the other is a hammer. The one with the hook is an arm very big and the one with the hammer is too short. And it has wings, one wing is little and the other is bigger." (Fig. 1 l)

xiv) *What is the lichen?* (A16; 4, 5 years old) Before the exploration of the garden outside the kindergarten: "It is the trunk of cork oak. It is brown and light brown. It has a light green things, also." "A trunk of cork oak. It has many crumbs!" "A lichen is a table. It gets up alone, but only at night when no one sees. It jumps at night, also." (Fig. 1 m) "[Lichens] are little things that are animals... they have four legs, two eyes, a mouth and three ears, so they hear very well!"

(i)

(j)

(k)

(l)

(m)

**Fig. 1** (continued)

The ateliers were accompanied by group discussions in class. The group discussions occurred within multiple purposes, individual expression, conversations within class, and sharing experiences among classes. Some of the ateliers were documented by video (www.montadomedia.com), which become available as an open tool to society. The children also actively collaborated in a radio programme dedicated to the project. The videos and the radio programme created the opportunity for children to be actors for transforming society.

**Fig. 2** *Painting with Nature* (A13) during the expedition to the Botanic Garden of the University of Coimbra (JBUC). "I have painted myself, a cork oak and the Sun. I also have painted a rainbow and firework because I like both very much!" (5-years old)

## 4.4 Consultation Process to the Children

To the question: "*Did you enjoy experiencing the biodiversity in the Mediterranean*?" all the children gave an affirmative answer. By detailing the biodiversity in the Mediterranean region among «Plants», «Animals», and «Mushrooms» (Fig. 3), it was possible to infer through the children preferences that plants were the most interesting element of analysis.

About the discovery processes and exploration/experiencing, the results stress the preference of the ateliers, the books, and the web (Fig. 4) by the children. However, there is clear evidence in the preference for the ateliers (Fig. 5) and thus their putative significance to engagement, appropriation of knowledge and attitude gains.

Table 5 reflects the preference of the ateliers by the children. Curiously, the preference for the atelier associated to the confection of the bread was rather discrete. This may be associated to when it was performed. The confection of the bread occurred in the first month of the project. However, children asked to repeat the atelier after the 9 months period. They confectioned bread and pizza to their families with flour of acorn and products of the Mediterranean forests.

The appropriation of knowledge regarding the biological understanding of biodiversity in the Mediterranean region was assessed through 17 closed questions, each one with four possible answers, and only one correct. To the question "*What is*

**Table 4** Appropriation of knowledge by the children, in understanding notions, concepts, and principles of biology and ecology associated to biodiversity and sustainable development

| Examples | Documentation |
|---|---|
| Abiotic interactions: soil, water | "We also put soil." "Well dome! My grandfather also put soil in plants." "Now we put the seeds." "And the water!" "And then they will grow." "Things [plants] were born!" "But not here." "We did not water…" "[Plants] They drink by the roots".<br>"…[We] may have a hose and make small holes to water slowly."<br>"We can make a small hole in a carboy and so the water flows very slowly." |
| Biotic interactions: food chains | *Dialogue between children about mammals* 4, 5, 6 years old<br>"We made a domino about mammals"<br>"First we saw [discussed] and then we decided to do a domino about mammals"<br>"We did two animals and what they eat."<br>"What eats the genet?" "Little mice!"<br>"And the lynx eats wild rabbits."<br>"The bats eat insects. When they are babies, the eat [drink] milk from Mom breast."<br>"The foxes are carnivorous and eat animals."<br>"I drew a boar! He eats mice and acorns."<br>"Acorns are for squirrels to eat [Squirrels et acorns]. And we too!"<br>"The pig also eats acorns."<br>"The boars eat truffles…"<br>"I have two mammals in my house!" |
| Biotic interactions: Mycorrhizas | *Role playing "there is an internet growing my under by feet!"* 3, 4, 5, 6 years old<br>"We intended to pretend that we were a network that exists in the forest. Some children are trees others are fungi. The fungi spread a network belowground. When there are mycorrhizas, born mushrooms." "We made a network of fungi with wool yarns, the mycelium, around trees." "…When the mushroom appears there is that network [mycelium]. When it rains, it may appear another mushroom." "The mushroom is near the roots. It's very good! The mushrooms [fungi] give water and minerals to the tree, and the tree gives sugar to mushrooms, at the mycorrhiza."<br>"We've been seeing mycorrhizas with magnifying glass. And below the mushroom there is the fungus because the mushroom is a fungus." |
| Biotic interactions: *Rhizobium* | *Exploration of plants through the construction of Mediterranean garden* (A16) 3, 4, 5, 6 years old. During the observation and description of the legumes pea and broad bean, "This is a *Rhizobium*. It is in the root." |

*a cork oak*?" 90.6 % of children answered "*A tree*". About the shape of cork oak leaves, 64.2 % children chose "*Green and peaks*"; the options: (1) Green and roundish; (2) Green and peaks; (3) Green and thin; (4) Green and rolled. About the acorns, 88.7 % children were able to associate acorns to the cork oak tree, but also other species of oaks. Also, 56.6 % of the children identified acorns as a fruit and

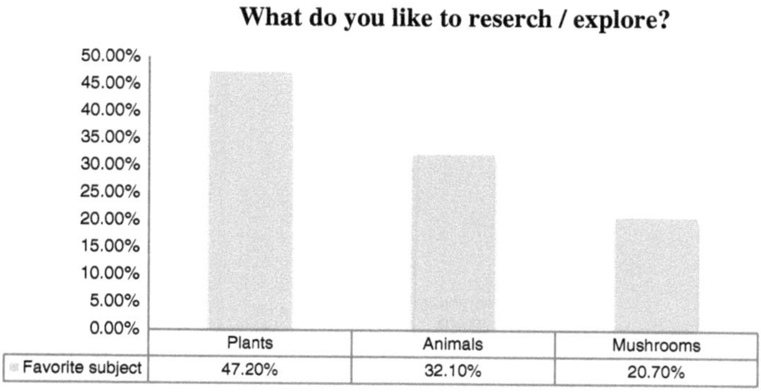


**Fig. 3** Preferences of the children in terms of biodiversity topics «Plants», «Animals», and «Mushrooms» (n = 53)

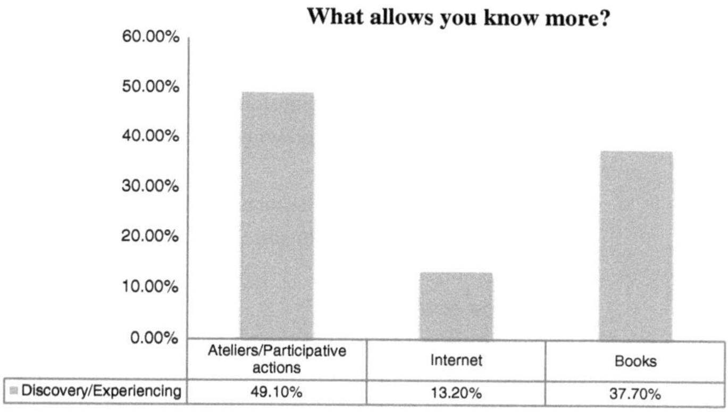


**Fig. 4** Perceptions of children in terms of information/appropriation of knowledge about biodiversity (n = 53)

67.9 % could recognize the bread as a product. The cork was mentioned by 79.2 % of children as a product likely to result from the cork oak trees.

To the question "*Which of these animals is not a bird*?" 71.7 % of children identified the "bat"; the options: (1) Black stork; (2) Great bustard; (3) Bat; (4) Iberian eagle, and 62.3 % recognize the ant as an insect quite current in the cork oak.

Regarding ecological interactions, to the question "*Which animal eats acorns*?" 83 % of the children chose the "*Pig*"; the options: (1) Bat; (2) Pig; (3) Bobcat; (4) Stork. The association between plant roots and fungi—mycorrhizas—41.5 % of the children indicated that mycorrhizas are in the soil; the association plant root and bacteria—*Rhizobium*—49.1 % of children could find the nodules on the roots of plants. The results are quite different on the relation to traits of plants. To the questions "*How plants drink water*?" and "*How plants breath*?" 92.5 % chose

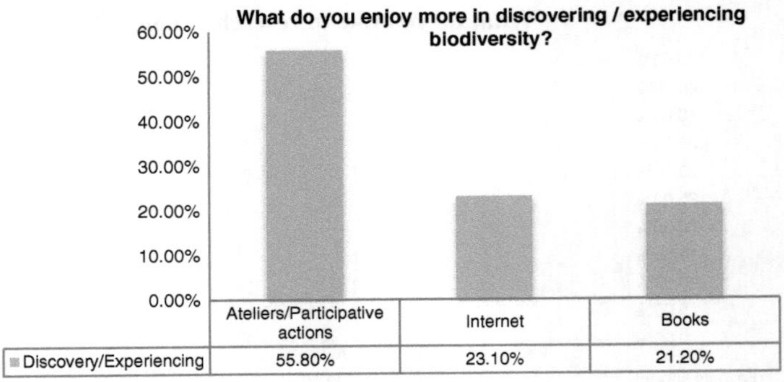


**Fig. 5** Perceptions of the children regarding pleasure in discovering/experiencing the biodiversity (n = 53)

**Table 5** Preferences of the children in relation to the Atelier

| Ateliers | % |
|---|---|
| Vertical garden | 1.9 |
| Discovering the acorn in the kitchen | 1.9 |
| Dance of Mediterranean forest | 18.9 |
| There is an internet growing my under by feet! | 7.5 |
| Garden to eat | 1.9 |
| Buildings from Mediterranean soil | 1.9 |
| Dinosaurs in the prehistoric forests | 1.9 |
| Painting with nature | 11.3 |
| How much it rain? | 3.8 |
| Risotto with wild mushrooms | 3.8 |
| Oks in the city! | 11.3 |
| Radio interview | 15.1 |
| Seeds of Mediterranean plants | 5.7 |
| Aromatics and teas | 1.9 |
| What is a mycorrhiza? | 7.5 |
| Where acorns came from? | 1.9 |
| Mediterranean garden | 1.9 |

through "*The root*" and 56.6 % chose through "*The leaves*", respectively. Moreover, 75.5 % of the children were able to indicate that efficient watering is associated to gradual procedures. Children were able to distinguish trees from other plants, plants cultivated and plants not cultivated, plants commonly used in dietary, plants that can be used to flavour food.

Notably, this participatory approach resulted in mutual learning about scientific concepts associated to biodiversity in the Mediterranean forests combining with multiple areas of proximal development of children. Moreover, this participatory approach promoted dialogues within multiple disciplines. Exceptionally, this experience at early childhood was crucial to construct the argument of the Animation "*Mediterranean Forest: Montado, as long as we live together*" (www.montadomedia.com).

Experiences at early childhood are not relief actions; they represent a key identity to cognitive, sensorial, and social development, and must be conceptualized as a dynamic and constructivist approach. This provides the challenge of how to be effective in BESD experience in terms of attitude gains and behaviours in a long-term perspective.

## 5 Concluding Remarks

The interdisciplinary and participatory research, resulting in active collaboration through mutual learning and constructivism, and involving science from the children's perspective, provided clear evidence of the engagement of children in understanding basic concepts of biology and ecology associated to biodiversity in Mediterranean region but also to ecosystems dynamics and to human well-being (e.g. dietary). The combination of science with children's explorations, questions, and affinities, revealed to promote curiosity and attitude gains in the context of BESD.

Children were critical actors in planning and developing the ateliers but also to extend the topics explored to their families. Children revealed distinct interests in different topics and actions. The interdisciplinary and participatory research in the Mediterranean forest experience highlights the influence of combining science with creativity and collective actions, such as painting or cooking.

The mutual learning between researchers, kindergarten teachers, children, and artists, throughout the plan programme was focal, to: (i) select the topics, (ii) adapt the science knowledge to children's development and interests; (iii) introduce local and regional level in terms of BESD. The dialogue between the multiple actors, through the regular meetings, was relevant to identify gaps between science and education, and to enhance participatory approaches from the children's perspective. Such participatory approach at early childhood that includes children's perspectives and interests corroborate the success of mutual learning processes and can be inspiring to different areas of knowledge, not only life sciences and environment.

The participatory approaches require extra time at kindergarten to construct daily programmes with specific expertise, but can be significantly improved by the dialogues within the academy. Finally, we substantiate the interdisciplinary and mutual learning actions in the higher education curricula for future professionals.

**Acknowledgements** Data in this paper were generated as part of a study entitled «Exploring and Experiencing Mediterranean» inserted in the project «Montado: From Biodiversity to Ecosystem Services», co-funded by COMPETE—Competitiveness Factors Operational Program (FEDER) and National Agency Ciência Viva, Lisbon, Portugal (FCOMP-01-0124-FEDER-016867). João Miranda and João Monteiro also acknowledge the financial support provided by the grants within FCOMP-01-0124-FEDER-016867. The authors gratefully acknowledge to: António Carmo Gouveia, Catarina Moura, Helena Castro, Helena Freitas, and Sofia Costa, the scientific team at

CFE-Centre for Functional Ecology, University of Coimbra; Carla Pacheco and Rui Dias, the scientific at CCVE-Centro Ciência Viva of Estremoz, Estremoz, Portugal; radio of university of Coimbra the support in the radio programme; media of the University of Coimbra for the edition of videos; Angela Filipe—sociologist—the engagement and collaboration in the atelier "risotto of wild mushrooms"; Marta Laranjeira—teacher of ballet—the engagement in the creative dance; and Mário Montenegro—theatre Marionet—the support in the programme meetings.

## References

Azul, A. M. (2009). *Mushrooms Paul of Madriz.* Portugal: Coimbra University Press. (in Portuguese).

Azul, A. M. (2011). Dialogue and collective action for sustainability of cork oak in Portugal. e-cadernos CES *10*, 70–90 (in Portuguese).

Azul, A. M., et al. (2008). Mycorrhiza in the Mediterranean Basin: safety teaching and training. *Journal of Biological Education, 42*(3), 54–61.

Azul, A. M., Reis, C. S., & Freitas, H. (2009). *Let's take care of Earth: Doing little can change a lot*. Portugal: Coimbra University Press. (in Portuguese).

Azul, A. M., Sousa, J. P., Agerer, R., Martín, M. P., & Freitas, H. (2010). Land use practices and ectomycorrhizal fungal communities from oak woodlands dominated by *Quercus suber* L. considering drought scenarios. *Mycorrhiza, 20*, 73–88.

Bach-Faig, A., Berry, E. M., Lairon, D., Reguant, J., Trichopoulou, A., Dernini, S., et al. (2011). Mediterranean diet pyramid today. Science and cultural updates. *Public Health Nutrition, 14* (12A), 2274–2284.

Dernini, S., & Berry, E. M. (2015). Mediterranean diet: From a healthy diet to a sustainable dietary pattern. *Frontiers in Nutrition, 2*(15), 1–7.

Dewey, J. (1989). The school and society. The University of Chicago Press.

Edwards, C., Gandini, L., & Forman, G. (2011). Hundred languages of children, the: The Reggio Emilia experience in transformation: The Reggio Emilia experience in transformation.

European Environment Agency. (2015). The European environment state and outlook 2015, synthesis report.

Freire, P. (1996). Pedagogia da Autonomia - Saberes Necessários à Prática Educativa Editora Paz e Terra. Coleção Saberes.

Hinds, J., & Sparks, P. (2008). Engaging with the natural environment: The role of affective connection and identity. *Journal of Environmental Psychology, 28*, 109–120.

IPCC. (2014). *Climate change 2014: Impacts, adaptation and vulnerability*. Cambridge, UK and New York, USA: Cambridge University Press.

Piaget, J. (1926). *The child's conception of the world*. London: Routledge & Kegan Paul Ltd.

Smidt, S. (2013). *Introducing Malaguzzi*. London, UK: Routledge.

UNESCO. (2007). Drivers and barriers for implementing learning for sustainable development in pre-school through upper secondary and teacher education. http://www.unesco.org/new/en/education/themes/leading-the-international-agenda/education-for-sustainable-development/education-for-sustainable-development/

UNESCO. (2008). Contribution of early childhood education to a sustainable society. http://www.unesco.org/new/en/education/themes/leading-the-international-agenda/education-for-sustainable-development/education-for-sustainable-development/

UNESCO. (2015). Education for sustainable development (ESD). At: http://www.unesco.org/new/en/education/themes/leading-the-international-agenda/education-for-sustainable-development/

Vygotsky, L. S. (1967). Play and its role in the mental development of the child. *Soviet Psychology, 5*(3), 6–18.

Wall, D. H., Nielsen, U. N., & Six, J. (2015). Soil biodiversity and human health Perspective. *Nature, 528*, 69–76.

Wells, A. M. (2013). Exploration and development of early childhood ecological education. *Master dissertation*, University of Coimbra, Portugal. Retrieved from https://estudogeral.sib.uc.pt/jspui/handle/10316/24707

Wells, A. M., Williams, B., & Azul, A. M. (2015). *Mommy, why I am bird?*. Portugal: Coimbra University Press.

World Health Organization and Secretariat of the Convention on Biological Diversity. (2015). Connecting global priorities: Biodiversity and human health. A state of knowledge review. https://www.cbd.int/health/SOK-biodiversity-en.pdf

# Part V
# Online Education to Biodiversity and Sustainability Awareness

# Engaging ODL Students with Biodiversity Issues: A South African Case Study on the Role of ESD

Rudi W. Pretorius, Mathilda E. Brand and Leslie R. Brown

**Abstract**

Biodiversity is regarded as a key asset in safeguarding the well-being of future generations. The threat to biodiversity through indiscretionary human activities is increasingly gaining attention, from local to global scales. Biodiversity forms part of the agenda for Education for Sustainable Development (ESD) through addressing inter-linkages between the various components and systems comprising the environment. This has been illustrated throughout the Decade of Education for Sustainable Development, which ended in 2014. As a result, significant progress has been made at various levels on the understanding of the impact of human consumption on biodiversity, together with the sensitization of students in terms of their potential roles to curb habitat and species loss as well as environmental degradation. This chapter utilises a case study approach to reflect on the way biodiversity is dealt with through the blended approach to ESD in the Diploma in Nature Conservation offered by the University of South Africa (UNISA). This blended approach is unique due to the combination of open distance learning (ODL), practical sessions, and work-integrated learning (WIL). A review of the flexibility of a blended approach to ODL, the challenges that were experienced, the means through which these were addressed and a future perspective, concludes this chapter.

R.W. Pretorius (✉)
Department of Geography, University of South Africa, Private Bag X6, Florida 1710, South Africa
e-mail: pretorw@unisa.ac.za

M.E. Brand · L.R. Brown
Applied Behavioural Ecology and Ecosystem Research Unit, Department of Environmental Sciences, University of South Africa, Private Bag X6, Florida 1710, South Africa

 
P. Castro et al. (eds.), *Biodiversity and Education for Sustainable Development*, World Sustainability Series, DOI 10.1007/978-3-319-32318-3_18

**Keywords**
Biodiversity education · Education for sustainable development · Nature conservation training · Open and distance learning · Blended learning · Work-integrated learning

# 1 Introduction

## 1.1 Linkages Between Biodiversity and ESD

The relationship between biodiversity and education for sustainable development (ESD) was established in 1992, when the Convention on Biological Diversity (CBD) was signed during the United Nations Conference on Environment and Development (UNEP). Despite several improvements, reports 18 years later during the launch of the International Biodiversity Year indicated that biodiversity loss was continuing at an unprecedented rate (Ki-Moon 2010). A variety of factors have been suggested to explain the lack of results in terms of the CBD. These include lacking political will, institutional weakness inadequate public education, to mention a few (COP6 Decision VI/26 2002). This fuelled the realisation that a turn-around strategy in terms of biodiversity loss should not only focus on technocratic solutions, but also need to incorporate communication, education and awareness (SCBD 2010). It is therefore not surprising that biodiversity and the need for its conservation is recognised as key theme in ESD (Collins-Figueroa 2012). ESD addresses biodiversity through a focus on linkages and interactions between the various subsystems comprising the environment, together with human activities related to agriculture, industry, urbanisation, livelihoods and more (Ramadoss and Poyya Moli 2011). Following the United Nations Decade of Education for Sustainable Development (UNDESD) which ended in 2014, it is now recognised that ESD constitutes a key element of educational quality, and is crucial to facilitate sustainable development (UNESCO 2014).

## 1.2 Biodiversity, ESD and the African Context

The African environment is notorious for challenges posed to people's livelihoods, associated with climate change, deforestation, resource exploitation, deterioration of ecosystems, and water quality issues. The impacts of these issues are frequently intensified because of wide-spread occurrence of poverty, food insecurity, instability, disease, drought, water provision, and sanitation problems (Togo 2009). Taking it a step further, the ultimate compounding factor as identified by

Togo (2009), is Africa's characteristic low capacity to respond to these challenges, associated with the fact that countries in Africa generally do not meet the human development threshold of 0.8 (Louw 2013). It has therefore been indicated that in the African context, the need for expanding access to education in general, but specifically to higher education, is huge and of an urgent nature (Barasa 2011). This challenge has to be considered against the background of Africa's vision to attain the Millennium Development Goals and the recognition of education as primary means by which this vision may be attained (African Union 2006). Taking the perspective of ESD, for which there is a crucial need to expand in scope in a relative short time period, the limited access to higher education in Africa is problematic. Achieving a turn-around in biodiversity loss is similarly dependant on matters such as capacity development, awareness-raising, engaging indigenous communities and mainstreaming biodiversity targets, to mention a few (Chandra and Idrisova 2011). Despite certain constraints, open and distance learning (ODL) has the ability to respond to many of these challenges, especially since ODL institutions have realised the need to transform to more engaging pedagogies (e.g. Unisa 2007).

## 1.3 Position Statement, Aim and Methodology

The position taken on biodiversity in this chapter, concurs with Van As et al. (2012), namely that biodiversity refers to the diversity of life on Earth in all forms at all levels of organisation within all ecosystems. Biodiversity conservation and teaching therefore have to focus on all habitats, ecosystems and biomes on Earth. The study programme under scrutiny in this chapter is the Diploma in Nature Conservation offered by the University of South Africa (Unisa), which is an ODL institution. This study programme has direct linkages with biodiversity conservation and ESD, and is well suited to the theme of the book. A case study format is used to review this study programme, with focus on links with biodiversity and ESD. The value of the case study approach is to allow contextualised description and analysis of the study programme as it is offered, with addition of critical reflection. Following the case study, the results of an assessment of the study programme is provided, firstly in terms of pivots to guide programs for biodiversity education (Kassas 2002), and secondly in terms of processes underpinning ESD (Tilbury 2011). This chapter contributes to the debate on access to sustainability related study programmes in higher education, while recognising that conservation science requires theoretical, practical and work related learning. The case study presented contributes to better understanding of the challenges faced by ESD, and how these challenges are addressed in the Diploma in Nature Conservation of Unisa through a blend of ODL, practical sessions and work-integrated learning (WIL).

## 2 Perspectives on Dealing with Biodiversity: The Higher Education Context

### 2.1 Review of Successes and Failures in Biodiversity Education

Navarro-Perez and Tidball (2012) refer to the results of several surveys on biodiversity awareness that have been conducted in a number of countries since the launch of the CBD in 1992. The results being reported confirm that levels of awareness of biodiversity issues by the general public as well as children still at school are generally low. In addition, it is clear that strategies (also in terms of education) that have been implemented to stimulate interest in and support of biodiversity restoration and conservation, have generally not succeeded in getting their message across very well. Despite these low levels of awareness being reported, national and international agreements concerning biodiversity have continued to gain relevance, resulting in frameworks to involve nations in the protection of biodiversity that have become commonplace, with organizations such as the International Union for the Conservation of Nature (IUCN) that has become well-known for contributions in this regard. There appears to be consensus that education has an important role to play in terms of transformation of attitudes towards nature, which should have positive spinoffs in terms of sustainability and biodiversity conservation (Ehrlich and Pringle 2008). Based on the research by Navarro-Perez and Tidball (2012), four challenges in terms of biodiversity education have to be attended to: (1) Defining an appropriate approach for biodiversity education; (2) Handling of a concept that is regarded as ill-defined; (3) Reaching different and broad audiences through a meaningful message; (4) Reconnection of people with nature.

### 2.2 Requirements for Biodiversity Education in Terms of ESD

In terms of ESD, consideration of interdependence between the environment and community concerns about societal, economic and cultural matters cannot be over-emphasised (Collins-Figueroa 2012). From this viewpoint, 'traditional' biodiversity education, with ecological focus on various plants and animals within their habitats, can benefit from the dimensions associated with ESD, such as critical reflection, clarification of values, envisioning of more sustainable futures and responding in terms of various modes of applied learning (Tilbury 2011). Although ESD has the potential to contribute to biodiversity conservation, it is not clear how to go about this in the best possible way. As indicated in the literature (Collins-Figueroa 2010), success seems to depend on the provision of opportunities for transformative learning in biodiversity education, in combination with hands-on engagement with multidisciplinary and interdisciplinary projects in real-life contexts, preferably including local environments. In South Africa, the Biodiversity

Act (No 10 of 2004) mandates higher education to develop the required capacity to manage the biological diversity of the country (South Africa 2004). As argued by Zietsman and Pretorius (2006), this capacity does not require specialists with a focus on only certain topics in specific fields, but with a holistic perspective on biodiversity within its cultural, social and economic context. This concurs with the view of Huntley (2003) namely that the qualifications of professionals working in conservation areas need to include not only conservation biology, but also embrace sociology, economics, marketing and politics.

### 2.3 Implications of the ODL Context

Since ODL does not require on-campus presence, students have the freedom to participate in the job market and to conduct their studies from locations of their choice and at times (to an extent) blended with their schedules (Taylor 2006). Instead of attending classes, interaction between participants in ODL (student-lecturer and student-student) is increasingly relying on various media facilitated through information and computer technologies (ICTs). The current trend is towards online presence through the internet, but which is limiting where computer literacy is inadequate (Wright et al. 2009) and where infrastructure presents a challenge, highlighted by Oyedemi (2012) for South African students and Unwin et al. (2010) for the broader African context. With many ODL institutions currently transforming to more engaging pedagogies (e.g. Unisa 2007), ODL programs are slowly but surely moving beyond the image of inferior, theory based learning experiences (Barasa 2011), towards provision of high quality qualifications in line with requirements of the job market and focus areas as ESD. However, catering for study fields as engineering, medicine and nature conservation, which require significant hands-on experience, remains challenging in ODL. This can be addressed by implementing variants of blended learning, defined as integration of face-to-face with online approaches to teaching and learning (De George-Walker and Keeffe 2010), and augmented with practical sessions and exposure to the world of work, as illustrated by the case study in the next section of this chapter.

## 3 Case Study: National Diploma in Nature Conservation Offered by Unisa

### 3.1 Contextual Setting: Unisa and ODL

The study programme that is presented and reviewed in this section of the chapter is associated with the Department of Environmental Sciences at Unisa, with the staff members who are involved also linked to the Applied Behavioural Ecology and Ecosystem Research Unit of Unisa. Although being regarded as a pioneer of distance education more than 100 years ago, Unisa has since grown significantly and is now a major player in the field of ODL (Tait 2008). By providing study opportunities to more than 400,000 students from South Africa, Africa and other parts of the world, a choice of study fields are offered at Unisa at the level of certificate to degree (Unisa 2015a). This places Unisa in the league of mega ODL institutions worldwide. The vision of Unisa is "*The* African University shaping futures in the service of humanity" (Unisa 2015b). Since 2007 the strategy at Unisa has been transformed to streamline interaction between students and the university, among other things with appropriate ICT solutions, including a move towards e-learning (Unisa 2007).

## 3.2 Background on Study Programme

The Diploma in Nature Conservation (previously labelled as the National Diploma in Nature Conservation) comprises 360 credits and is pitched at Level 6 of the National Qualifications Framework of South Africa (for comparison, a three year B-degree is pitched at Level 7). Work-integrated learning (WIL) amounts to almost 25 % of these credits. This study programme was recently revised, and currently in transition between the previous and new curriculum. The rationale for this study programme is to provide input to the improvement of professional competence in the field of nature conservation on a regional and national level in Southern Africa. As a result, the curriculum has been compiled with the aim to deliver conservation officers that will be able to provide support to conservators and resource managers and/or actively participate in management related functions in natural resource management areas. In this way people that can participate in ecosystems management, environmental education, and ecotourism, are supplied to the conservation industry. This also supports sustainable utilisation of natural resources and the conservation of biodiversity in Southern Africa, and feeds into economic development. A last, but very important aim is to provide accessible opportunities for individual, career and entrepreneurial development to people from all social and cultural groups, thereby contributing to social and economic transformation.

## 3.3 Teaching and Learning Approach

In order to inform biodiversity and its management, it is important for conservation science to engage with students in different ways. Different modes of teaching should therefore be applied in order to enhance learning. This is supported by Finn et al. (2002), suggesting that ecological experiments, in addition to theoretical teaching, would greatly enhance the environmental consciousness of students. A trend to take into account is that new tertiary students in biological related fields generally have a lower appreciation and understanding of nature than students during past decades (Ens et al. 2014). This may be one of the reasons for the detachment of humans from nature, and an aspect to take note of during curriculum design, teaching and learning. The Diploma in Nature Conservation has been

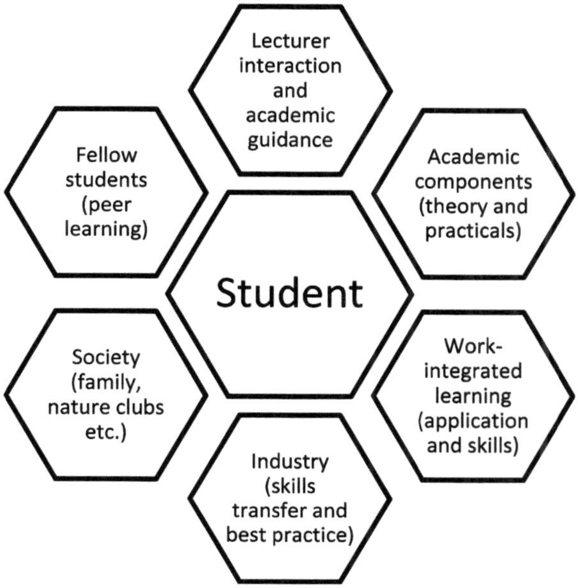


**Fig. 1** Components of the Unisa programme in nature conservation

developed with sustainable *veld* and game management in mind, while focussing on biodiversity conservation. Stakeholders that were consulted before and during curriculum development include representatives from the conservation industry, government departments, NGOs and tertiary institutions. A blended approach to facilitate teaching and learning in this ODL study programme, was deemed most suitable. The study components that have been blended include theoretical work, practical work and work-integrated learning. Concerning teaching approach, both ODL and face-to-face contact sessions are utilised. The different components comprising this study programme and how they come together to create the learning experience for students, are shown in Fig. 1.

## 3.4 Theoretical Component of Study Programme

The theoretical component of the study programme forms the basis to expose students to different aspects of biodiversity, within the context of sustainable *veld* and game management. The focus is distinctively South African. Four major areas are covered, namely Zoology, Botany, Ecology, and Resource Management. A matrix approach during curriculum development ensured that each area that is covered theoretically, forms part of the practical sessions as well. In this way integration between theory and practical modules is ensured. Table 1 provides a synopsis of the themes and topics that are covered. Basic terminology and ecological principles are dealt with during the first year of study, population dynamics and interactions during the second year and ecosystems and biodiversity management during the third year.

**Table 1** Themes utilised in the nature conservation programme

| Theme | Examples of topics covered |
|---|---|
| Plants | Classification, identification, anatomy, morphology, physiology, vegetation condition assessment |
| Animals | Classification, identification, anatomy, morphology, physiology, invertebrates and vertebrates, population dynamics, genetics |
| Ecology | Principles, processes, interactions, biogeochemical cycles, succession, freshwater, terrestrial and marine ecosystems, climate change |
| Resource management | Wildlife monitoring and census techniques, stocking rates, species selection; supplementary feeding |
| Technical aspects | Fences; water provision, soil science; soil erosion (prevention and rehabilitation), fire as a management tool, invasive plant control techniques, management of damage causing animals |
| General topics | Communication and nature interpretation, environmental education, general administration, legal and security aspects, history of conservation, relevant conventions and treaties, environmental citizenship, use of technology |

Examples of supportive modules that are included are Soil Science, Conservation Legislation, Communication and Fundamentals of Conservation. Basic technological skills form part of the curriculum and are practiced in most of the modules. The design of the study material for all modules promotes student engagement through active learning. The activities that are included aim to ensure that students do not only read the study material, but have to participate in the learning process as well. These activities may be for self-assessment or else form part of the formative assessment strategy in the form of assignments that have to be submitted and contribute towards the year mark portion of final marks.

## 3.5 Practical Component of Study Programme

A practical module is linked to each of the three levels of study so that the integration of the theory modules at each level is reinforced by means of hands-on field work in an appropriate nature-based environment. Examples of skills and techniques taught during these contact sessions include the following:

- Basic plant and animal identification keys;
- Game management (e.g. animal monitoring, census techniques, human-animal conflict);
- Technical aspects (e.g. fences, water provision, soil erosion, game capture);
- Vegetation management (e.g. *veld* condition assessment, use of fire, control of alien invasive plants; wetland delineation);
- Basic communication skills (oral and/or poster presentation, environmental education, nature guiding);
- Aquatic ecosystem (freshwater and marine), ecology, and management techniques.

All venues for practical sessions are selected to meet specific objectives. This is achieved through collaboration with industry partners and ensures that students are exposed to best practice. Practical activities are planned to emphasise the integration and interconnectivity of all components of the study programme. This increases the awareness of students of the importance of biodiversity management for sustainability. During the duration of the study programme, soft skills are also integrated in assessments and activities. Examples of these include critical thinking, working in a team, organising and managing activities, collecting and organising information, using science and technology effectively, reflection on learning and responsible environmental citizenship.

## 3.6 Work-Integrated Learning (WIL)

The final part of the nature conservation study programme is devoted to work-integrated learning (WIL), during which students are exposed to real-life situations in the natural environment. Students are required to complete six modules, consisting of a total of 30 activities (Unisa 2015c). These activities are linked to all the modules of the theoretical component of the study programme and include prescribed and elective topics. The prescribed topics are specifically set up so that the students acquire the necessary skills to be able to apply the principles of *veld* and game management. These activities are completed with assistance of a mentor and students have to submit a detailed report for assessment.

The final module serves as capstone and requires students to compile an ecological management plan for a specific protected area. Students, who are employed or have access to a specific protected area, may use real data to compile the ecological management plan according to the vision and mission of that area. However, if students are not employed or do not have access to a protected area, they are required to do a vegetation survey in a natural area chosen by themselves. This information is then used to provide these students with their own unique simulation, in which they use the vegetation information and all its associated data to compile an ecological management plan for the protected area (Unisa 2015d).

## 3.7 Reflection on Student Experience

Feedback received from students indicates that they value the role of study groups as part of their learning experience. These groups are formed voluntary once students have registered for a module and post their contact information in the discussion forums on myUnisa, the virtual learning platform used by Unisa. This form of interaction is encouraged as it forms part of Unisa's teaching and learning strategy and facilitates peer learning. The members of study groups are inclined to attend the same practical contact sessions, which they generally seem to regard as highlights in their study calendar. The different venues of the contact sessions offer opportunities to some students who have not visited such areas before (e.g. Maropeng and Mogale's Gate situated in the Cradle of Humankind—UNESCO World Heritage Site; or the Marakele National Park).

The challenges experienced by students do not necessarily have to do with the study programme or modules, but with the context of studying at an ODL institution. Some are, in fact, studying full-time, lack an income and therefore financing their studies is a challenge. Although this does not specifically relate to the context of conservation, it impacts on student's learning experience. Technological issues in terms of using the virtual learning platform present a further challenge. Some students work in remote nature conservation areas, where infrastructure to access the internet is lacking. As Unisa is increasingly relying on technology, issues like these pose a real challenge. Although such access is provided at the Unisa learning centres in various major cities and towns in South Africa, this does not always provide a practical solution to the problems students are experiencing.

### 3.8 Reflection on Lecturer's Experience

Teaching nature conservation to students from different backgrounds is challenging, but also rewarding. One of the greatest rewards is to experience the change in attitude and behaviour of students as they make progress through their studies. After engaging with the basics of sustainability and conservation, they are eager to implement recycling programmes, organise river clean-ups and become passionate about environmental education and biodiversity conservation in their respective communities. From a teaching perspective, the use of technology to provide additional resources complementary to the study material, presents a positive challenge. Teaching students to access relevant information to ensure that they get the best learning experience is very satisfying. The dedication shown by some students to conservation and sustainable use of resources provides proof of the success being achieved.

That being said—no study programme will ever be so successful that there are no challenges. Since the programme is offered in collaboration with the conservation industry, continuous reflection on the needs of graduates is required, in order to keep up with the challenges of an ever changing environment. Another major challenge is to ensure that students are actively involved in the learning process—which is achieved through regular communication and by supplying motivation. In this regard it indeed appears as if the teaching strategies and the blended approach of the nature conservation programme reduce the 'distance' in distance education. In terms of communication, and with the majority of students having English as second and even third of fourth language, the challenge is to find suitable approaches to reduce the impact of language barriers on the teaching and learning process.

# 4 Role of the Unisa Study Programme in ESD: Critical Review

In this section the Unisa study programme in nature conservation is critically reviewed in the context of firstly the approach taken to biodiversity education, and secondly the extent to which this approach relates to ESD. In terms of the approach to biodiversity education, the review is based on the alignment between the Unisa programme and the five pivots to guide programs for biodiversity education as identified by Kassas (2002), often cited in recent literature as benchmark (e.g. Navarro-Perez and Tidball 2012; Saito 2013). To determine the extent to which the approach to biodiversity education followed at Unisa relates to ESD, the benchmark is provided by the processes underpinning ESD as identified by Tilbury (2011), recognised for her leading role during the UNDESD. In addition, both parts of this review address an issue that is not well researched yet, namely the challenges experienced when exposing students from differing contexts and with varying needs to biodiversity education, thus adding to the body of knowledge and experience in this regard.

## 4.1 Approach to Biodiversity Education

Scope in terms of view of biodiversity: Since ‘biodiversity’ has different meanings to different stakeholders, study programmes in higher education need to be tailored to take this into account. However, since the Unisa programme has been designed with vocational training in mind, the scope had to be narrowed down. Sustainable use and management of biodiversity as resource thus emerged as golden thread running through the programme. The programme furthermore aims to strike a balance between the aims of vocational training of nature conservation technicians (which are very specific), and the aims associated with human development and life-long learning (which are more general).

Guiding perspectives: The Unisa programme focuses on conservation of biodiversity in the Southern African context. As a result course materials are designed with a view to engage students to all facets of the Southern African environment as an integrated whole, including the socio-economic and socio-political environments. Students need to be aware of the needs and challenges associated with this region and how to deal with these (e.g. lack of sufficient useable water, poaching, cultural diversity, food security, intensified development, etc.). To achieve this, the programme aims to get students to learn (from theory modules) by doing and implementing (through practical sessions and WIL).

Matching of actors, stakeholders and aims: For biodiversity education to be effective, the aims of actors (lecturers and students) need to be matched to those of stakeholders (industry, government, etc.). For the Unisa programme this match is achieved through an Advisory Committee (comprising of representatives from the conservation industry, academics and educational consultants). This committee provides advice on changing needs and trends which need to be incorporated in the curriculum. Linked to the guiding perspective and scope, the curriculum and course materials are then adapted accordingly to ensure that at successful completion of the programme, students will meet the set requirements.

Themes/sites utilised in the programme: This pivot refers to themes/sites for biodiversity education such as a garden attached to the campus, or accessible wetlands, river banks, or nature reserves. Selection of these will relate to the guiding perspective, associated aims and required stakeholders-actor match. In the case of the Unisa programme, the practical sessions offer some students the opportunity to visit places they would else never have been to (e.g. Maropeng, situated in the Cradle of Humankind—a World Heritage Site). Similarly the course dealing with marine environments presents the first opportunity for many students to experience the ocean and to observe/identify organisms usually just seen in textbooks/media.

Assimilation of programme: It is important that the study programme, in the way that it has been designed and implemented, produces its target and that all actors have undertaken their roles as envisaged. Means to evaluate this aspect therefore need to be developed and put into place. Since the Unisa programme has specifically been designed to deliver students that meet the requirements set by the conservation industry, the programme is deemed successful if students are indeed taken up in positions in the conservation industry. Constant feedback from industry (through ad hoc discussions and formally through the Advisory Committee) ensures that the programme is updated to meet the needs of the industry.

## 4.2 Relationship with ESD

Inclusion of collaboration and dialogue: ESD emphasizes engagement in collaboration and dialogue to increase capacity and to facilitate learning towards sustainable development. In the Unisa programme, this aspect features through participation in discussion forums on myUnisa, and in addition to that, especially during attendance of and participation in the practical sessions. On a different level, feedback from industry (through ad hoc discussions and through the Advisory Committee) ensures the relevance of the programme. In addition, collaborative research with industry creates new information, which is transferred to students so that they are exposed to best practice in conservation and biodiversity management.

Utilisation of active and participatory learning: Regarding a suitable pedagogical approach, it is generally accepted that active and participatory learning forms the core of ESD, although supporting empirical evidence is still largely lacking. The blended model of teaching and learning, as implemented in the Unisa programme, provides for theoretical learning as well as practical application. This ensures that students do not only study theory from texts, but perform activities on their own (e.g. assignments and projects) and participate in group work as well (practical sessions and WIL). Utilisation of the myUnisa virtual study platform provides further opportunities to support active and participatory learning.

Curriculum focussed on 'whole system': This requirement involves steps to be taken to ensure engagement between ESD focused study programmes and the 'whole system' in which such programmes are based and through which they operate. Engagement with the system as a whole is regarded as vital to make progress in terms of learning for sustainable development. In the context of the Unisa programme, the role of the Advisory Committee is to ensure that that a 'whole system' approach is adhered to. On a different level, the matrix approach to curriculum development in this programme ensures continuity and the focus to remain on biodiversity and the sustainable management thereof in the South African context.

Stimulation of innovation and transformative learning: Although acquiring knowledge, values and theories about sustainable development is important, ESD is also about changing mindsets in terms of sustainable futures, whereas learning processes and practices that are in place, mostly do not take this transformative view into account. In the Unisa programme, this transformation is encouraged through interaction with nature and aspects of biodiversity that is required though-out the programme. The change in attitude and thinking is proven by the increased number of students employed in the conservation industry and whose progress through the ranks of various organisations has proven to be beneficial to the programme.

## 5 Conclusion

### 5.1 Potential of ODL for Biodiversity Education

Using the Diploma in Nature Conservation of Unisa as example, this chapter highlighted that ODL has a huge potential to fulfil in terms of biodiversity education, linked to ESD. This is especially true in the African context, where the threat to biodiversity is characterised by continent specific challenges, including lacking capacity to deal with the problem, together with issues limiting access to higher education. However, courses that focus only on theory, and do not provide hands-on, practical engagement with nature, stand little chance of success. To this end addition of real time practical sessions in nature to enhance student's understanding of natural processes and biodiversity, proved to be successful for the Unisa study programme. The final component of the blended teaching and learning model being used comprises work integrated learning, which exposes students to authentic learning in real world contexts.

### 5.2 Biodiversity in the Context of ESD

Consideration of the Unisa study programme in nature conservation suggests a synergy and integration between biodiversity education and ESD through collaboration and dialogue on various levels: between students, between students and lecturers, and between lecturers and stakeholders. This results in the curriculum to maintain its relevance for biodiversity issues in local contexts, to show responsiveness to the needs of a range of stakeholders and to stay in line with industry requirements. Lecturers are furthermore empowered to facilitate active learning within a participatory context, which peaks during the practical sessions when both lecturers and students have opportunity to enact their practice through relevant ways. The project, theme-oriented approach encapsulated by the WIL component allows 'whole system' learning and facilitates the multi-perspective approach to studying biodiversity within ESD.

### 5.3 Future Perspective

The future of transformative learning in ESD requires more emphasis on integration between environmental, economic, societal and cultural issues, while incorporating the processes of active learning, participation and collaboration in teaching and learning to a greater extent than currently the case. To this end, a theme as biodiversity can be used effectively to bridge the gap between human and physical sciences and to facilitate the move towards approaches that emphasize interdisciplinarity. Although the role played by ICTs is becoming increasingly important in the facilitation of ESD, consideration of themes such as biodiversity illustrate that teaching and learning in some fields will always be in need of practical, hands-on and/or work-based experiences, where direct guidance/support by physically present mentors/peers are required, which cannot be replaced by online learning.

## References

African Union. (2006). Second Decade of Education for Africa (2006–2015): Plan of Action. Retrieved from http://www.unesco.org/new/fileadmin/MULTIMEDIA/FIELD/Dakar/pdf/AU%20SECOND%20DECADE%20ON%20EDUCTAION%202006-2015.pdf

Barasa, F. (2011). Promoting open and distance learning in Africa: A critical reflection on rhetoric, real and ideal. Retrieved from http://events.aau.org/userfiles/file/corevip11/papers/fred_barasa_Promoting_ODL.pdf

Chandra, A., & Idrisova, A. (2011). Convention on biological diversity: A review of national challenges and opportunities for implementation. *Biodiversity Conservation, 20*, 3295–3316.

Collins-Figueroa, M. (Ed.). (2010). *Learning in and beyond the classroom: Biodiversity initiatives in teacher education*. Kingston: Joint Board of Teacher Education and Jamaica Environment Trust.

Collins-Figueroa, M. (2012). Biodiversity and education for sustainable development in teacher education programmes of four Jamaican educational institutions. *Journal of Education for Sustainable Development, 6*(2), 253–267.

COP 6 Decision VI/26. (2002). *Strategic plan for the convention on biological diversity*. Retrieved from https://www.cbd.int/decision/cop/default.shtml?id=7200

De-George-Walker, L., & Keeffe, M. (2010). Self-determined blended learning: A case study of blended learning design". *Higher Education Research and Development, 29*(1), 1–13.

Ehrlich, P. R., & Pringle, R. M. (2008). Where does biodiversity go from here? A grim business-as-usual forecast and a hopeful portfolio of partial solutions. *The National Academy of Sciences, 105*(1), 11579–11586.

Ens, E., Burns, E., Russes-Smith, J., Sparrow, B., & Wardle, G. (2014). The cultural imperative: Broadening the vision of long-term ecological monitoring to enhance environmental policy and management outcomes. In E. Burns, D. Lindenmayer, A. Lowe & N. Thurgate (Eds), *Biodiversity and environmental change: Monitoring, challenges and direction* (pp. 83–107). Collingwood: CSIRO.

Finn, H., Maxwell, M., & Calver, M. (2002). Why does experimentation matter in teaching ecology? *Journal of Biological Education, 36*, 158–162.

Huntley, B. J. (2003). Developing the capacity to manage protected areas—South African experience. In G. I. Cowan, J. Yawitch & M. Swift. (Eds.), *Strategic innovations in biodiversity conservation—The South African experience*. Pretoria: Department of Environmental Affairs and Tourism.

Kassas, M. (2002). Environmental education: Biodiversity. *The Environmentalist, 22*, 345–351.

Ki-Moon, B. (2010). Secretary-general's remarks at the event to launch the international biodiversity year at the American Museum of Natural History. Retrieved from www.un.org/sg/STATEMENTS/index.asp?nid=4386#

Louw, W. (2013). Green curriculum: Sustainable learning at a higher education institution. *The International Review of Research in Open and Distributed Learning, 14*(1).

Navarro-Perez, M., & Tidball, K. G. (2012). Challenges of biodiversity education: A review of education strategies for biodiversity education. *International Electronic Journal of Environmental Education, 2*(1), 13–30.

Oyedemi, T. K. (2012). Digital inequalities and implications for social inequalities: A study of internet penetration amongst university students in South Africa. *Telematics and Informatics, 29*(3), 302–313.

Ramadoss, A., & Poyya Moli, G. (2011). Biodiversity conservation through environmental education for sustainable development—A case study from Puducherry, India. *International Electronic Journal of Environmental Education, 1*(2), 97–111.

Saito, C. H. (2013). Environmental education and biodiversity concern: Beyond the ecological literacy. *American Journal of Agricultural and Biological Sciences, 8*(1), 12–27.

SCBD—Secretariat of the CBD. (2010). *Global biodiversity outlook 3*. Montreal: CBD. Retrieved from http://www.cbd.int/gbo3/ebookCBD

South Africa. (2004). *National environmental management: Biodiversity Act*, No 107 of 1998. Pretoria: Government Printer.

Tait, A. (2008). What are open universities for? *Open Learning: The Journal of Open, Distance and e-Learning, 23*(2), 85–93.

Taylor, D. (2006). *Issues, trends and challenges in distance education: An international perspective*. Retrieved from http://ejournal.narotama.ac.id/files/Issues,%20Trends,%20and%20Challenges%20in%20Distance%20Education%20%20An%20International%20Perspective.pdf

Tilbury, D. (2011). *Education for sustainable development: An expert review of processes and learning*. Paris: UNESCO.

Togo, M. (2009). *A systems approach to mainstreaming environment and sustainability in universities: The case of Rhodes University, South Africa* (Ph.D. thesis). Rhodes: Rhodes University.

UNESCO—United Nations Educational, Scientific and Cultural Organisation. (2014). *Roadmap for Implementing the global action programme on education for sustainable development*. Paris: UNESCO.

Unisa—University of South Africa. (2007). *Self-evaluation portfolio for the commonwealth of learning trail audit*. Pretoria: Unisa.

Unisa. (2015a). *Unisa online—about: Facts and figures*. Retrieved from http://www.unisa.ac.za/Default.asp?Cmd=ViewContent&ContentID=15888

Unisa. (2015b). *Unisa strategic plan: 2016–2030*. Pretoria: Unisa.

Unisa. (2015c). *Nature conservation application IA (NCA2603) tutorial letter 101/2015*. Pretoria: Unisa.

Unisa. (2015d). *Nature conservation application IVA (NCA3701) tutorial letter 101/2015*. Pretoria: Unisa.

Unwin, T., Kleessen, B., Hollow, D., Williams, J. B., Oloo, L. M., Alwala, J., Mutimucuio, I., Eduardo, F., Muianga, X. (2010). Digital learning management systems in Africa: Myths and Realities, *Open Learning, 25*(1), 5–23

Van As, J., Du Preez, J., Brown, L., & Smit, N. (2012). *The story of life and the environment: An African perspective*. Cape Town: Struik.

Wright, C. R., Dhanarajan, G., & Reju, S. A. (2009). Recurring issues encountered by distance educators in developing and emerging nations. *International Review of Research in Open and Distance Learning, 10*(1), 1–25.

Zietsman, S., & Pretorius, R. W. (2006). Learning programmes for environmental sustainability: A different approach to curriculum design. *South African Journal of Higher Education, 20*(5), 691–702.

# Online Teaching for Biodiversity Conservation

Ulisses Miranda Azeiteiro and Paula Bacelar-Nicolau

**Abstract**

The aim of this work was to present the Biodiversity and Conservation teaching contents, teaching and learning methodologies and their results/effectiveness through e-learning in higher education (HE) at the Universidade Aberta (UAb), Portugal. Two curricular units covering the thematic of Biodiversity Conservation are taught in the three cycle degree system at UAb: *Biodiversity and Conservation* (compulsory CU, 6 ECTS, integrated in the curricular plan of the *major* in Environmental Sciences) and *Biodiversity, Geodiversity and Conservation* (optional CU, 6 ECTS, integrated in the curricular plan of the Master degree in Environmental Citizenship and Participation). Based on the teaching experience of nine editions of the e-learning courses on Biodiversity Conservation, particularly on the students perceptions and on the number of completed dissertations on the topic of Biodiversity, we consider that their main objectives were fulfilled, i.e. knowledge acquisitions about fundamentals on biodiversity conservation, interpretation, and applications. Some improvements are needed: foster the competencies needed among teachers to use e-learning in a meaningful way; materials suitable for use on e-learning programmes and the need to create synergies and interfaces between scientific knowledge and traditional knowledge (essential in this area of Biodiversity conservation) and also the lead role that e-learning can play establishing and developing a broader awareness on biodiversity conservation.

U.M. Azeiteiro (✉) · P. Bacelar-Nicolau
Department of Science and Technology,
Universidade Aberta and Centre for Functional Ecology,
University of Coimbra, Coimbra, Portugal
e-mail: paula.nicolau@uab.pt



P. Castro et al. (eds.), *Biodiversity and Education for Sustainable Development*,
World Sustainability Series, DOI 10.1007/978-3-319-32318-3_19

**Keywords**
E-learning · Biodiversity · Bachelor · Master

# 1 Introduction

E-learning in higher education can be of great relevance in effective life-long learning education for sustainable development in a population of students who are simultaneously full time employees (Azeiteiro et al. 2014, 2015; Martinho et al. 2014, 2016). Azeiteiro et al. (2015) assess the Education of Sustainable Development effectiveness through e-learning in higher education in a case study at the Universidade Aberta, the Portuguese Distance Learning University. The expectations and experience of students who enrolled in more than one programme in environmental and/or sustainability science offered by Universidade Aberta were analysed in six dimensions: general expectations; learning quality; teaching resources, pedagogical tools, and evaluation; acquired competences in education for sustainable development; satisfaction and interactions; and reasons to pursue enrolment in a new programme at Universidade Aberta. Authors conclude that the surveyed students felt that they attained a high level of motivation and satisfaction, and had reached an effective learning outcome of knowledge, competences, values, attitudes and behaviour in environment and/or sustainability sciences. Azeiteiro et al. (2015) results are in line with previous works (Bacelar-Nicolau et al. 2009, 2012) and let us conclude that formal e-learning programmes can provide an effective alternative to face-to-face training (see also Moura et al. 2010), allowing students to pursue their studies, in a flexible, collaborative and interactive way, whilst holding down full time jobs (Bacelar-Nicolau et al. 2015). In the environmental and sustainability science fields, online programmes may produce the same level of student performance as in face-to-face courses, with comparable learning outcomes, and with high levels of satisfaction (Azeiteiro et al. 2015; Bacelar-Nicolau et al. 2009, 2012).

The teacher's education mission, contents and goals for biodiversity conservation can be achieved through transdisciplinary approaches, innovative teaching methods and e-learning (see for this purpose the work from Huettmann (2015). The aim of this work was to present the Biodiversity and Conservation teaching contents, teaching and learning methodologies and their results/effectiveness through e-learning in higher education (HE) at the Universidade Aberta, Portugal.

# 2 Biodiversity Conservation Teaching and Education for Sustainable Development

Biodiversity and Conservation teaching is essential for Education for Sustainable Development (ESD). The central reason for this is what we define as *Ecosystem Services* (Millennium Ecosystem Assessment [MEA] 2005). Raising awareness of the critical role that biodiversity plays in ensuring environmental sustainability, economic prosperity, and social and cultural well-being is essential for ESD. It is necessary, in order to support sustainable development.

The 2006 Communication from the European Commission—'Halting Biodiversity Loss by 2010—and Beyond: Sustaining ecosystem services for human well-being'—underlined the importance of biodiversity protection as a pre-requisite for sustainable development (SD), and set out a detailed Biodiversity Action Plan to achieve this. At the moment, European Union Biodiversity Targets for 2020 ("Halting the loss of biodiversity and the degradation of ecosystem services in the EU by 2020, and restoring them in so far as feasible, while stepping up the EU contribution to averting global biodiversity") and The 2050 EU Biodiversity Vision ("By 2050, European Union biodiversity and the ecosystem services it provides—its natural capital—are protected, valued and appropriately restored for biodiversity's intrinsic value and for their essential contribution to human well-being and economic prosperity, and so that catastrophic changes caused by the loss of biodiversity are avoided") emphasises the importance of the Biodiversity Conservation.

However, there is little public perception of which *taxa* are most important in terms of their total biomass, biodiversity or the ecosystem services they perform. Such awareness is important for conservation, as without appreciation of their value and conservation status, species are unlikely to receive adequate conservation protection (Snaddon et al. 2008). Biodiversity and ecosystem services appear to be under-represented in teaching environmental sustainability and these are essential components in developing an understanding of natural capital. More holistic approaches between nature, society and culture are needed (Alves et al. 2013). In our analysis, we consider that it is essential to deconstruct the social and environmental development, giving priority to the structural aspects that have a close influence on social, cultural and environmental factors, and particularly to the agency of individuals (demonstrating the plural knowledge they convey), and to the spaces of empowerment and participatory democracy that they require (Alves et al. 2013).

In this area of Biodiversity, Conservation, and Ecosystem services, scientific knowledge must relate with personal, local and cultural diversity.

## 3 E-learning and Online Teaching

Technology has had a major influence on the ways that information is generated and accessed, and increasingly, on teaching and learning in higher education (see Dlouha et al. 2013). A growing body of research has described the benefits (flexibility, interaction, teaching presence, collaborative learning, and a great sense of community and interaction among peers, as well as with teachers, is privileged by online students promoting the existence of a learning community) of online technologies in this context (see Azeiteiro et al. 2014, 2015, Halog and Dishman 2014; Leal Filho 2014; Sibbel 2014; Moura et al. 2010).

Through open learning environments, control is given to the learner through processes that allow for self-guidance and tools that facilitate individual and collaborative exploration of concepts (Barth and Burandt 2013) supported by three principles: *Self-directed learning*; *Collaborative learning* and *Problem Orientated Learning*. These principles are put in place in Online Distance Learning in the Modelo Pedagógico® (Pereira et al. 2008).

## 4 Case Study: The Universidade Aberta (The Portuguese Open Distance University)

Established in 1988, Universidade Aberta (UAb) is the only Distance Learning public higher education institution in Portugal. All pedagogical offers at UAb are integrated in the Bologna European process and are organised according to the European Credit Transfer and Accumulation System (ECTS). All programmes are taught in e-learning since 2008, and the open source Moodle (http://eLearning.uab.pt/) is used as the course management system. UAb is a reference HE European institution in the area of online and advanced e-learning and learning, using the most advanced information and communication technologies, and is acknowledged by its virtual pedagogical model (Pereira et al. 2008). UAb has more than 12,000 students and offers more than 40 degrees and several life-long study programmes. All pedagogical offer is directed to an adult public (over 21 years old), who are mostly working-students seeking professional development. The UAb was awarded the international Prize of European Foundation for Quality in e-learning and certification of The UNIQUE Quality Label for the use of ICT in HE. It was also awarded the 1st Level of Excellency of the European Foundation for Quality Management, and for the Diploma Supplement Label, by the European Commission. UAb was also considered as a key player in the HE system both in Portugal and in Europe (European Commission/EACEA/Eurydice 2014) (from Azeiteiro et al. 2015).

The virtual pedagogical model at UAb is based on four major principles: (i) student-centred learning, (ii) flexibility (access to learning without pressure of time and space, with primacy for asynchronous communication), (iii) interaction (student-teacher, student-student and student-learning resources, and also socially contextualised), and (iv) digital inclusion (Pereira et al. 2008).

A core aspect of the pedagogical model is the Plan of Curricular Unit—for undergraduate courses, or the Learning Contract—for postgraduate courses (Figs. 1 and 3), which guide the teaching and learning process in each curricular unit (CU). These documents, developed by the teacher of each CU, are structured into topics, outcomes, competences to be acquired or developed, list of diversified learning materials, learning methodology, e-activities, assessment, and timetable for developing learning activities. Besides the details that are discussed below, the two documents differ by the fact that the Plan of Curricular Unit is more static, in nature and content, while the Learning Contract is regarded as an agreement between

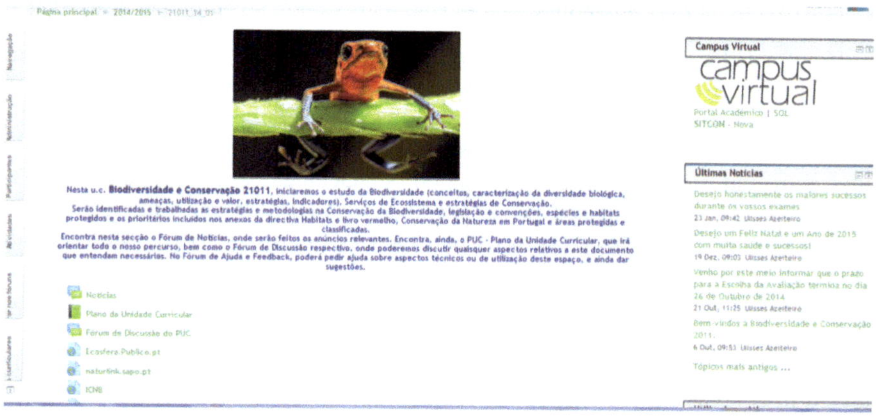


**Fig. 1** Main page of the undergraduation CU on Biodiversity and Conservation, with emphasis to the Plan of Curricular Unit ("Plano de Unidade Curricular") and to the fora—News ("Notícias") and General Forum ("Fórum de Discussão do PUC")

students and teacher, who may contain some degree on openness to debate e.g. deadline adjustments. The virtual pedagogical model privileges continuous assessment (through the e-activities), which weigh 40 % (in undergraduate courses) or a minimum of 60 % (in postgraduate courses) of the final mark. Face-to-face final exams are only mandatory in undergraduate courses. Different learning strategies are developed depending on the study cycles: from mainly individual assignments (in undergraduate courses) to more collaborative and cooperative learning where peer dimension interaction becomes essential (in MSc and PhD).

A two week online introductory module is available for students before the start of the curricular year, which is intended to familiarise the students with the virtual environment and the e-learning tools, as well as promoting the acquisition of online communication and online social skills. All the courses have virtual class sites, a coordination site for student matters, a coordination site for the teacher's team, an online secretariat, and a virtual "café".

## 5 Education for Sustainable Development Through E-learning at UAb

In line with the global requirement for an education for sustainable development, the UAb has developed a three cycle degree system, according to the reform of the "Bologna process", which aims to actively promote education for sustainable development, along with an increase in transdisciplinarity across subjects, and student cooperation:

- first cycle—Bachelor degree in Environmental Sciences (180 ECTS);
- second cycle—Master degree in Participation and Environmental Citizenship (100 ECTS); and

- third cycle—PhD degree in Social Sustainability and Development (180 ECTS).

The Bachelor degree in Environmental Sciences is a b-learning programme, which began in the academic year 2007. The general purpose of the course is to promote and develop a set of professional skills and competences within the Environmental Sciences, and includes three optional minors: Natural Heritage, Environmental Health, and Environmental Management and Sustainability. The first two years are composed of 20 mandatory curricular units of Science and Environmental Technology, Biological Sciences, Earth Science, Mathematics, Chemistry, Physics, and Legal Sciences. With the exception of two curricular units (Fieldwork I and II, that include, each, a face-to-face week), the undergraduate course follows UAb's pedagogical model in its virtual class regime.

The Master degree in Participation and Environmental Citizenship was initially offered in 2005 as a b-learning programme (all CU are offered in e-learning, except for one which includes a one day face-to-face session). The main purpose of this programme is to develop competences in the area of Sustainable Development, covering both the environmental and the social aspects.

Lastly, the PhD degree in Social Sustainability and Development was developed as a natural sequence, due to the increasing demand for an advanced specialisation and research in sustainable development science, both from the professionals connected to the tertiary sector and public sector, as well as educators, managers in business and international organisations, whose area of activity is socio-environmental sustainability. This PhD aims to contribute to forming a generation of leading individuals capable of understanding the different meanings and implications of changes in global, social, and human systems and who choose the path of sustainability in implementing policies on the basis of this understanding.

Thus, although the three cycle degree system is not fully specialized in SD, it transversally integrates SD into the courses of each programme and increases focus on SD, primarily at the Master and PhD levels covering all aspects of SD.

The evaluation of sustainability programmes at UAb started with the distance learning experience of curricular units of Ecology, Environmental Education and Environmental Sciences (Caeiro et al. 2001, 2004; Caeiro and Azeiteiro 2004, 2007). The transition from the open and distance education to the e-learning, and namely the e-learning for the environment, was described in terms of new challenges for e-learning in environmental sciences by Martinho et al. (2010), Bacelar-Nicolau et al. (2007, 2009), Amador et al. (2008), and Caeiro et al. (2008). In summary, the studies of Bacelar-Nicolau et al. (2007, 2009, 2012), Martinho et al. (2010), Moura et al. (2010) showed: (i) a very high level of student motivation and satisfaction with the online programmes (changing their attitudes about environmental domains and contributing to others' changing attitudes and behaviours), and (ii) that online students were highly satisfied with their courses (in terms of general expectations and acquired competences) perceiving environmental online education as effective as face-to-face education. Also Martinho et al. (2014) indicated that the 1st cycle e-learning programme in Environmental Sciences of UAb, was globally adjusted to employability requirements, through the acquisition and/or development of competences and skills, according to its graduates. Based on the results of a questionnaire survey applied to the master's students for 3 academic years, starting in 2007, Martinho et al. (2010) established a link between the knowledge acquired and changing attitudes and behaviours towards sustainability. However, the works of Moura et al. (2010) and Oliveira et al. (2012) reinforce the need for more laboratory and practical classes (b-learning). Also Amador et al. (2015) revealed a lack of consistency between the theoretical framework of ESD during the curricular year and the implementation of *praxis* during the dissertation. The new virtual technologies are important but not sufficient, because they do not encourage the development of key learning skills, attitudes and values towards environmental conservation and sustainability development to the same level as face-to-face fieldwork (Oliveira et al. 2012).

## 6 Curricular Units on Biodiversity Conservation

Two curricular units covering the thematic of Biodiversity Conservation are taught in the three cycle degree system at UAb: *Biodiversity and Conservation* (compulsory CU, 6 ECTS, integrated in the curricular plan of the *major* in Environmental Sciences; Table 1; Figs. 1 and 2) and *Biodiversity, Geodiversity and Conservation* (optional CU, 6 ECTS, integrated in the curricular plan of the Master degree; Table 2; Figs. 3, 4, and 5).

**Table 1** Characterisation of biodiversity and conservation, curricular unit of the 1st cycle major in environmental sciences

| *Curricular unit* |
|---|
| 6 ECTS<br>Mandatory<br>Online teaching contact: 26 h contact (in 156 h, total study hours of student) |
| *Learning outcomes* |
| • It is intended that at the end of this CU, students should be able to:<br>• Know, understand and apply the Concepts about Biodiversity and Ecosystem Services<br>• Describe and explain the major argumentation about Species and Habitats Conservation<br>• Describe and explain the major argumentation about Biodiversity Conservation<br>• Explain the threats to Biodiversity<br>• Know Species and Habitats Conservation Priorities<br>• Know Species Conservation Strategies<br>• Know the Instruments and Nature and Species Conservation Policies in Portugal and European Union<br>• Recognise that Biodiversity is the very basis of the Territory Management Plans and Economic Activities |
| *Syllabus* |
| • Biodiversity and Ecosystem Services. The Study and Biodiversity Conservation (Concepts, Quantification, Threats, Valuation, Ecosystem Services and Conservation) |

(continued)

**Table 1** (continued)

| |
|---|
| *Curricular unit* |
| • Biodiversity: Indicators, Strategies and Instruments (the European Union Strategies, Indicators, the Portuguese Situation)<br>• Conservation (Nature Conservation Strategies, Legislation and Conventions, Protected Areas, Red Lists, Natura 2000 as the centrepiece of EU nature and biodiversity policy, Birds Sites and Habitats Sites, biogeographical regions) |
| *Teaching-learning methodologies (including evaluation)* |
| According to the pedagogical virtual model of UAb and its four major principles (see Sect. 3)<br>Continuous assessment (with teacher's formative feedback) is privileged: 2 or 3 digital written assignments (e-folios) during the semester (40 %) and a face-to-face final exam (p-folio) at the end of the semester (60 %). In due time, students can alternatively choose to perform one final face-to-face exam (100 %)<br>Students gradually practice their knowledge and competences acquisition in a series of self-evaluation formative activities<br>Students who choose final evaluation have access to the same study materials as those who choose continuous evaluation |

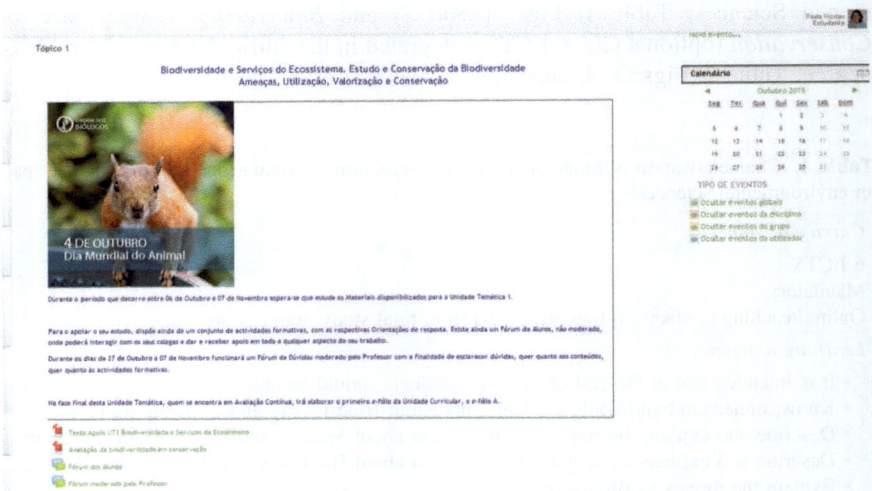


**Fig. 2** View of the topic on Biodiversity and Ecosystem Services (undergraduate CU on Biodiversity and Conservation), with emphasis to the topic's study guidelines, learning materials and asynchronous fora—Student Forum and Teacher moderated Forum

**Table 2** Characterisation of biodiversity, geodiversity and conservation, curricular unit of the master degree in environmental citizenship and participation

| |
|---|
| *Curricular unit* |
| 6 ECTS<br>Optional<br>Online teaching contact: 26 h contact (in 156 h, total study hours of student) |
| *Learning outcomes* |
| It is intended that at the end of this CU, students should be able to:<br>• Describe and debate the concepts and values of biodiversity<br>• Identify threats and destructive agents of biodiversity and conservation issues globally<br>• Recognize the values of geodiversity<br>• Articulate goals and set priorities for nature conservation<br>• Know national, European and international strategies for the conservation of nature<br>• Develop assignments of analytical and exploratory nature, in the scientific area of the CU, presented as written assignment or poster |
| *Syllabus* |
| • Geodiversity: Concepts, Value and threats<br>• Biodiversity: Concepts, Value and threats<br>• National strategies, European and international nature conservation (Biodiversity and Geodiversity)<br>• Urban Biodiversity<br>• Biodiversity and Climate Change |
| *Teaching-learning methodologies (including evaluation)* |
| According to the pedagogical virtual model of UAb and its four major principles (see Sect. 3)<br>In this CU only continuous evaluation is used (100 %). The evaluation activities (research, individual and collaborative work and debates with peer-review dimension) are designed for students to acquire and consolidate the acquired knowledge. The learning materials, course e-activities in the form of written assignments (individual and collaborative) and collaborative discussions allow students to achieve course competencies and objectives of knowledge acquisition, comprehension and practical application of the acquired consolidated knowledge<br>The continuous evaluation includes several tasks/strategies and pedagogical instruments e.g. blogs, individual and group projects, essays, Problem Based Learning, Case Studies, participation in the discussions, research reports and online tests). Individual evaluation activities can contemplate the elaboration of short papers or a project with online presentation, discussion and defence |

**Fig. 3** Main page of the postgraduation CU on Biodiversity, Geodiversity and Conservation. Emphasis is given to the Learning Contract ("Contrato de Aprendizagem"), asynchronous fora—News ("Notícias"), General Forum ("Forum Geral") and Help ("Ajuda"), link to online Library ("Biblioteca on line"), and to the student's blogs of this CU, which serve as porfolio, reflexion log and sharing intrument

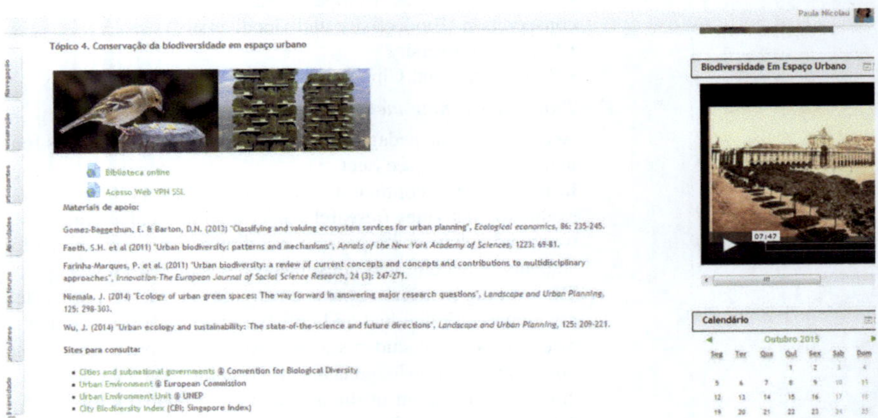


**Fig. 4** View of the topic on Biodiversity conservation in urban areas (postgraduation CU on Biodiversity, Geodiversity and Conservation), with emphasis to the diversified learning materials and link to online Library ("Biblioteca on line") for further individual research

## 7 The Msc Dissertation Research Topics in Biodiversity

Since its beginning, 34 % (17 in a total of 50) of the dissertations were concluded, within the thematic of Biodiversity: Education, teaching textbooks and biodiversity teaching (n = 2); Interactive Teaching Materials about Biodiversity Conservation and Protected Areas and Conservation Areas (n = 2); Biodiversity Literacy (n = 5);

**Fig. 5** View of the topic Online Seminar: Projects on Biodiversity and Geodiversity Conservation (of the postgraduation CU on Biodiversity, Geodiversity and Conservation). Emphasis is given to the Call for papers guidelines, individual research through online library, student-teacher fora for individual tutoring on preparing the scientific posters, e-Seminar forum for presentation and debate (including peer-review)

Biodiversity and Ecosystem Services (n = 2); Environmental governance for biodiversity (n = 1); Evaluation of Life projects (n = 1); Importance of fish biodiversity for the management of fisheries (n = 2); Perceptions and Values of Biodiversity (n = 1); Ethics and Biodiversity (n = 1). Taking into consideration that the Biodiversity and Geodiversity Conservation 2nd cycle CU is an optional CU, the research project choice on the topic biodiversity denotes the importance attributed by the students to the topic.

## 8 Biodiversity Teaching in Education for Sustainable Development: Some Conclusions

The main purpose of the above mentioned two e-learning curricular units is to provide graduate and postgraduate students the contents and competences on the field of Biodiversity Conservation. More specifically, the 1st cycle course introduces the basic contents and the 2nd cycle course provides higher competencies.

Based on the numbers of students who have completed the curricular units (namely progressing from the first cycle to the second cycle choosing the 2nd cycle Biodiversity CU), we consider that the main objectives have been fulfilled by far. After completing the CUs, students acquire an important knowledge about the topic's fundamentals, interpretation and applications. The results obtained along the nine editions of the e-learning CU in Biodiversity Conservation point to a very positive experience, both for the students and for the teachers, as also indicated by the number of completed dissertations on biodiversity thematic (see topic 7).

Learning, as a process emerging from a relational network of ideas, conceptions and representations is based on this social process of building experiences by the community. Due to its complex nature, this process is not limited to the transfer of information but rather expands with the construction of knowledge (Aires et al.

2014) and in this context of Biodiversity and Ecological and traditional knowledge e-learning has a decisive role to play.

Still, improvements are needed to the effectiveness of e-learning. As already stated by Leal Filho (2014) some of the improvements are: further fostering the competencies needed among teachers to use e-learning in a meaningful way and the conceptualization, elaboration and delivery of more suitable materials for use on e-learning programmes. The need to create synergies and interfaces between scientific knowledge and traditional knowledge (essential in this area of Biodiversity Conservation), and also the lead role that e-learning can play establishing and developing a broader awareness on environment and socio-environmental issues in biodiversity conservation are essential to reach the Biodiversity Conservation targets. Hence, these are some interesting challenges to improve biodiversity conservation teaching and learning, particularly in the e-learning context, thus further contributing for ESD in the future.

Other initiatives substantiate our teaching path on biodiversity Conservation. Zahariev and Mihnev (2015) and Mihnev and Raycheva (2010) showed that e-learning provides an accessible option for remote collection, publication and verification of information of different nature and sources relating to the topic biodiversity. These authors also reported on the effectiveness of an e-learning course as a form of self-study that complements traditional teaching and learning forms.

## References

Aires, L., Dias, P., Azevedo, J., Rebollo, M. Á., & Garcia-Pérez, R. (2014). Education, digital inclusion and sustainable online communities. In U. M. Azeiteiro, W. Leal Filho, & S. Caeiro (Eds.), *E-learning and education for sustainability* in the series Umweltbildung, Umweltkommunikation und Nachhaltigkeit—Environmental Education, Communication and Sustainability, Peter Lang (pp. 263–273). ISSN 1434-3819 ISBN 978-3-631-62693-1 (Print) E-ISBN 978-3-653-02460-9 (E-Book). doi:10.3726/978-3-653-02460-9 © Peter Lang GmbH International Academic Publishers Frankfurt am Main 2014.

Alves, F., Leal Filho, W., Araujo, M. J., & Aeiteiro, U.M. (2013). Crossing borders and linking plural knowledge—Biodiversity conservation, ecosystem services and human well-being. *International Journal of Innovation and Sustainable Development (IJISD), 7*(2), 111–125 doi: 10.1504/IJISD.2013.053323

Amador, F., Caeiro, S., Oliveira, C., Caetano, F., Bacelar-Nicolau, P., Azeiteiro, U. M., et al. (2008). Curricular development on higher education and sustainability science: A b-learning course in environmental science. In: *Proceedings 5th EMSU Conference, Environmental Management for Sustainable Universities 2008*. A New Knowledge Culture. Barcelona, Spain, 15–17 October, pp. 1e12. Available at: www.emsu.org

Amador, F., Martinho, A. P., Bacelar-Nicolau, P., Caeiro, S., & Oliveira, C. P. (2015). "Education for sustainable development in higher education: Evaluating coherence between theory and praxis. *Journal Assessment & Evaluation in Higher Education*. doi:10.1080/02602938.2015.1054783

Azeiteiro, U. M., Bacelar-Nicolau, P., Caetano, F., & Caeiro, S. (2015). Education for sustainable development through e-learning in higher education: Experiences from Portugal. *Journal of Cleaner Production, 106*, 308–319. doi:10.1016/j.jclepro.2014.11.056

Azeiteiro, U. M., Leal Filho, W., & Caeiro, S. (Eds.) (2014). *E-learning and education for sustainability* in the series Umweltbildung, Umweltkommunikation und Nachhaltigkeit—Environmental Education, Communication and Sustainability, Peter Lang. ISSN 1434-3819 ISBN 978-3-631-62693-1 (Print) E-ISBN 978-3-653-02460-9 (E-Book). doi:10.3726/978-3-653-02460-9 © Peter Lang GmbH International Academic Publishers Frankfurt am Main 2014.

Bacelar-Nicolau, P., Amador, F., Martinho, A., Azeiteiro, U. M., & Caeiro, S. (2007). E-learning in environmental science post-graduation: A case study at Universidade Aberta, Portugal. In: *EADTU's 20th Anniversary Conference* (p. 19). European Association of Distance Teaching Universities. 8e9 November, Lisbon, Portugal.

Bacelar-Nicolau, P., Caeiro, S., Martinho, A., Azeiteiro, U. M., & Amador, F. (2009). E-Learning for the environment. The Universidade Aberta (Portuguese Open Distance University) experience in the Environmental Sciences Post-Graduate Courses. *International Journal of Sustainability in Higher Education, 10*(4): 354–367. doi:10.1108/14676370910990701

Bacelar-Nicolau, P., Caeiro, S., Martinho, A. P., & Azeiteiro, U. M. (2015). Attitudes, barriers and motivators as factors for sustainability of higher education e-learning programmes at Universidade Aberta, Portugal. In Leal Filho, W., Brandli, L., Kuznetsova, O., & Paco, A. (Eds.), *Integrative Approaches to Sustainable Development at University Level: Making the links in the series* "World Sustainable Development Series", Springer (pp. 567–582). 2015, XII, 731 p. 120 illus., 58 illus. in color. ISBN 978-3-319-10689-2, pp 567–582. doi:10.1007/978-3-319-10690-8_39, http://www.springer.com/gp/book/9783319106892

Bacelar-Nicolau, P., Martinho, A. P., Amador, F., Caeiro, S., & Azeiteiro, U. M. (2012). Online learning for sustainability: The student perception in an environmental science post-graduation. In F. Gonçalves, R. Pereira, W. Leal Filho, & U. M. Azeiteiro (Eds.), *Contributions to the UN decade of education for sustainable development* in the series Umweltbildung, Umweltkommunikation und Nachhaltigkeit—Environmental Education, Communication and Sustainability, Peter Lang, 33: 281–294.

Barth, M., & Burandt, S. (2013). Adding the "e-" to learning for sustainable development: Challenges and innovation. *Sustainability, 5*(6), 2609–2622.

Caeiro, S., & Azeiteiro, U. M. (2004). Distance learning of environmental education in Portugal "O Ensino a Distancia da Educaçao Ambiental em Portugal: um caso de estudo". In: U. M. Azeiteiro, M. J. Pereira, W. Leal-Filho, S. Caeiro, P. Bacelar-Nicolau, F. Morgado, & F. Gonçalves (Eds.), *Global trends on environmental education*, Discursos Língua, Cultura e Sociedade, n especial. Universidade Aberta, Lisboa, pp. 97e111.

Caeiro, S., & Azeiteiro, U. M. (2007). Evaluation of the distance learning teaching of environmental education in Portugal "Avaliação do Ensino a Distancia da Educação Ambiental em Portugal". In: F. Gonçalves, R. Pereira, U. M. Azeiteiro, M. J. Pereira (Eds.), *Pratical activities in science and environmental education* "Actividades Praticas em Ciencia e Educaçao Ambiental.", Coleção Horizontes Pedagogicos 133 Ediç∼oes Piaget, Lisboa, pp. 439e455.

Caeiro, S., Carapeto, C., & Azeiteiro, U. M. (2001). Distance learning teaching of ecology and environmental education "O Ensino a Distancia da Ecologia e Educação Ambiental". Rev. Biol. 19, 107e114.

Caeiro, S., Costa, F. P., Moura, A. P., Martinho, A., Bacelar-Nicolau, P., Ramos, M. R., et al. (2008). New challenges for e-learning in Post- Graduation in exact and technological sciences. In: *CD-ROM Proceedings EDEN 2008 Annual Conference*. EDEN European Distance and E-learning Network, New Learning Cultures. 11–14 June, Lisbon, pp. 1e6.

Caeiro, S., Martinho, A. P., Azeiteiro, U. M., Carapeto, C. (2004). Open and distance learning in the environmental area "O ensino aberto e a distância na area das ciencias ambientais". In: F. Santana, G. Martinho, I. Peres, J. Farinha, J. J. Melo, L. Amaral (Eds.), *CD-Rom Proceedings da 8ª Conferencia Nacional de Ambiente*. 27 e 29 de Outubro, Lisboa, pp. 1e8.

Dlouha, J., Machackova-Henderson, L., & Dlouhy, J. (2013). Learning networks with involvement of higher education institutions. *Journal of Cleaner Production, 49*, 95–104.

Huettmann, F. (2015). Teaching (Tropical) biodiversity with international field schools: A flexible success model in a time of "wireless" globalization. In Huettmann, Falk (Ed.), *Central*

*American biodiversity conservation, ecology, and a sustainable future* (pp 215–245). doi:10.1007/978-1-4939-2208-6_9, ISBN 978-1-4939-2208-6

Halog, A., & Dishman, G. (2014). Developing e learning material for teaching industrial ecology and environmental sustainability. In U. M. Azeiteiro, W. Leal Filho, & S. Caeiro (Eds.) *E-learning and education for sustainability* in the series Umweltbildung, Umweltkommunikation und Nachhaltigkeit—Environmental Education, Communication and Sustainability, Peter Lang (pp. 83–103). ISSN 1434-3819 ISBN 978-3-631-62693-1 (Print) E-ISBN 978-3-653-02460-9 (E-Book). doi:10.3726/978-3-653-02460-9 © Peter Lang GmbH International Academic Publishers Frankfurt am Main 2014.

Leal Filho, W. (2014). E-learning for sustainable development: The way ahead. In U. M. Azeiteiro, W. Leal Filho, & S. Caeiro, (Eds.), *E-learning and education for sustainability* in the series Umweltbildung, Umweltkommunikation und Nachhaltigkeit—Environmental Education, Communication and Sustainability, Peter Lang (pp. 275–277). ISSN 1434-3819 ISBN 978-3-631-62693-1 (Print) E-ISBN 978-3-653-02460-9 (E-Book) DOI 10.3726/978-3-653-02460-9 © Peter Lang GmbH International Academic Publishers Frankfurt am Main 2014.

Martinho, A. P., Bacelar-Nicolau, P., Caeiro, S., Amador, F., & Oliveira, C. (2010). Environmental citizenship and participation. The role of education programs. In: *Knowledge Collaboration & Learning for Sustainable Innovation ERSCP-EMSU Conference*, October 25–29, Delft, The Netherlands, pp. 1e17.

Martinho, A. P., Caeiro, S., Caetano, F., Azeiteiro, U. M., & Bacelar-Nicolau, P. (2014). Environmental professional training and employability: The e-learning degree program in environmental sciences. In U. M. Azeiteiro, W. Leal Filho, & S. Caeiro (Eds.), *E-learning and education for sustainability* in the series Umweltbildung, Umweltkommunikation und Nachhaltigkeit—Environmental Education, Communication and Sustainability, Peter Lang. ISSN 1434-3819 ISBN 978-3-631-62693-1 (Print) E-ISBN 978-3-653-02460-9 (E-Book). doi:10.3726/978-3-653-02460-9 © Peter Lang GmbH International Academic Publishers Frankfurt am Main 2014

Martinho, A. P., Vaz-Fernandes, P., Oliveira, C. P., Bacelar-Nicolau, P., Azeiteiro, U. M., & Caeiro, S. (2016). Training and employability, competences from an e-learning undergraduate programme in environmental sciences. In W. Leal & P. Pace (Eds.), *Teaching education for sustainable development at university level* in the series "World Sustainable Development Series", published with Springer (in press).

Mihnev, P., & Raycheva, N. (2010). An assignment-based e learning course on the use of KeyToNature e-keys. In P. Nimis, & V. Lebbe (Eds.), *Tools for identifying biodiversity: Progress and problem*. Proceedings of the International Congress Paris, September 20–22, 2010. EUT—Edizioni Università di Triest, Italy, pp. 355–360.

Millennium Ecosystem Assessment. (2005). *Ecosystems and human well-being: Biodiversity synthesis*. Washington, D.C.: World Resources Institute.

Moura, A. P. M., Cunha, L. M., Azeiteiro, U. M., Aires, L., & de Almeida, M. D. V. (2010). Food consumer science post-graduate courses: Comparison of face-to-face versus online delivery systems. *British Food Journal, 112*(5), 544–556.

Oliveira, C. P., Martinho, A. P., Caeiro, S., Amador, F., & Caetano, F. (2012). Field activities within an undergraduate program of environmental sciences: A b-learning case study. In F. Gonçalves, R. Pereira, W. Leal Filho, & U. M. Azeiteiro (Eds.), *Contributions to sustainability* (pp. 255e280) in the series Umweltbildung, Umweltkommunikation und Nachhaltigkeit e Environmental Education, Communication and Sustainability Peter Lang.

Pereira, A., Mendes, A. Q., Morgado, L., Amante, L., & Bidarra, J. (2008). *Universidade Aberta's pedagogical model for distance education* © (p. 109). Lisbon: Universidade Aberta.

Snaddon, J. L., Turner, E. C., & Foster, W. A. (2008). Children's perceptions of rainforest biodiversity: Which animals have the lion's share of environmental awareness? *PLoS ONE, 3* (7): e2579. doi:10.1371/journal.pone.0002579

Sibbel, A. (2014). An experience in developing and implementing blended learning for sustainability. In U. M. Azeiteiro, W. Leal Filho, & S. Caeiro (Eds.), *E-learning and education for sustainability* (pp. 15–27) in the series Umweltbildung, Umweltkommunikation und Nachhaltigkeit—Environmental Education, Communication and Sustainability, Peter Lang. ISSN 1434-3819 ISBN 978-3-631-62693-1 (Print) E-ISBN 978-3-653-02460-9 (E-Book). doi:10.3726/978-3-653-02460-9 © Peter Lang GmbH International Academic Publishers Frankfurt am Main 2014.

Zahariev, D., & Mihnev, P. (2015). E-Learning in support to educational field trips – a subject and platform independent curriculum and teacher training model, based on practice. In *ICERI2015: 8th International Conference of Education, Research and Innovation*, (pp. 1314–1324), IATED Publications.

# Knowledge Dissemination and Best Practice Transfer on Biosafety, Biosecurity and Biorisk Management Through a Sustainable and Effective Education and Awareness System

Carola Argiolas, Veronica Baldo and Maurizio Martellini

**Abstract**

Education is the first step to create a society that respects the others and the environment and that works to design and build a peaceful future. An effective and sustainable education system could rely on a tailored methodology that synergizes self-evaluation, gap-analysis, and train-the-trainers methods. This combination allows gathering information about real needs and expectations of training targets, elaborating a made-to-measure educational program and training future educators on topics of interest, making education and awareness system sustainable. Using these methodologies, we set up a knowledge development and transfer of best practice system on biosafety/biosecurity/biorisk management, in order to spread awareness and know-how on these topics. Twenty-two countries, in four different regions (South East Europe, South East Asia, North and West Africa) were involved in the project titled "Knowledge development and transfer of best practice on biosafety, biosecurity and biorisk management". National Experts (NEs) from each country have been trained by intensive and e-learning courses to improve learning efficacy, to raise awareness and to foster networking as well as best practice sharing within each region. Consequently, the trained NEs disseminated the knowledge in their own countries, tailoring the courses on their local needs and expectations, amplifying the educational impact of the project. Here, we review the methodologies applied to develop a sustainable education and awareness system and the training contents related to biosafety/biosecurity/biorisk management.

C. Argiolas · V. Baldo · M. Martellini (✉)
Insubria Center on International Security, University of Insubria, Como, Italy
e-mail: maurizio.martellini@uninsubra.it



P. Castro et al. (eds.), *Biodiversity and Education for Sustainable Development*,
World Sustainability Series, DOI 10.1007/978-3-319-32318-3_20

## 1 Introduction

The impact of an incident involving a biological substance can be enormous. Hospitals, research institutes and industrial facilities must combine their efforts to strengthen the global preparedness and response against incidents happened accidentally or intentionally. Working with such biological agents requires bio-safety,[1] bio-security[2] and biorisk-management.

To this end, it is necessary to build a responsible conduct in the life science and to achieve this goal, education and awareness raising constitute the main pillars. A sustainable education system together with awareness raising activities are indeed the first steps to promote a safety culture and create a society that respects the others and the environment and that works to design and build a peaceful future (National Research Council [NRC], 2010).

The importance of these two pillars has also been highlighted in the Report of the BWC Meeting of States Parties in 2008 (United Nations [UN] 2008; Australia et al. 2011), where the States Parties recognized the value of education and awareness programs:

(i) Explaining the risks associated with the potential misuse of the biological sciences and biotechnology;
(ii) Covering the moral and ethical obligations incumbent on those using the biological sciences;
(iii) Providing guidance on the types of activities which could be contrary to the aims of the Convention and relevant national laws and regulations and international law;
(iv) Being supported by accessible teaching materials, train-the-trainer programmes, seminars, workshops, publications, and audio-visual materials;
(v) Addressing leading scientists and those with responsibility for oversight of research or for evaluation of projects or publications at a senior level, as well as future generations of scientists, with the aim of building a culture of responsibility;
(vi) Being integrated into existing efforts at the international, regional and national levels.

Without any doubt, the interest in biosecurity education has increased recently but, as showed by an investigation conducted by the Landau Network-Fondazione Volta and the University of Bradford, it appears that this interest has not manifest in significant concrete action at the level of the life science academic community. Indeed, it becomes apparent that very little exists in terms of biosecurity related

[1]In this paper, biosafety is used to "to describe the containment principles, technologies and practices that are implemented to prevent unintentional exposure to pathogens and toxins, or their accidental release" (WHO, LBM, 2003).

[2]In this paper, biosecurity refers to "to institutional and personal security measures designed to prevent the loss, theft, misuse, diversion or intentional release of pathogens and toxins" (WHO, LBM, 2003).

education. Some exceptional existing cases could be considered as examples or models, but we are far from achieving the necessary levels of biosecurity related education for life sciences students to be able to generate a culture of responsibility (Mancini and Revill, 2008). Furthermore, there is also, in general, a lack of awareness of the risk of malevolent misuse of the biological sciences as demonstrated also by Dando and Rapper who conducted 90 interactive seminars with more than 2500 life scientists in 13 different countries (Dando and Rappert, 2005).

Last but not least, alongside these intentional misuse concerns, safety risks arising from the unintentional exposure to pathogens and toxins of humans, animals and plants are of great concern, especially with the increasing number of advanced research laboratories dealing with higher-level pathogens (Butler, 2009). Scientists often recognize the biorisks related to pathogens they are researching accidentally infecting people and animals or contaminating the environment outside the laboratories, rather than recognizing the risk of theft or intentional malicious use of bioagents for biocrime and terrorism (Gaudioso, 2006). Moreover, in her survey of Asian life scientists, Gaudioso described how the main means to manage biorisk management in laboratories are biosafety operations manual, institutional biosafety committee, biosafety training procedures, and laboratory management plan, while biosecurity issues do not figure as predominantly in most biorisk assessment, and simple biosecurity measures are utilized to protect laboratories containing pathogens or toxins (Gaudioso, 2006). Therefore, in this context, biosafety education for scientists working with highly pathogenic microorganism and toxins is based on WHO and CDC (Center for Disease Control and Prevention) manuals and, in some cases, on national or institutional regulations. Biosafety training and dissemination of good practices are mostly responsibility of the institutes where the pathogens are used and, as it happens for biosecurity principles, they are very seldom integrated in future scientists education as topics discussed during university courses.

Of course there is no "one-size-fits-all" approach dealing with bio-related issues, as different aspects should be taken into consideration that differs from case to case, namely local and regional needs and priorities, past and present national efforts implemented (such as training organized, regulations, certification/accreditation for laboratories, general level of knowledge and awareness on these issues, etc.), structure of academic curricula, and different linguistic and cultural backgrounds (Minehata et al. 2013). Therefore, a multidisciplinary and holistic approach should be applied in order to address and cover all the aspects related to biosafety and biosecurity.

This paper intends to describe the project titled "*Knowledge development and transfer of best practice on biosafety, biosecurity and biorisk management*", and implemented by the Insubria Center on International Security in the framework of the EU CBRN Centres of Excellence (CoE) Initiative jointly implemented by the European Commission (EC) and the United Nations Interregional Crime and Justice Research Institute (UNICRI). In particular, the methodology used, the training contents, the project's added value and the lessons learned are highlighted hereafter.

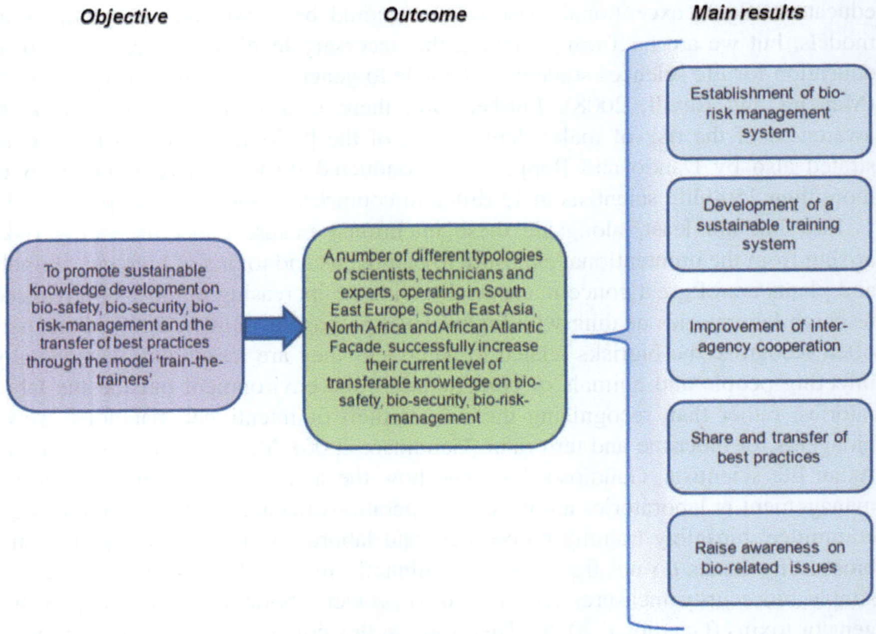


**Fig. 1** Schematic representation of the project objective, outcome and main results

The strategic objective of the project was to promote sustainable knowledge development in four geographical regions (South East Europe, the Southern Caucasus, Moldova and Ukraine,[3] South East Asia,[4] North Africa[5] and African Atlantic Façade[6]) on bio-safety, bio-security and biorisk management, via subsequent train-the-trainer phases in order to deliver to the regions sustainable capacity (Fig. 1). In particular the project planned to:

- Develop holistic and multidisciplinary approach to bio-related issues;
- Strengthen regional and international cooperation on knowledge sharing and best practice exchange;
- Raise awareness on bio-related issues and support synergies among scientific communities and institutions.

---

[3]In this region, the participating countries were Albania, Armenia, Croatia, Georgia, Moldova, Republic of Macedonia, Serbia and Ukraine.

[4]In this region, the participating countries were Cambodia, Brunei Darussalam, Laos, Myanmar, The Philippines, Singapore, Thailand and Vietnam.

[5]In this region, the participating countries were Algeria, Tunisia, Libya.

[6]In this region, the participating countries are Gabon, Mauritania, Morocco.

## 2 Methodology

The methodology used to implement the project was aimed at guaranteeing the highest level of ownership and sustainability of the actions carried out. An effective education system should rely on a tailored methodology that synergizes *self-evaluation, gap-analysis*, and *train-the-trainers* methods. This combination allows gathering information about real needs and expectations of training targets, elaborating a made-to-measure educational program and training of future educators on topics of interest, making education and awareness system sustainable (Fig. 2).

First activities implemented were the self-evaluation, by disseminating to the local stakeholders in the four regions a survey questionnaire in order to understand local situation and local needs, and consequently gap analysis. The results of these activities showed a big discrepancies among the countries (some more technological advanced than others) and showed that regions have differentiated needs from an educational point of view. Differences emerged not only among countries of the same region, but even within each country, between main cities and the periphery or rural areas.

In order to take into consideration these results, the train-the-trainers model, applied to this project to develop knowledge and build a sustainable capacity in bio-related issues, was common among the above mentioned regions, yet flexible enough to be tailored to local needs. The model was inspired to EU levels of risk management, and designed in cooperation with international institutions and European partners. The development of a sustainable knowledge was achieved by

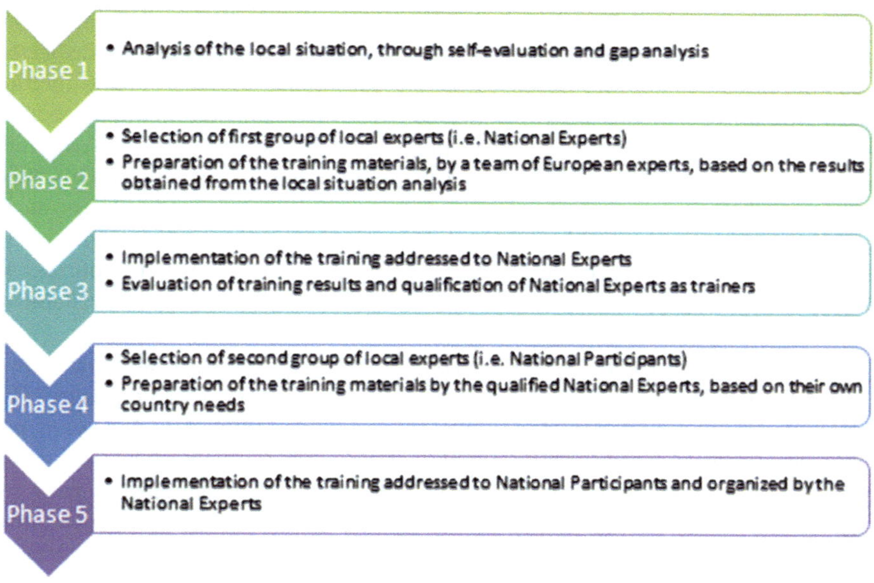


**Fig. 2** Project main phases

training, first, groups of local experts, i.e. National Experts (NEs), coming from the 22 participating countries, in the concepts of biosecurity, biosafety and biorisk management system and, then, by supporting the NEs, who successfully completed the training sessions and were qualified to become trainers in their own countries, in implementing the training locally addressed to second groups of local experts (hereafter referred to as "National Participants"), so that NEs become trainers locally.

The NEs and NPs were selected according to precise profiles, to make sure they had the necessary scientific, technical, educational and language (only for NEs) qualifications: a list of criteria were indeed proposed in each country, taking into account professional expertise, academic curriculum, age, geographical representation and female participation. The challenge in this phase of selection was to make sure to select experts really interested in further disseminating and delivering of acquired capacities at national level. The successful completion of training by NEs was evaluated against predetermined SMART (Specific Measurable Attainable Results-Focused Time-Focused) learning objectives using a range of assessment methods, including assignments, case studies and situational analysis. In total, the project trained 60 National Experts, and 410 National Participants, for a total of 470 trainers trained. These 470 trainers were doctors and technicians from hospitals, research institutes and industrial facilities so as officials from different ministries such as Health, Agriculture, Environment, Defense, Science and Technology, Internal affairs, Trade and, among the authorities, from Customs and Civil Protection in a multidisciplinary perspective. During the training sessions, the different background of the experts indeed helped and fostered the exchange of ideas and lessons learned, based on their own professional experience.

The training methodology was diversified and consisted of e-learning phases and intensive courses. The e-learning phases were delivered through a dedicated Learning Support Platform, housing different educational materials and e-libraries and based on open-source software (Moodle). In particular, the training modules were set up with presentations, assessment tests and interactive tools, such as forum and chat, to foster the discussion among the experts and the trainers. Open source papers, guidelines, and manuals were uploaded in the libraries, used as a repository of the documents and useful to deepen the topics discussed in the courses. This e-learning phase was conceived as a preparatory step for the intensive courses. At the end of the project, the Learning Support Platform remained as a tool available to National Experts and National Participants (Fig. 3).

Intensive courses (both theoretical and practical) were from one week up to two weeks duration and were structured through active learning methodologies. The intensive courses were a mix of frontal lessons, delivered through power point, and video presentations, brainstorming, analysis of case studies, open debates and group exercises. To animate the sessions, laboratory visits and demonstrations were organized as well. The interactive approach guaranteed the exchange of ideas, best practice and share of expertise, especially where the gap between the background and the level of knowledge on bio-related issues was significant. The heterogeneity

**Fig. 3** E-platform course page

fostered a wide sharing of past experiences by the most skillful participants, coming from countries that are at a higher level in terms of awareness on biosafety and biosecurity issues, to the less experienced attendants.

The trainings of National Participants organized by the qualified NEs included only theoretical-practical intensive courses aimed at the common learning objectives, previously mentioned. In general, these training sessions addressed the management from the technical but also policy, legal, and "teaching methods" points of view related to biosafety, biosecurity and biorisk management.

## 3 Training Content

The content of the training modules addressed to NEs was conceived to have a common basis but at the same time to incorporate specific local requests, at regional level: it was based on identified transferable best practices, as well as on the further development of the lessons learned by past similar training efforts, and the common learning objectives, which were tailored and specified on the needs of each region.

Approaching the topics of biosafety, biosecurity and biorisk management a multidisciplinary, omni-comprehensive and holistic approach is pivotal to understand and apply good practices to safeguard security and to guarantee safety in all the aspects involving biological threats. In order to cover all the topics the training addressed to NEs was divided in 5 modules: Module 0 "*Core training specifications for biosafety professionals (content from CWA 16335)*", Module 1 "*Teaching and Assessment methods for bio-risk management training*", Module 2 "*Legal, ethical, environmental aspects*", Module 3 "*Laboratory Biorisk Management*", Module 4 "*Accreditation, implementation and the CWA Standards*".

In detail:

- *Module 0 "Core training specifications for biosafety professionals (content from CWA 16335)"*: was based on the CEN Workshop Agreement (CWA) 16335:2011 Bio-safety Professional Competences. It was conceived as a comprehensive introductory module and as a preparatory phase of training useful to provide the participants with a common and crosscutting background of knowledge required to attend successfully the following modules and to overcome the gaps existing among countries in the same region, emerged during the *gap-analysis* and *self-evaluation* phases of the project. In particular: the risks associated with biological agents and other hazards; concepts of containment and its limitations, the most important types of containment and their installation, validation, certification and maintenance; main elements of infection control, disinfection, decontamination and sterilization; biological waste management plans and requirements on transport, import and export, labelling and means of transport for different biological agents.

- *Module 1 "Teaching and Assessment methods for bio-risk management training"*:
    - *Teaching and Training methodologies* consisted of four different subtopics, namely "Adult as Learners", "Facilitation versus Teaching", "Active Learning" and "Course Design". This course was designed to help participants develop an in-depth understanding of facilitation techniques focused on adult training, concepts of active learning and active learning strategies for teaching bio-risk management. The training was designed to provide the attendants with the ability to distinguish among various approaches for adult training, to compare different methods for training in bio-risk management, to apply active learning methodologies in training and to distinguish among various types of assessment (Fig. 4).
    - *E-learning platform:* this theoretical and practical course was aimed to make experts more familiar with the e-learning platform. It illustrated how to use

**Fig. 4** Moments of active learning

the platform as a student (i.e. updating personal profile, downloading course presentations, filling and uploading assessment tests and communicating through chats and forums) and as a teacher (showing how to create new courses and profiles, upload files and grade assessment tests).

- *Background elements of Quantitative Microbial Risk and Assessment and presentation of toolkit* focused on the biosafety and the biosecurity program in laboratories. To point out the synergy and potential conflicts between biosafety and biosecurity both programs were discussed in detail. The experts were provided with international guidelines (WHO, BMBL, CWA Laboratory Biorisk management) and tools like the biosecurity toolkit in order to think over biosecurity management in their own organization and to identify gaps in their biosecurity management system. The aim of this lecture was to show the importance of doing a risk assessment and how to create awareness. As an example of risk assessment, the basic steps of Quantitative Microbial Risk Assessment (QMRA) for food and water were explained: hazard identification, exposure assessment and risk characterization, including also variability and uncertainty.

– *Module 2 "Legal, ethical, environmental aspects"*

- *Legal framework for Biosafety and Biosecurity* was aimed to review the basic legal concepts and ideas relating to biosafety and biosecurity and legislative provisions. During the training, participants were asked not only to compare and contrast the legislative frameworks in their respective countries but also to exchange ideas on best practices and areas where further regulation may be required.
- *Environmental Biosafety and Biosecurity* focused on general environmental concepts (biodiversity, resilience, and sustainability), environmental impact assessment, and environmental management (Environmental Protection). In particular, biosafety, biosecurity and biorisk analysis, Environmental Impact Assessment (EIA), Environment Risk Assessment (ERA), Environment Management Systems (EMS) were presented and discussed. Moreover, the Invasive Alien Species (IAS) topic was developed elucidating the principle of invasion ecology, explanation of the invasion process, and determinants of invasiveness and invisibility. It was deeply analyzed the theme of biodiversity and the importance of its protection for earth and earth inhabitants health.
- *Ethical/dual use aspects* discussed the ethical questions related to the dual use dilemmas in science and, in particular, it focused on the possibility to use the new emerging technologies, like nanotechnology, technologies based on atomic fission, chemistry and synthetic biology, for good as well as for bad purposes. The question of the moral responsibility of scientists and on their accountability in dual use dilemmas was deepened and some case studies were discussed.

– *Module 3 "Laboratory Biorisk Management"*

- *Laboratory Biorisk Management* was divided in different subtopics. The first one focused on natural occurring bio emergency. In particular, the social, economical and demographical impacts of SARS outbreak (2003) were analyzed in detail. Moreover, the recent Ebola outbreak (2014) effects in society, economy and demography in the three most affected countries (Liberia, Guinea and Sierra Leone) and in the rest of the word have been discussed. The second course reviewed the classification of infective microorganism by risk group, the characteristics of microorganisms belonging to different groups, the bio containment levels that have to be applied for each group and the diagnostic tools used to identify viruses and bacteria. Moreover, this course aimed to refresh and improve knowledge about safety in clinical diagnostic Biosafety Level Laboratories 2, 3 and 4 (BSL2, BSL3 and BSL4) of L. Sacco University Hospital based in Milan (Italy), to encourage laboratory workers to think over safety issues while they are working and to stimulate the birth of a safety culture in biological laboratory. Guidelines to work safely in a contained laboratory were discussed and their implementations presented in details, analysing laboratory building, personal protective equipments and good laboratory practices, as well as how to face an accident or an emergency. Experts had also the possibility to visit both the cellular and molecular laboratories of a BSL2, BSL3 and BSL4 facilities. The laboratory visit and practical exercise consisted in donning the BSL3 and BSL4 suit and all the required PPE appropriately, in getting in the BSL3 and BSL4, where laboratory workers showed equipment and tools and explained in detail how to work safely in containment laboratories, using biosafety cabinets and appropriate protocols especially for decontamination of outgoing material (Fig. 5).
- *Biological Risk Assessment Methodologies/Measurements and analysis of biorisk management system performance* were mainly the Laboratory Biorisk Management through the AMP model (Assessment, Mitigation,

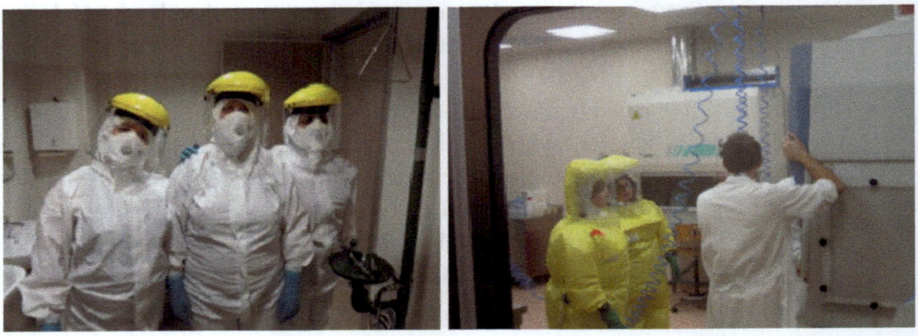

**Fig. 5** BSL3 and BSL4 laboratory visits

Performance). In particular, BioRAM, software for risk assessment developed by Sandia, and "What if" analysis, FMEA, FTA, HAZOP study were presented in detail as tools that can be used to assess biological risk correlated to a microbiological laboratory. The measures to mitigate the risk were highlighted as well, together with the methods to evaluate the performance.

– *Module 4 "Accreditation, implementation and the CWA Standards"*

- *Accreditation, implementation and the CWA/ISO Standards and A wide introduction to GMOs and GMO regulations* was a discussion and experience sharing among the experts from different countries. In particular, discussions focused on regulations already present in each countries, ISO, Standards and guidelines implemented and required for laboratory equipment and PPE. Moreover, Certification, Guidelines and Manuals used in Biosafety and Biorisk Management have been deeply analyzed, paying a particular attention to CWA15793 and ISO15190. The second part of this module was added to fulfil a request by the participants, emerged during module 1 and 2 attendance, and focused on Genetically Modified Organisms (GMOs) regulations and guidelines already present in each single country for the use of GMOs.

Concerning the National Participants' training, National Experts, according to needs and specificities of their own countries, chose among the modules described above the ones that were more appropriate and interesting for their audience, tailoring the courses on local situations.

## 4 Project's Added Value

Notwithstanding a number of trainings programmes already organised in the past and conducted in the four regions considered, the present project provided a significant added value in the bio-sector, based on the following four factors (Fig. 6):

(a) Regional and inter-regional coordinated approach;
(b) Demand-side and participatory approach;
(c) Holistic and multidisciplinary approach;
(d) Establishment of a sustainable training method.

(a)
*Regional and inter-regional coordinated approach*

In line with the Centres of Excellence initiative, the project supported the development of capacities in a regional coordinated manner. The project developed and spread a common-ground of knowledge and expertise about bio-safety, bio-security and bio-risk management in the targeted regions. The training content, along with the involvement of stakeholders from different institutions and countries, contributed to stimulate networking, cooperation and strengthen synergies among

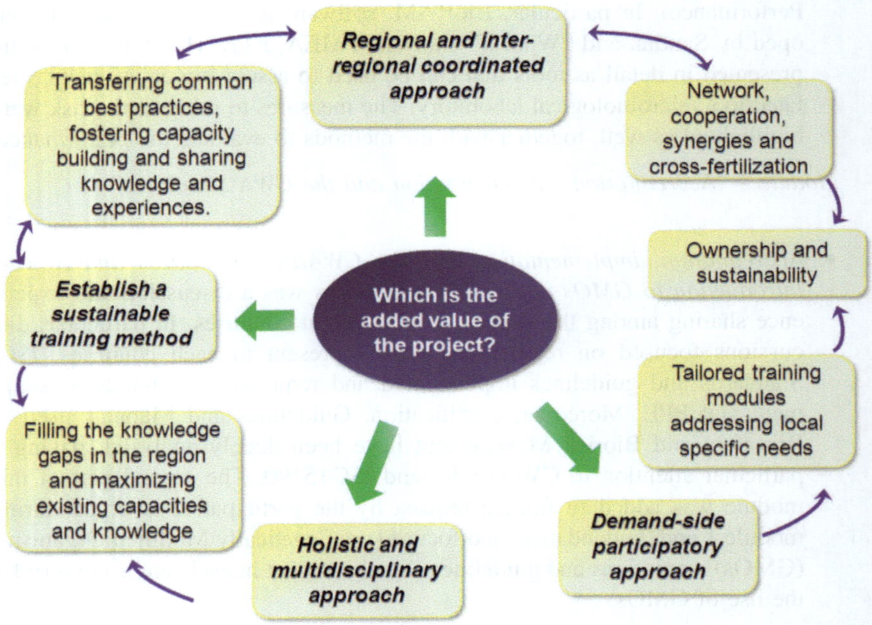


**Fig. 6** Project's added value

the scientific communities and institutions, thus facilitating knowledge sharing and transfer of lessons learned at regional level. In order to maximise the networking and exchange of knowledge, the project relied on already established regional networks of local partners, well grounded in the targeted areas: this allowed to build on what has been already achieved and implemented and to further strengthen the regional networking, cooperation and mutual trust. The project promoted cross-fertilisation of ideas also across different regions. ICIS supported and foster exchange of knowledge, lessons learned and best practices beyond the regional boundaries, particularly through the e-learning platform.

(b) *Demand-side and participatory approach*

The regional and inter-regional coordinated approach did not prevent to keep into duly account also local specificities. In practical terms, training modules were tailored on the specific needs of each region and, as far as possible, country. As previously described, the assessment of local specific training needs was carried out according to a participatory demand-side approach. Such a method better addressed local specific needs, keeping into account their past experiences, the existing level of knowledge, the difference among institutes operating in bio-related issues (either private or public), thus ensuring both the ownership and sustainability of the project. This entailed, for example, the possibility to translate key training materials in local languages, especially in those countries where English is not widespread. The participatory approach was also reflected in the ‘train-the-trainer’ method, which transforms the trained people from passive users to active actors in the process of knowledge exchange, putting the grounds to deliver training beyond the time span of the project.

(c) *Holistic and multidisciplinary approach*

This project addressed a wide variety of bio-related issues spanning from bio-related international and national legal aspects, bio-related environmental and ecotoxicological aspects, to bio-ethical aspects, accreditation, implementation and the CWA standards and laboratory bio-risk management. Through such a multi-disciplinary approach, the project aimed at filling the knowledge gaps in the region and at maximising existing capacities and knowledge in different fields.

(d) *Establishment of a sustainable training method*

The sustainability of the project’s training method was ensured firstly by the train-the-trainers model, not new in most of the countries involved, which was complemented by a modern methodological tool for training, i.e. a web based e-learning platform. The platform constituted an interactive tool favoring the active participation of both trainers and trainees (e.g. through a Question/Answer section, virtual classrooms, live sessions, etc.) within the same region, but also across different regions. Through the platform, experts from different regions had the possibility to interact and confront each other on common experiences, so as to increase inter-regional cooperation and networking and the exchange of knowledge. Moreover, by making training materials available on-line before the start of intensive training courses in presence, the e-learning platform guaranteed a preliminary common ground of knowledge among all participants to the training courses in preparation for the subsequent intensive course, so as to maximize the effectiveness of training. The most important feature of the e-learning platform is that it remains available as a tool to experts, to continue using and updating it. In this way, the e-platform represents a concrete legacy of the project, which will continue to favor information sharing and will put the basis for the potential organization of further intensive courses (both theoretical and practical) ensuring, in this way, the long-term sustainability of the project, by generating further positive effects in the regions and also increasing the number of beneficiaries beyond the project’s direct users.

## 5 Lessons Learned

Basing on this experience, some lessons learned can be highlighted so that these could be useful for future similar projects, aiming at knowledge development, also not necessarily in the bio-field. It is of outmost importance to:

- Apply and foster real participatory approaches: this methodology requires time, patience and understanding of mutual needs and institutional structures. To guarantee ownership and commitment, it is necessary to devote efforts and energy especially at the beginning of the project;

- Transfer the importance of a multidisciplinary approach and the importance of bringing together experts with different backgrounds;
- Stimulate the use of simple and, where possible, open source tools that don't imply financial costs to be supported also after the end of the project, tailored on the users' real capacities and needs; it is important to evaluate the opportunity to use IT tools and adapt them to the local situation, taking into consideration the difficulties some countries may face with the internet connection;
- Create tailored didactic tools and training sessions, considering local needs and priorities, the cultural background, the importance of the human factor and interaction among experts with different skills and experiences;
- Produce/translate didactic materials in local languages, which is also a very important element in terms of sustainability and impact, since these can be shared and distributed also to non-English -speaking experts;
- Enhance awareness raising on bio-related issues and support synergies among the scientific community and institutions within each country;
- Create an international network of experts, fostering knowledge sharing and best practice exchange within and among regions;
- Last but not least, remain constantly engaged in the biosecurity and biosafety issues and on the biorisk evaluation and metrics aspects, due to the continuous development of the life sciences.

## References

Australia, et al. (2011). Revised: Possible approaches to education and awareness-raising among life scientists BWC/CONF.VII/WP.20/Rev.1.

Butler, D. (2009). European biosafety labs set to grow. *Nature, 462*, 146–147.

Dando, M., & Rappert, B. (2005) Codes of conduct for the life sciences: Some insights from UK Academia. *Bradford Briefing Papers, 16*.

Gaudioso, J. (2006). *A Survey of Asian Life Scientist: The State of Biosciences, Laboratory Biosecurity, and Biosafety in Asia*. California: Sandia National Laboratories.

Mancini, G., & Revill, J. (2008). *Fostering the Biosecurity Norm: Biosecurity Education for the Next Generation of Life Scientists*. Bradford: University of Bradford.

Minehata, M., Sture, J., Shinomiya, N., Withby, S., & Dando, M. (2013). Promoting education of dual-use issues for life scientists: a comprehensive approach. *Journal of Disaster Research, 8* (4), 674–685.

National Research Council. (2010). *Challenges and Opportunities for Education About Dual-Use Issues in the Life Sciences*. National Academies Press.

United Nations. (2008). Report of the Meeting of States Parties, BWC7MPS72008/5.

WHO. (2003). *Laboratory Biosafety Manual*.

MIX
Papier aus verantwortungsvollen Quellen
Paper from responsible sources
FSC® C105338

If you have any concerns about our products,
you can contact us on
**ProductSafety@springernature.com**

In case Publisher is established outside the EU,
the EU authorized representative is:
**Springer Nature Customer Service Center GmbH**
**Europaplatz 3, 69115 Heidelberg, Germany**

Printed by Libri Plureos GmbH
in Hamburg, Germany